PERSISTENT POLLUTANTS IN MARINE ECOSYSTEMS

Earlier Volumes in the SETAC Special Publications Series

BERGMAN *et al.*
Environmental Hazard Assessment of Effluents

BODEK
Environmental Inorganic Chemistry: Properties, Processes and Estimation Methods

CAIRNS
Multispecies Toxicity Testing

DICKSON *et al.*
Fate and Effects of Sediment-Bound Chemicals in Aquatic Systems

HUGGELT *et al.*
Biomarkers: Biochemical, Physiological and Histological Markers of Anthropogenic Stress (*Published by Lewis Publishers, Chelsea, Michigan*)

KENDALL
The Population Ecology and Wildlife Toxicology of Agricultural Pesticide Use: A Modeling Initiative for Avian Species
(*Published by Lewis Publishers, Chelsea, Michigan*)

Other Related Pergamon Publications

OPEN UNIVERSITY OCEANOGRAPHY SERIES
The Ocean Basins: Their Structure and Evolution

Seawater: Its Composition, Properties and Behaviour

Ocean Circulation

Waves, Tides and Shallow-Water Processes

Ocean Chemistry and Deep-Sea Sediments

Case Studies in Oceanography and Marine Affairs

Related Journals

Biochemical Systematics and Ecology

Chemosphere

Current Advances in Ecological Sciences

Environment International

Environmental Toxicology and Chemistry

Marine Pollution Bulletin

Water Research

PERSISTENT POLLUTANTS IN MARINE ECOSYSTEMS

Edited by

Colin H. Walker

School of Animal and Microbial Sciences, University of Reading

and

David R. Livingstone

Cellular and Molecular Biology Group, Plymouth Marine Laboratory

SETAC Special Publications Series

Volume Editor

Robert L. Lipnick

Office of Pollution Prevention and Toxics, US Environmental Protection Agency

Series Editor

Thomas W. La Point

Institute of Wildlife and Environmental Toxicology, Clemson University

PERGAMON PRESS

OXFORD · NEW YORK · SEOUL · TOKYO

U.K. Pergamon Press Ltd, Headington Hill Hall, Oxford OX3 0BW, England
U.S.A. Pergamon Press, Inc., 660 White Plains Road, Tarrytown, New York 10591–5153, USA
KOREA Pergamon Press Korea, KPO Box 315, Seoul 110–603, Korea
JAPAN Pergamon Press Japan, Tsunashima Building Annex, 3–20–12 Yushima, Bunkyo-ku, Tokyo 113, Japan

---

First edition 1992

Library of Congress Cataloging-in-Publication Data
Persistent pollutants in marine ecosystems / edited by Colin H. Walker and David R. Livingstone.
p. cm. — (SETAC special publications series)
1. Persistent pollutants—Environmental aspects—Congresses. 2. Persistent pollutants—Bioaccumulation—Congresses. 3. Marine fauna—Effect of water pollution on—Congresses. 4. Marine pollution—Congresses. I. Walker, C. H. (Colin Harold), 1936–. II. Livingstone, David R. III. Series.
QH545.P39P47 1992 574.5′2636—dc20 92–6423

British Library Cataloguing in Publication Data
A catalogue record for this book is available from the British Library.

ISBN 0 08 041874 0

Printed in Great Britain by B.P.C.C. Wheatons Ltd., Exeter

Contents

PART III FISH

PART IV MARINE MAMMALS

PART V FISH-EATING BIRDS

PART VI OVERVIEW

Foreword

It has been my pleasure to serve as SETAC Special Publications Associate Editor in the development of this volume with Drs. Colin Walker and David Livingstone, both of whom are highly regarded for their research in understanding the effects of persistent bioaccumulative pollutants within the marine aquatic environment. This book evolved from a session chaired by Dr. Walker at the 10th Annual Meeting of the Society of Environmental Toxicology and Chemistry in Toronto, Canada. Following a very favorable reception of the papers at this session, the Editors were able to recruit a broad group of recognized specialists to prepare the chapters that follow. Differences in style and length reflect the fact that most represent expansion of the original Symposium presentations.

This book represents the latest in the Special Publication Series of the Society of Environmental Toxicology and Chemistry (SETAC). This Series was established to provide authoritative works covering a broad range of disciplines in environmental toxicology, environmental chemistry and hazard and risk assessment, with works selected based upon approval of the SETAC Technical Committee and SETAC Board of Directors.

In addition to the individual chapter reviews, the overall structure and content of the book was assessed by both a hired independent reviewer and by this SETAC Associate Editor. Support for the independent review by Dr. Daniel Osborn, Institute of Terrestrial Ecology, UK, from the SETAC Foundation for Environmental Education (Mr Rodney Parrish, Executive Director) is gratefully acknowledged.

This work provides a foundation for understanding the behavior of hydrophobic persistent organic compounds in the environment and the subtle factors involving uptake, metabolism and excretion that control their ultimate body burdens in marine organisms. Fish and other aquatic organisms are able to bioconcentrate such substances by enormous factors, leading to both acute and chronic effects to fish, and to the biomagnification of such chemicals within the tissues of birds and mammals, including humans, with cascading results. The theoretical principles set forth in this book, as they continue to be expanded, will provide a logical basis for prediction of the relative risks posed by persistent hydrophobic chemicals released into the environment.

ROBERT L. LIPNICK
U.S. Environmental Protection Agency
Washington, DC

Preface

A familiar image of marine pollution is one of oil spills producing polluted beaches and oiled seabirds. In a number of well-publicized incidents, tankers have been wrecked, releasing large quantities of crude oil into the marine environment. Recently, there has been a massive release of oil in the Persian Gulf, during the Gulf War. The ecological damage caused by such disasters is immediately apparent when dead and dying seabirds are washed up on the shore; the harmful effects of pollution are clearly visible to the naked eye.

This is not the case with the majority of the persistent pollutants, principally lipophilic organic compounds, which are the subject of the present book. These chemicals are widely distributed throughout the seas and oceans of the world. They are, by definition, substances which have long half-lives in sediments, water and marine organisms. They travel considerable distances, carried by the movements of water, air and migrating animals. They also move through food chains, often reaching relatively high concentrations in predators occupying the highest trophic levels. Their effects can be insidious and long term, unlike the immediate effects of oil pollution. Often these effects are very difficult to establish because of the sheer complexity of the situation – the number of compounds is very large.

The elimination of foreign compounds (xenobiotics) depends on the processes of metabolism and excretion, and the compounds are persistent when these processes fail to operate adequately. This may be for various reasons – the processes are ineffective, the compounds are bound to other molecules such that they are not available for metabolism, or the rates of metabolism and elimination simply cannot cope with the rates of uptake. It is thought that these processes have evolved to aid the detoxication of naturally occurring lipophilic xenobiotics. They vary greatly between species and different groups of organism, e.g. marine invertebrates compared to marine vertebrates. Consequently, particular chemicals may be persistent in some species but not in others, e.g. many hydrocarbons are strongly persistent in marine invertebrates but usually much less so in marine vertebrates. In contrast, certain other chemicals, e.g. many polychlorobiphenyls, are persistent in all animals that have been studied. This book gives particular emphasis to the latter group of compounds, which are liable to undergo strong bioaccumulation at the top of the food pyramid. However, compounds which are only persistent in certain species are also considered.

The chapters of the book are arranged in a sequence that essentially follows the trophic levels of marine food chains. To emphasize this point, the book is arranged into five parts each of which deals mainly (although not exclusively) with one group of organisms. Thus, invertebrates of the lower trophic levels are considered in Chapters 1 and 2, paying particular attention to the direct uptake of pollutants from water and sediments. Chapters 3 and 4 deal mainly with invertebrates higher in the food chain which receive pollutants largely from their prey, the latter chapter being concerned with the modelling of processes of accumulation. Chapter 5 focuses on fish, with some consideration of marine mammals, whereas essentially the reverse is the case in Chapters 6 and 7 which deal with marine mammals and their food sources. Chapters 8–10 cover fish-eating birds. Thus, Chapters 6–10 deal largely with vertebrates at the top of the food chain. Chapter 11 provides an overview which pulls together and builds on the findings of all the previous chapters.

The primary concern is with the persistent pollutants in organisms of the different trophic levels – the concentrations they reach in tissues and the processes which determine these concentrations. Particular attention is paid to metabolism by the different species of the marine food chain, and the role of metabolism in detoxication and activation. Where possible, toxic effects of pollutants upon individuals and populations are reviewed. The concluding chapter focuses on the major issues of processes of uptake, metabolism and bioaccumulation. Particular attention is paid to the comparative metabolism of xenobiotics, the relative toxicities of parent compounds and metabolites, and evolutionary aspects. The overall picture gives cause for considerable concern. Not only is the range of compounds very large, but certain compounds can reach substantial levels in predators at the top of food chains. Much more is known of the levels of compounds in organisms than of their effects. Nevertheless, there are already clear indications of toxic effects that persistent pollutants have had, or are having, in the marine environment. More work is urgently required to investigate the ecological effects of these compounds, singly and in combination, in highly polluted areas, so that appropriate action can be taken to control marine pollution.

List of Abbreviations

AAF	2-acetylaminofluorene
AH	aromatic hydrocarbon
AHH	aryl hydrocarbon hydroxylase
BaA	benzo[a]anthracene
BaP	benzo[a]pyrene
BAF	bioaccumulation factor
BCF	bioconcentration factor
BPH	benzo[a]pyrene hydroxylase
CB	chlorinated biphenyl
CB-138	2,2′,3,4,4′,5-hexachlorinated biphenyl
CB-153	2,2′,4,4′,5,5′-hexachlorinated biphenyl
CB-180	2,2′,3,4,4′,5,5′-heptachlorinated biphenyl
CH	chlorinated hydrocarbon
DCBP	dichlorobiphenyl
DDD	1,1-dichloro2,2-*bis*(*p*-chlorophenyl)ethane
DDE	1,1-dichloro-2,2-*bis*(*p*-chlorophenyl)ethylene
DDT	1,1,1-trichloro-2,2-*bis*(*p*-chlorophenyl)ethane
2DGC-ECD	two-dimensional gas chromatography with electron capture detection
ECOD	7-ethoxycoumarin *O*-deethylase
EROD	7-ethoxyresorufin *O*-deethylase
FDA	Food and Drug Administration
GC	gas chromatographic or gas chromatography
GC-ECD	gas chromatography with electron capture detection
GC-MS	gas chromatography-mass spectrometry
GSH	glutathione
HAH	4–7 ring aromatic hydrocarbon
HCB	hexachlorobenzene
HCBP	hexachlorobiphenyl
HCH	hexachlorocyclohexane
γ-HCH	γ-hexachlorocyclohexane (lindane)

HEOD	1,2,3,4,10,10-hexachloro-1,3,3a,5,6,7,8,8a-octahydro-6,7-epoxy-exo-1,4-endo-5,8,-dimethanonaphthalene
HEOM	1,2,3,4,9,9-hexachloro-1,4,4a,5,6,7,8,8a-octahydro-6,7-epoxy-1,4-methanonaphthalene
HMO	hepatic microsomal monooxygenase
HxCDD	hexachlorodibenzo-*p*-dioxin
LAH	1-3 ring aromatic hydrocarbon
K_{ow}	octanol/water partition coefficient
3-MC	3-methylcholanthrene
MCBP	monochlorobiphenyl
MCOD	7-methoxycoumarin *O*-demethylase
MFO	mixed-function oxygenase (oxidase)
OCS	octachlorostyrene
PAH	polynuclear aromatic hydrocarbon
PB	phenobarbital
PBB	polybrominated biphenyl
PCB	polychlorinated biphenyl
PCBP	pentachlorobiphenyl
PCDD	polychlorinated dibenzo-*p*-dioxin
PCDF	polychlorinated dibenzofuran
PCP	pentachlorophenol
PnCDD	pentachlorodibenzo-*p*-dioxin
PnCDF	pentachlorodibenzofuran
ppt	parts per trillion
PROD	7-pentoxyresorufin *O*-deethylase
QSAR	quantitative structure–activity relationship
RI/FS	Remedial Investigation/Feasibility Study
$t_{1/2}$	depuration half-life
TBT	*bis*(tributyl)tin oxide
TCBP	trichlorobiphenyl
4,5,6-TCG	4,5,6-trichloroguajacol
2,4,5-TCP	2,4,5-trichlorophenol
TeCBP	tetrachlorobiphenyl
TCDD	tetrachlorodibenzo-*p*-dioxin
TEF	toxic equivalency factor
TSR	tissue to sediment concentration ratio

Part I

Invertebrates of Lower Trophic Levels

CHAPTER 1

Persistent Pollutants in Marine Invertebrates

DAVID R. LIVINGSTONE

NERC Plymouth Marine Laboratory, Citadel Hill, Plymouth PL1 2PB, UK

INTRODUCTION

For the last fifty to a hundred years in particular, marine organisms have been faced with an increasing input of a wide variety of potentially toxic chemicals. The accumulation and persistence of many such compounds is marked in marine invertebrates. Determining persistence of foreign compounds (xenobiotics) in organisms are processes of uptake, metabolism, retention and release of parent compound and metabolites. Persistence can occur in the absence or presence of a continuing input of the chemical: in the latter case, the relative rates of the different processes will be critical, e.g. uptake compared to metabolism and elimination. This chapter reviews these aspects, and the data presented are treated in a quantitative manner in order to understand the tissue levels of xenobiotics accumulating in organisms in the field. Emphasis is placed on identifying the similarities and differences between the major marine invertebrate phylogenetic groups and, to some extent, on examination of the extent to which the invertebrate picture fits the paradigm of mammalian toxicology. The data are largely for marine species, but information on freshwater ones is included where relevant. This chapter represents a synthesis and extension of views presented in a recent comprehensive review on organic xenobiotic metabolism in marine invertebrates (Livingstone, 1991a).

BIOACCUMULATION: TISSUE LEVELS, UPTAKE, DEPURATION

Tissue Levels of Contaminants in the Field

Typical tissue concentrations for various major groups of chemicals for invertebrate groups from the field are given in Table 1.1. Concentrations are higher in more polluted environments and for tissues containing more lipid. Many other contaminants have been detected, e.g. chlordane compounds in the eight-armed squid *Gonatopsis borealis*, antarctic krill *Euphausia superba* and various crab species (Lee *et al.*, 1982; Kawano *et al.*, 1988), and polychlorinated dibenzo-*p*-dioxins (PCDDs) and dibenzofurans (PCDFs) in the mussel *Mytilus edulis* (Miyata *et al.*, 1987, 1989). Field levels of contaminants, such as polynuclear aromatic hydrocarbons (PAHs), polychlorinated biphenyls (PCBs) and/or chlorinated hydrocarbons, are observed to vary in molluscs, crustaceans and echinoderms with season (Satsmadjis and Voutsinou-Taliadouri, 1983; Mix, 1984; Knickmeyer and Steinhart, 1988; Hummel *et al.*, 1990; Den Besten *et al.*, 1990) and food intake (Knickmeyer and Steinhart, 1988; Van der Oost *et al.*, 1988). Seasonal changes are partly, though not totally, related to changes in lipid levels (Den Besten *et al.*, 1990; Hummel *et al.*, 1990).

Field bioconcentration factors (BCFs; ratio of xenobiotic concentration in animal to that in external medium), on a wet weight basis, for PAHs have variously been recorded ranging from $\times 600$ to $\times 1200$ for sponge species, the crab *Cancer pagurus* and the starfish *Asterias rubens* (Grahl-Nielsen *et al.*, 1978; Sieben *et al.*, 1983). The BCFs for petroleum hydrocarbon mixtures for *M. edulis* were approximately constant at 2×10^5 for seawater concentrations of 1 to 400 ppb, with maximum observed tissue hydrocarbon concentrations of 30 mg per g of body lipid (Burns and Smith, 1981). Maximal limits of accumulation have been indicated for other xenobiotics, e.g. residues of 1,1,1-trichloro-2,2-*bis*(*p*-chlorophenyl) ethane (DDT) taken up by various detritivore freshwater crustaceans and molluscs (Webber *et al.*, 1989). The bioaccumulating capability of bivalve and other molluscs, for contaminants such as PAHs, is generally regarded as being greater than that of other comparable marine invertebrates, such as crustacea (Neff, 1979), and for this and other reasons (e.g. because they are sessile, filter-feeders, ubiquitous and robust) they have been used world-wide as sentinels in mussel watch pollution monitoring programmes (Livingstone, 1991a). Time courses of contaminant uptake by bivalves have also been monitored following oil spills (Boehm *et al.*, 1982), creosote spills (De Leon *et al.*, 1988), oil terminal operations (Widdows *et al.*, 1987) and dredging activities (Rice and White, 1987). Maximal levels of contaminants accumulated, however, particularly in acute situations such as spills, are often similar in different organisms (Table 1.1), and differences in routes of uptake, species physiology, etc., make exact comparisons between organisms difficult, e.g. levels of PCBs and DDT were greater in the blue crab *Callinectes sapidus* than in the American oyster *Crassostrea virginica* at sites in South Carolina, USA (Marcus and Renfrow, 1990); levels of unsubstituted PAHs were similar in the polychaete *Nereis succinea* and the clam *Macoma balthica* at particular sites in Chesapeake Bay, USA (Foster and Wright, 1988); and the uptake of PCBs following dredging was greater in fathead minnows *Pimephales promelas* than in fingernail clams *Sphaerium striatinum* (Rice and White, 1987).

Uptake and Depuration

A plethora of information exists on the uptake and depuration of xenobiotics by marine invertebrates, the processes being dependent on a number of variables, including bioavailability and route of uptake, time course of exposure and various other physicochemical and biological factors (Barron, 1990; Livingstone, 1991a). Lipophilic organic compounds are readily taken up from water: provided the external xenobiotic concentration is effectively constant, the process usually involves an initial high linear rate of uptake, followed by the eventual reaching of a maximal tissue equilibrium concentration. Depuration is normally exponential, with the pattern of loss often being resolved into an initial faster phase followed by a longer-term slower phase. Exceptions to these generalizations occur, viz. biphasic or multiphasic uptake, e.g. picric and picramic acids taken up by *C. virginica* (Burton *et al.*, 1984); tissue equilibrium concentrations apparently not being reached even after long periods of exposure, e.g. various bivalves exposed to hydrocarbons for up to 6 months (Clement *et al.*, 1980); and single phase elimination, e.g. anthracene loss from the clam *Rangia cuneata* (Jovanovich and Marion, 1987) and 2,4,5-trichlorophenol (2,4,5-TCP) from intermolt (stage C) grass shrimp *Palaemonetes pugio* (Ranga Rao *et al.*, 1981).

The uptake and depuration of xenobiotics is thought to be essentially a passive process, involving movement and equilibrium of the chemical between aqueous (external) and biotic (organism) compartments (Burns and Smith, 1981). In some cases active excretion has been suggested, e.g. naphthalene lost from the gills and kidneys of *M. edulis* (Widdows *et al.*, 1983), whereas in others it has been clearly shown not to be the case, e.g. blocking of the nephrophores and anus of the crab *Hemigrapsus nudus* had no affect on the clearance of naphthalene (Laurén and Rice, 1985). Faecal material and urine can also make a contribution to elimination, e.g. for petroleum compounds lost from zooplankton (Corner *et al.*, 1976) and sulfadimethoxine from the lobster

TABLE 1.1 *Concentrations of Classes of Pollutants in Tissues of Major Marine Invertebrate Groups from the Field*

Pollutant	Phylum	Species	Tissue	Field site	Concentration[a] ($\mu g\ g^{-1}$ wet wt)	Reference
Total PAH[b]	Porifera	*Halichondria panicea*	whole	oil spill	13.7	Grahl-Nielsen *et al.* (1978)
Total PAH	Coelenterata	*Teclia felina*	whole	oil spill	18.0	Grahl-Nielsen *et al.* (1978)
Total PAH	Annelida	*Nepthys incisa*	whole	New York Bight, USA	0.06	McElroy *et al.* (1989)
Total PAH	Mollusca	*Mytilus edulis*, other bivalves	whole	commerce	0.5 – 1.3	Mix (1984)
		M. edulis	whole	industrial, Norwegian fjord	0.5 – 3.1	Klungsøyr *et al.* (1988)
		M. edulis *Modiolus modiolus*	whole	industrial, Norwegian fjord	0.7 – 29.9	Bjorseth *et al.* (1979)
		Mytilus galloprovincialis	whole	industrial, French coast	60.6	Claisse (1989)
Total PAH	Arthropoda	*Carcinus maenas*	whole	industrial, Norwegian fjord	0.03 – 0.05	Klungsøyr *et al.* (1988)
		Homarus americanus	muscle	coal-coking plant	0.32 – 2.04	Uthe and Musial (1986)
			digestive gland		15.9 – 60.1	
		Uca pugnax	whole	oil spill	18.3 – 56	Burns (1976)
Total PAH	Echinodermata	*Asterias rubens*	whole	urban, industrial	0.07 – 0.09	Knutzen and Sortland (1982)
		A. rubens	whole	oil spill	7.0	Grahl-Nielsen *et al.* (1978)
Total hydrocarbons	Mollusca	*Anomalocardia brasiliana*	whole	Todos os Santos	0.1 – 42	Porte *et al.* (1990)
		M. edulis	whole	Australian coast	117 – 375	Burns and Smith (1981)

Total PCB	Mollusca	*Crassostrea virginica*	whole	Gulf of Mexico, USA	0.002 – 0.8	Sericano *et al*. (1990)
		M. galloprovincialis	whole	Mediterranean coast	0.02 – 1.63	Geyer *et al*. (1984)
		M. edulis	whole	New Bedford Harbor, USA	3.4	Farrington *et al*. (1983)
		Sphaerium striatinum	whole	industrial river, Michigan, USA	0.16 – 13.9	Rice and White (1987)
Total PCB	Arthropoda	*C. maenas*	whole	industrial, Norwegian fjord	0.03 – 0.1	Klungsøyr *et al*. (1988)
		Callinectes sapidus	muscle	industrial, S. Carolina, USA	0.2 – 7.7	Marcus and Mathews (1987)
		Carcinus mediterraneus	whole	Mediterranean Sea	0.3 – 3.6	Geyer *et al* (1984)
Total DDT	Mollusca	*C. virginica*	whole	Gulf of Mexico, USA	0.001 – 0.079	Sericano *et al*. (1990)
		M. californianus	whole	industrial, S. California, USA	0.02 – 1.05	Young *et al*. (1976)
		M. californianus	whole	urban, industrial, San Francisco Bay, USA (1986 data)	0.01 – 4.49	Phillips and Spies (1988)
Total DDT	Arthropoda	*C. sapidus*	muscle	industrial, S. Carolina, USA	0.005 – 0.17	Marcus and Renfrow (1990)
	Mixed mollusca/ crustaceans	–	whole	industrial	25.0 – 31.6	Webber *et al*. (1989)
Total (non-DDT) pesticides	Mollusca	*C. virginica*	whole	Gulf of Mexico, USA	0.001 – 0.046	Sericano *et al*. (1990)
Tributyltin	Mollusca	mussel species	whole	various	0.04 – 3.42	Page (1989)

[a] Dry weight to wet weight conversion factor of × 0.2 used. For other details of calculations see Livingstone (1991a).
[b] Variously 2–6 rings.

Homarus americanus (James and Barron, 1988). Multiphasic uptake or elimination has usually been interpreted in terms of equilibration of the xenobiotic into, or out of, multiple compartments within the organism (Stegeman and Teal, 1973); another possible explanation is the saturation of different rate-limiting steps at different concentrations of the xenobiotic. Such phenomena presumably could be tissue-, cellular-, subcellular- or molecular-based. For example, the patterns of uptake of picric and picramic acids by *C. virginica* (see before) could be the result of covalent adduct formation via metabolism of their nitro-groups. Uptake can also take place from sediment/abiotic particles and food sources, although the rates are usually somewhat lower than from water, e.g. benzo[a]anthracene (BaA) taken up by the polychaete *Nereis virens* (McElroy, 1990).

The distribution and disposition of a xenobiotic in an organism is largely determined by the route of intake and tissue lipid levels, e.g. in *C. sapidus* exposed to BaP (Lee *et al.*, 1976) or *bis*(tributyltin)oxide (Lee, 1986) the gills were an initial site of bioaccumulation when uptake was from the water, but the stomach was more important when input was through the food. Similarly, in bivalve molluscs, the gills may be an initial site of uptake (Knezovich and Crosby, 1985), but in both molluscs and crustaceans the lipid-rich hepatopancreas or digestive gland is a major site of bioaccumulation (Livingstone, 1991a). Patterns of bioaccumulation and depuration are also affected, directly, by reproductive condition and/or seasonality, temperature, tidal cycle, nutritional state and species feeding habits.

Key factors in persistence are the external concentration, hydrophobicity and chemical structure of the xenobiotic and the exposure time (Livingstone, 1991a). The initial linear rate of uptake and final tissue equilibrium concentration of the chemical increase with increasing concentration in the external medium. The tendency to bioaccumulate increases with increasing hydrophobicity of the chemical and with increasing levels of tissue lipids. For many xenobiotics taken up by bivalve molluscs and crustaceans, linear relationships have been observed between equilibrium BCFs and chemical hydrophobicity measured in terms of either octanol/water partition coefficient (K_{ow}) (linearity observed for log K_{ow} of from about 2 to 6) or water solubility (Ernst, 1977; Geyer *et al.*, 1982; Zaroogian *et al.*, 1985; Hawker and Connell, 1986; Ferraro *et al.*, 1990). Thus compounds of high hydrophobicity such as PCBs and DDT show high BCF values of up to × 50,000 (Geyer *et al.*, 1982). Similarly, the half-life of depuration ($t_{1/2}$) generally increases with increasing hydrophobicity of the chemical, e.g. about 3 hours for 2,4,5-TCP eliminated from *P. pugio* (Ranga Rao *et al.*, 1981) compared to 46 days for DDT from *H. americanus* (Guarino *et al.*, 1974), and can increase with increasing tissue lipid levels, viz.

for various PAHs eliminated from the amphipod *Pontoporeia hoyi* (Landrum, 1988). Deviations from these relationships also occur, particularly for very hydrophobic compounds ($K_{ow} > 6$) which show BCFs and depuration rates less than expected (Hawker and Connell, 1985; Gobias *et al.*, 1986; Donkin and Widdows, 1990; also see Chapter 11).

For compounds of similar hydrophobicities, bioaccumulation can be affected by chemical structure, e.g. PAHs were bioaccumulated less from sediments than were PCBs by *M. edulis* (Pruell *et al.*, 1986), and the BCF for alkyl-substituted benzenes taken up by stage V spot shrimp *Pandalus platyceros* appeared to be related to the degree of alkylation (Sanborn and Malins, 1980). Similarly, for depuration, PCBs are generally more persistent than PAHs, and elimination is dependent on both the degree of chlorination and the structure of the congener. Depuration can also be markedly affected by the duration of previous exposure to the xenobiotic. Short-term exposure results in rapid and complete or almost complete elimination, whereas longer-term exposure is followed by slower and often incomplete elimination. Thus, the $t_{1/2}$ of depuration increases with increasing time of exposure, and the relationship is particularly evident for the elimination of petroleum hydrocarbons from molluscs (see later). Information on what particular aspects of chemical structure, in addition to general hydrophobicity, affect the persistence of xenobiotics in marine invertebrates is limited, but clearly indicated are chemical reactivity and molecular size and shape. Chemical reactivity can increase persistence through the formation of macromolecular adducts and, somewhat paradoxically, of polar metabolites and conjugates, if the elimination of the latter two from the animal is slower than the elimination of the parent compound (see later). Non-metabolism can cause persistence if the parent compound is not eliminated from the animal, as is the case for a number of PCB congeners in many invertebrates and vertebrates (Chapters 6, 7 and 11). Large molecular size and globularity (non-planarity) have been shown to restrict the movement of chemicals across membranes in model systems and fish (see Chapter 11) and the same is indicated for marine invertebrates, e.g. for PCBs (see later).

METABOLISM OF CONTAMINANTS

Biotransformation Enzymes

The metabolism of organic xenobiotics is usually divided into phase I and phase II (Gibson and Skett, 1986; Livingstone, 1991a). Phase I enzymes catalyse virtually every reaction that a compound can undergo (oxidation, reduction, hydration, hydrolysis, etc) and introduce a

TABLE 1.2 *Biotransformation Enzymes in Major Marine Invertebrate Groups*[a]

Enzyme[b]	Polychaeta	Mollusca	Crustacea	Echinodermata
PHASE I				
MFO system				
Cytochrome *P*-450	82	73 ± 10	328 ±70	63 ± 12
MFO activities:				
Aldrin epoxidase	–	8.1 ± 4.2	11.3	–
Benzo[a]pyrene hydroxylase	8.5 ± 6.8	21.2 ± 3.4	40 ± 31	8.4 ± 5.5
O-dealkylase activity	–	0.6–5	4.9–29	–
N-dealkylase activity	–	100–666	15–225	–
Flavoprotein monooxygenase	–	0.1–1	–	+
Epoxide hydratase	+	0.1–13	2.2–22	+
Azo/nitroreductases	–	0.2–2.9	0.03	+
Esterases	+	+	4.5	+
Phosphatases	+	+	4.0	+
Organophosphate acid anhydrase	–	20–100	–	–
β-glucuronidases	+	62,000	+	–
Sulphatases	–	+	+	220
Deacetylase	–	+	–	–
PHASE II				
Glutathione *S*-transferases				
epoxide substrates	–	0.2–16	0.2–3.3	–
other substrates	+	21–10,700	135–5,400	11–52
UDP-glucuronyltransferase	+	7.6–38	0.01	+
UDP-glucosyltransferase	–	+	0.2–1.2	–
Sulphotransferase	+	+	0.53	+
Amino acid conjugase	–	–	+	–
Formylase	–	+	–	–
Methylase	–	+	+	–
Acetylase	–	+	+	+

[a] Data from Den Besten *et al.* (1990) and Livingstone (1990, 1991a) which contain details of tissues, subcellular fractions and enzyme substrates.
[b] Enzyme content/activity in pmol mg^{-1} protein (*P*-450), pmol min^{-1} mg^{-1} protein (MFO activities) or nmol min^{-1} mg^{-1} protein (all others).
Single value, range or mean ± SEM given; –, no information available; +, enzyme activity indicated from bacterial mutagenicity or *in vivo* metabolic studies.

functional group, e.g. –OH, $–NH_2$, –COOH, into the chemical. Phase II enzymes utilize a variety of endogenous molecules (glutathione, sugars, amino acids, sulphate, etc.) to attach polar moieties to the functional groups to produce water-soluble compounds that generally account for the bulk of the excreted products.

Most types of enzymes responsible for phase I and II metabolism which have been identified in mammals are present, or appear to be present, in marine invertebrates (Table 1.2).

Not surprisingly, activities are generally lower than in vertebrates, but nevertheless are in the same proportion to the activities of non-biotransformation enzymes, i.e. the "detoxication potential" is that which would be predicted from the general level of enzymic or metabolic activity, e.g. in molluscs and crustaceans, the mixed-function oxygenase

(MFO) system compared to the enzymes of intermediary metabolism (Moore *et al.*, 1989; Livingstone, 1990). Least is known of the lower invertebrate phyla, such as Porifera and Coelenterata, but key enzymes such as the MFO system, flavoprotein monooxygenase and glutathione *S*-transferase are indicated (Livingstone, 1990, 1991a). Activities of biotransformation enzymes are highest in digestive tissues, which generally are also important sites of contaminant uptake and bioaccumulation, viz. intestines (Polychaeta), digestive gland (Mollusca), hepatopancreas (Crustacea) and pyloric caeca (Asteriodea). Some enzymes appear to have a wider tissue distribution, e.g. glutathione *S*-transferases, compared to the MFO system. Multiple forms of enzymes (isoenzymes) exist, e.g. cytochrome *P*-450 and glutathione *S*-transferase, but possibly in fewer numbers than in vertebrates (Livingstone, 1990). Endogenous functions are evident for a number of the biotransformation enzymes (Livingstone, 1991a).

In Vivo Metabolism; Qualitative Aspects

Studies on the metabolism of some 68 xenobiotics by 63 species from the Porifera, Coelenterata, Platyhelminthes, Polychaeta, Mollusca, Crustacea and Echinodermata have been described (Table 1.3).

The action of most types of phase I and II enzymes are evident, resulting in the formation of polar metabolites, conjugates and macromolecular adducts. Multiple metabolites are generally produced, e.g. *o*-toluidine was converted to 2-nitrosotoluene, *N*-methyl-*o*-toluidine, *N*-formyl-*o*-toluidine and small amounts of *N*-hydroxy-*o*-toluidine in the common mussel *Mytilus edulis* (Knezovich and Crosby, 1985). Sulphated conjugate and glucoside formation possibly predominate over glucuronide formation, e.g. metabolism of *p*-nitrophenol derivatives by various crustacea (Foster and Crosby, 1986, 1987) and 3-hydroxy-BaP by the spiny lobster *Panulirus argus* (Schell and James, 1989), but the overall database for this conclusion is limited. Glutathione *S*-conjugates have not been isolated, but are indicated from mercapturic acid pathway and other studies, e.g. mercapturic acids were detected or indicated for the metabolism of octachlorostyrene by *M. edulis* (Bauer *et al.*, 1989) and 2,6-dimethylnaphthalene by the sea urchin *Strongylocentrotus droebachiensis* (Malins and Roubal, 1982). Other pathways identified in various species include dearylation (methyl parathion metabolism), dechlorination (DDT, pentachlorophenol), amino acid conjugation (phenylacetic acid) and oxon formation (fenitrothion) (Feroz *et al.*, 1990; Livingstone, 1991a). Species differences in the metabolism of the same xenobiotic are evident, e.g. the major route of metabolism of *p*-nitroanisole by the starfish *Strongylocentrotus purpuratus* was reduction and subsequent *N*-acetylation (Crosby *et al.*, 1979) compared to mainly

TABLE 1.3 In Vivo *Metabolism of Organic Xenobiotics by Marine Invertebrates: List of Compounds Studied (see Livingstone, 1991a for Original References and Other Details, Including Calculations of* In Vivo *Rates of Metabolism)*[a]

Chemical	Animal group[b] (number of species in parenthesis)
Hexadecane	Crust (1)
Dotriacontane	Crust (1)
Naphthalene	Cnid (2); Moll (3); Crust (6)
Methylnaphthalene	Moll (1); Crust (1)
2,6-dimethylnaphthalene	Echin (1)
Anthracene	Crust (2)
Phenanthrene	Poly (1); Moll (1); Crust (1)
Benz[a]anthracene	Poly (1)
Fluorene	Crust (1)
3-methylcholanthrene	Cnid (2)
Benzo[a]pyrene	Por (1); Cnid (2); Poly (1); Moll (3); Crust (7)
Chrysene	Poly (1)
Chlorinated paraffins	Moll (1)
1-naphthol	Moll (1); Crust (1)
Biphenyl	Moll (2); Crust (2); Echin (1)
4-monochlorobiphenyl	Poly (1)
2,2′-dichlorobiphenyl	Poly (1)
2,5,4′-trichlorobiphenyl	Poly (1)
2,4,2′,4′-tetrachlorobiphenyl	Poly (1)
2,4,6,2′,4′-pentachlorobiphenyl	Poly (1)
2,4,5,2′,4′,5′-hexachlorobiphenyl	Poly (1)
Aldrin	Cnid (1); Plat (1); Ann (1); Moll (4); Crust (6)
Chlordane	Moll (1)
Dieldrin	Moll (1)
1,1-*bis*(4-chlorophenyl)-2,2,2-trichloroethane (DDT)	Plat (4); Ann (2); Mol (1); Crust (4)
1,2,3,4,5,6-γ-hexachlorocyclohexane (lindane)	Poly (1)
Aniline	Moll (2); Crust (1)
2,6-diethylaniline	Moll (1); Crust (1)
2-acetylaminofluorene	Moll (1)
2-aminofluorene	Moll (1)
Anisole	Moll (1); Crust (1)
Antipyrine	Moll (1)
Benzidine	Moll (1); Crust (1)
Benzoic acid	Moll (1); Crust (1)
Chlorobenzene	Moll (1); Crust (1)
Cypermethrin	Ann (1)
Dibutylphthalate	Moll (1); Crust (1)
Dimethylphthalate	Crust (1)
Di(2-ethylhexyl)phthalate	Moll (1); Crust (1)
Ethoxychlor	Moll (2)
Fenitrothion	Moll (2); Crust (3)
Hexachlorobenzene	Moll (1); Crust (1)
Methoxychlor	Moll (2)
Nitrobenzene	Moll (1); Crust (1)
p-nitroanisole	Moll (4); Crust (3); Echin (3)
4-nitro-m -cresol	Crust (3)

TABLE 1.3 *Continued*

Chemical	Animal group[b] (number of species in parenthesis)
p-nitrophenetole	Moll (2)
p-nitrophenol	Echin (2)
p-nitrotoluene	Echin (1)
Parathion	Ann (1)
Methylparathion	Crust (3)
Pentachlorophenol	Moll (3); Crust (2)
2,4,6-trinitrophenol (picric acid)	Moll (1)
2-amino-4,6-dinitrophenol (picramic acid)	Moll (1)
Phthalic anhydride	Moll (1); Crust (1)
Phenylacetic acid	Crust (1)
2,4-dichlorophenoxyacetic acid	Crust (1)
2,4,5-trichlorophenoxyacetic acid	Crust (1)
bis(4-chlorophenyl)acetic acid	Crust (1)
3,5,6-trichloro-2-pyridinol	Moll (1); Crust (1)
Trifuralin	Moll (1); Crust (1)
p-toluidine	Moll (1); Echin (2)
o-toluidine	Moll (2)
N-acetyl-*o*-toluidine	Moll (1)
Sulfadimethoxine	Crust (1)
Vinyl chloride	Moll (1); Crust (2)
Bis(tributyl)tin oxide	Moll (1); Crust (2)
Phenylmercuric acetate	Moll (1)

[a] Includes some non-marine and terrestrial invertebrates; in some cases no metabolism was detected (see Livingstone, 1991a).
[b] Por: Porifera; Cnid: Cnidaria; Plat: free living Platyhelminthes; Poly: Polychaeta; Ann: Oligochaeta and Huridinae; Moll: Mollusca; Crust: Crustacea; Echin: Echinodermata.

oxidative pathways in the starfish *Pisaster ochraceus* and the sea cucumber *Cucumaria miniata* (Landrum and Crosby, 1981a); biphenyl was converted primarily to 2-hydroxybiphenyl by the whelk *Buccinum undatum* (Meyer and Bakke, 1979) and to 4-hydroxybiphenyl by the snail *Lymnaea stagnalis* (Livingstone, 1991a).

In Vivo Metabolism: Quantitative Aspects

Using literature data, whole body tissue concentrations of parent compound and rates of metabolism have been calculated for the biotransformation of a large number of xenobiotics by various marine and freshwater species (Table 1.3). Most information was available for molluscs and crustaceans, and less for polychaetes and echinoderms. The exposure times for calculating the rates of metabolism varied, but were mostly 48 hours or less: information about the dependence of metabolism on exposure time is limited, but rates were approximately linear for 48 hours for the metabolism of naphthalene by the crab

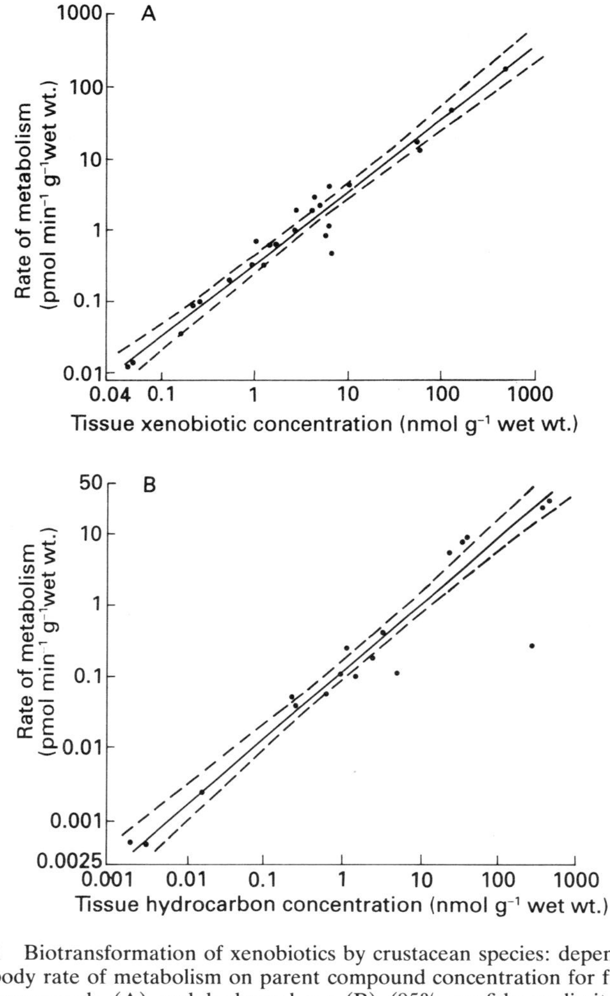

FIG. 1.1 Biotransformation of xenobiotics by crustacean species: dependence of whole body rate of metabolism on parent compound concentration for functional group compounds (A) and hydrocarbons (B) (95% confidence limits for the regression lines shown) [from Livingstone (1991a) with permission]. A: pooled data for metabolism of variously aniline, anisole, benzoic acid, chlorobenzene, nitrobenzene, *p*-nitroanisole, 4-nitro-*m*-cresol, methyl parathion, phthalic anhydride, phthalate esters, hexachlorobenzene, pentachlorophenol, 2,6-diethylaniline, 3,5,6-trichloropyridinol and *bis*(tributyl)tin oxide by *Sicyonia ingentis, Macrobrachium rosenbergii, Procambrus clarkii, Daphnia magna* and *Penaeus aztecus*; corr. coeff. = 0.97 (n = 24). B: pooled data for metabolism of variously naphthalene, anthracene, phenanthrene, hexadecane, fluorene and benzo[a]pyrene by *Uca pugnax, Hemigrapsus nudus, Callinectes sapidus, Panulirus argus* (hepatopancreas only), *Nephrops norvegicus, Calanus helgolandicus, Pandalus platyceros, Hyalela azteca, Rhepoxynius abronius* and *Eohaustorius washingtonianus*; corr. coeff. = 0.98 (n = 17); also shown, but not included in the regression, is the metabolism of benzo[a]pyrene (hepatopancreas only) by *Homarus americanus* (★).

Uca pugnax (Burns, 1976) and BaP (incorporation into protein) by the amphipod *Rhepoxynius abronius* (Reichert *et al.*, 1985). For studies using radiolabelled xenobiotic, "unextractable radioactivity" was interpreted as macromolecular adduct formation, as opposed to being simply tightly bound parent compound, and was included in the rate calculations (inclusion of this factor increased rates of xenobiotic metabolism, but did not affect the overall picture; see below). The data were pooled to examine the dependence of rate of metabolism on tissue xenobiotic concentration for different groups of xenobiotic and animals. A distinction was made between hydrocarbons and other xenobiotics already containing functional groups (termed functional group compounds) such as aromatic amines, nitroaromatics, phenols and others. Whereas hydrocarbons require initial metabolism by the MFO system, functional group commands can be acted on immediately by many different phase I and II enzymes.

Despite the variety of species and chemicals pooled in the regressions, and the other variables and assumptions involved in the individual calculations of rate and tissue concentration (see Livingstone, 1991a), rates of metabolism linearly increased with tissue xenobiotic (substrate) concentration, over up to six orders of magnitude of the two parameters, for the metabolism of functional group compounds by crustacea and molluscs (Figs. 1.1A and 1.2A) and of hydrocarbons by crustacea (Fig. 1.1B).

From fewer data, the same relationship was observed or indicated for the metabolism of "hydrocarbons" by molluscs (hydrocarbon metabolism is discussed in more detail in the next section) (Fig. 1.2B), BaA by *N. virens* and nitroaromatics by the echinoderm *S. purpuratus* (Table 1.4). With the exception of hydrocarbon metabolism by molluscs, the correlation coefficients for the regressions were all 0.96 or greater (Table 1.4).

Increased rates of metabolism with increased tissue xenobiotic concentrations have also been seen for other single compounds and species, viz. sulphated pentachlorophenol formation by *M. edulis* (Ernst, 1979a), phthalate esters by *C. virginica* and the brown shrimp *Penaeus aztecus* (Wofford *et al.*, 1981) and BaP by the amphipods *Eohaustorius washingtonianus* and *R. abronius* (Reichert *et al.*, 1985; Varanasi *et al.*, 1985) (see Livingstone, 1991a for calculations).

Functional group compounds are metabolized faster than hydrocarbons by both crustaceans and molluscs (Figs. 1.1 and 1.2), and hydrocarbon metabolism is indicated to decrease in the order crustaceans $\geqslant$ polychaete (*N. virens*) $>$ molluscs. This can be illustrated by use of the regression equations to compare rates of metabolism at a fixed tissue concentration of xenobiotic, viz. for a latter concentration of 10 nmol g^{-1} wet weight, mean rates in pmol min^{-1} g^{-1} wet weight are: functional

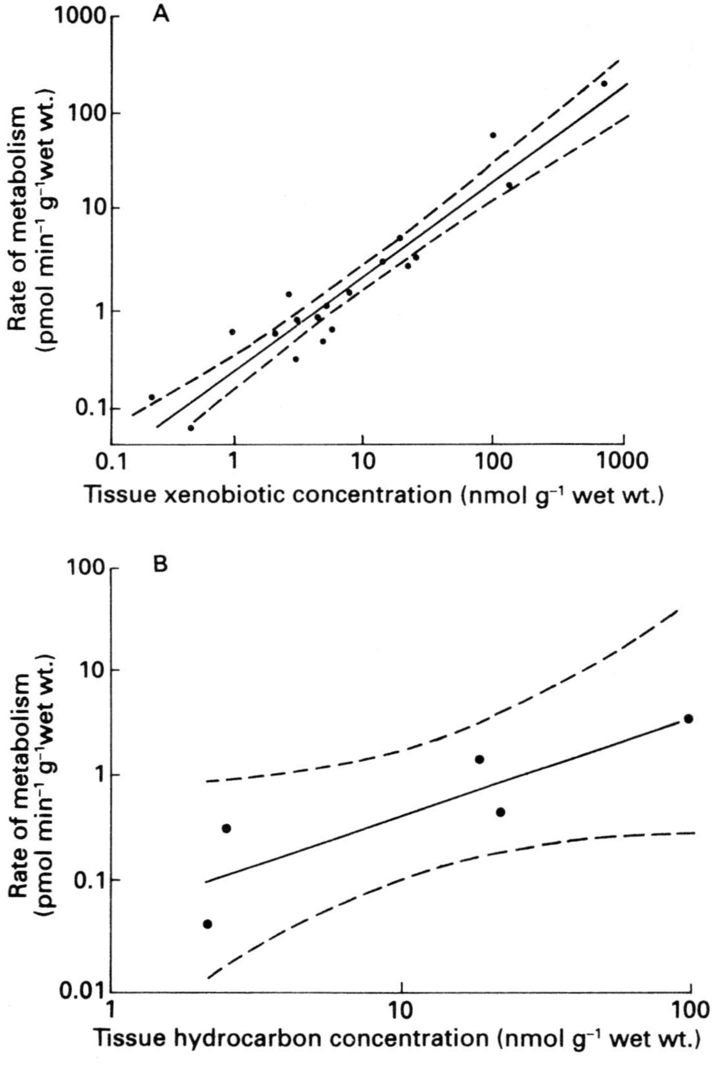

FIG. 1.2 Biotransformation of xenobiotics by molluscan species: dependence of whole body rate of metabolism on parent compound concentration for functional group compounds (A) and "hydrocarbons" (B) (95% confidence limits for the regression lines shown) [Fig. 1A from Livingstone (1991a) with permission]. A: pooled data for metabolism of variously aniline, diethylaniline, *o*-toluidine, benzidine, benzoic acid, nitrobenzene, picric acid, picramic acid, anisole, *p*-nitroanisole, chlorobenzene, pentachlorophenol, aldrin, trichloropyridinol, phthalic anhydride, phthalic esters and phenylmercuric acetate by *Mytilus edulis, Crassostrea gigas, Crassostrea virginica, Anondonta* sp., *Physa* sp., *Lymnaea pallustris, Helisoma campanulata* and *Cryptochiton stelleri*; corr. coeff. = 0.96 (n = 20). B: pooled data for metabolism of variously naphthalene, benzo[a]pyrene, chlorinated hydrocarbons and hexachlorobenzene by *M. edulis, Macoma inquinata, Ostrea edulis* and *Physa* sp.; corr. coeff. = 0.87 (n = 5).

TABLE 1.4 *Comparative Rates of Xenobiotic Metabolism by Marine and Other Invertebrates Calculated for a Theoretical Tissue Xenobiotic Concentration of 10 nmol g^{-1} wet weight*[a]

Type of compound	Number of chemicals in regression	Animal group	Number of species in regression	Regression equation[b] type	intercept (S.E.[c])	slope (S.E.[c])	corr. coeff. (*n*)	Rate of metabolism (pmol min^{-1} g^{-1} wet wt) (S.E.[c])
Functional group xenobiotics	17	Crustacea	5	Log_{10}	−0.46 (0.05)	1.01 (0.05)	0.97 (24)	3.55 (2.82–4.47)
	18	Mollusca	8	Log_{10}	−0.51 (0.08)	0.94 (0.07)	0.96 (20)	2.69 (1.91–3.80)
Hydrocarbons	6	Crustacea	10	Log_{10}	−0.90 (0.07)	0.93 (0.04)	0.98 (17)	1.07 (0.83–1.38)
	4	Mollusca	4	Log_{10}	−1.31 (0.37)	0.92 (0.30)	0.87 (5)	0.41 (0.09–1.91)
Benz[a]anthracene	1	Polychaeta[d]	1	Linear	0.038 (0.054)	0.077 (0.018)	0.95 (4)	0.81 ± 0.23
Nitroaromatics[e]	2	Echinodermata[f]	1	Linear	−5.55 (1.46)	1.75 (0.06)	0.99 (4)	11.95 ± 2.06

[a] Taken from Livingstone (1991a): for details of xenobiotics and species see Figs 1.1 and 1.2.

[b] Log_{10} type: log_{10} rate = intercept + (slope × log_{10} tissue xenobiotic concentration). Linear type: rate = intercept + (slope × tissue xenobiotic concentration); rate of metabolism and tissue concentration of xenobiotic in, respectively, pmol min^{-1} g^{-1} wet weight and nmol g^{-1} wet weight.

[c] Standard error, given in parenthesis as ± S.E. (linear type) or range for 2 S.E. (log_{10} type).

[d] *Nereis virens*.

[e] *p*-nitroanisole and *p*-toluidine.

[f] *Strongylocentrotus purpuratus*.

group compounds, 3.6 (crustacea) and 2.7 (molluscs); hydrocarbons, 1.1 (crustaceans), 0.8 (polychaete) and 0.4 (molluscs) (Table 1.4). The indicated lower rate of metabolism of hydrocarbons compared to functional group compounds is presumably due either to the MFO system being rate-limiting (catalyses epoxidation of C=C and hydroxylation of C-H in hydrocarbons), or to several enzymes being able to act at once on the functional group compounds. Although (with the exception of the MFO system) comparative biotransformation enzyme data on crustaceans and molluscs is limited (Table 1.2), the generally lower levels of xenobiotic metabolism in molluscs compared to crustaceans is consistent with their lower levels of general (non-biotransformation) enzymic activities (Moore *et al.*, 1989). Not surprisingly, maximum observed rates of *in vivo* metabolism are significantly less than calculated whole body maximal enzyme activities, e.g. for molluscs, activities of the flavoprotein monooxygenase system, *N*-dealkylases, reductases and phase II enzymes range from 346 to 34,200 pmol min^{-1} g^{-1} compared to *in vivo* metabolism of functional group compounds of up to 100 pmol min^{-1} g^{-1} (Livingstone, 1991a). It is interesting to note that in none of the pooled examples of Table 1.4 has the maximum rate of *in vivo* metabolism been reached (Figs. 1.1 and 1.2), indicating a greater potential for xenobiotic metabolism with increased tissue substrate concentration. Alternatively, the tissue concentrations of xenobiotics may become lethal before such maximum rates of metabolism are reached. A relatively high rate of metabolism was obtained for the biotransformation of nitroaromatics by the echinoderm *S. purpuratus* (Table 1.4), but this is for a specific group of xenobiotics.

The relationships between rates of metabolism and tissue xenobiotic concentration, as described in Figs. 1.1 and 1.2 and Table 1.4, are very generalized and obviously must be used with caution. Exceptions, falling well outside the 95% confidence limits of the regressions, will exist with respect to particular compounds and species, e.g. the lobster *Homarus americanus* metabolized BaP about ×100 slower than did *P. argus* (Foureman *et al.*, 1978; Little *et al.*, 1985; see star on Fig. 1.1B). Less dramatic differences, generally falling within the 95% confidence limits, are also seen, e.g. *R. abronius* metabolized BaP faster than the other amphipod species *E. washingtonianus* (Varanasi *et al.*, 1985), and the red swamp crayfish *Procambrus clarkii* metabolized *p*-nitroanisole faster than the ridgeback prawn *Sicyonia ingentis* (Foster and Crosby, 1986, 1987; Livingstone, 1991a). However, the relationships clearly demonstrate that rates of *in vivo* metabolism of xenobiotics are dependent on their tissue concentrations, and that, for example, statements on comparative rates of xenobiotic metabolism of different species, without reference to the tissue xenobiotic concentrations, can be somewhat misleading. Furthermore, the relationships could have application in modelling studies (Chapter 11), at least in giving first approximations of rates of xenobiotic metabolism. Further experimental studies would obviously refine and extend the application of such relationships.

Metabolites have been found in many tissue types but generally are present in highest levels in those tissues containing the highest activities of biotransformation enzymes, e.g. hepatopancreas of crustaceans and digestive gland of molluscs (Livingstone, 1991a). Whole body rates of xenobiotic metabolism will depend not only on tissue levels of enzyme activity, their properties and regulation (inducibility), and the availability of the substrate to them, but also on the percentage composition of the different tissues, e.g. generally, the pyloric caeca makes up a much bigger proportion of the total tissues in starfish than does the digestive gland in mussels.

BIOACCUMULATION: INTEGRATED ASPECTS

Rate of Uptake Versus Rate of Metabolism

Initial linear rate of uptake and final tissue equilibrium concentration of a xenobiotic both increase with increasing exposure concentration of the chemical in the water column. The former two parameters are obviously related, and for the uptake of hydrocarbons the relationship is seen for either pooled data or for studies on single species and chemicals, viz. for crustacea (Fig. 1.3) and molluscs (Fig. 1.4). Not surprisingly, the correlation coefficients of the regressions are lower for the pooled data, reflecting the differences in species physiology and other variables in the different studies.

The final tissue xenobiotic concentration will determine the maximum rate of metabolism of the xenobiotic (see before). Therefore, using the regression equations of Figs. 1.3 and 1.4 (tissue concentration vs. uptake) and Figs. 1.1B and 1.2B (tissue concentration vs. metabolism) it is possible to compare, at a fixed concentration of hydrocarbon, rate of uptake with rate of metabolism (Table 1.5).

Comparing the two processes, it can be seen that rate of uptake markedly exceeds rate of metabolism, particularly in molluscs, so accounting for the bioaccumulation of hydrocarbons and the overall process being describable in terms of a simple lipid/water equilibrium model (Burns and Smith, 1981; Hawker and Connell, 1986; Ferraro *et al.*, 1990). Strictly speaking, the rates of uptake as measured or calculated are rates of bioaccumulation (uptake minus simultaneous depuration), but this would mean that the former are underestimates which would not affect the argument. A similar imbalance in uptake

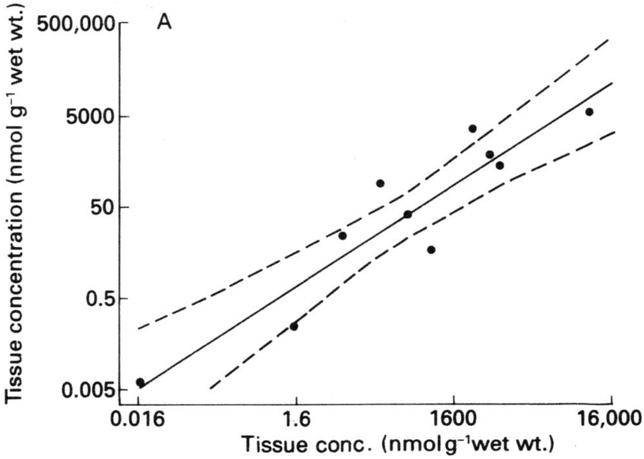

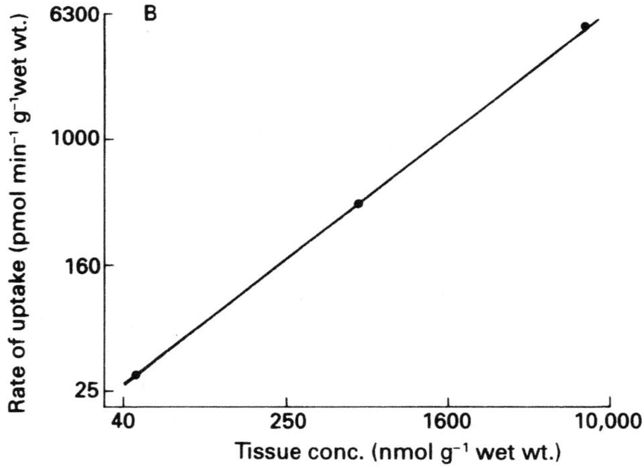

FIG. 1.3 Relationship between rate of uptake and tissue equilibrium concentration for various hydrocarbons bioaccumulated by crustacean species. A: pooled literature data from table 17 of Livingstone (1991a) on uptake of naphthalene, anthracene and benzo[a]pyrene by variously *Callinectes sapidus, Hemigrapsus nudus, Calanus helgolandicus, Eurytemora affinis, Pandalus platyceros, Penaeus aztecus* and *Hyallela azteca* (95% confidence limits for the regression line shown, corr. coeff. $= 0.94, n = 10$); B: calculated from Corner *et al.* (1976) for the uptake of naphthalene by *C. helgolandicus*.

and metabolism is indicated for other xenobiotics, e.g. rate of metabolism was 6.4% of rate of uptake for *M. edulis* exposed to *o*-toluidine; similar differences are indicated for the bioaccumulation of *p*-toluidine and *p*-nitroanisole by *S. purpuratus* [see Livingstone (1991a)].

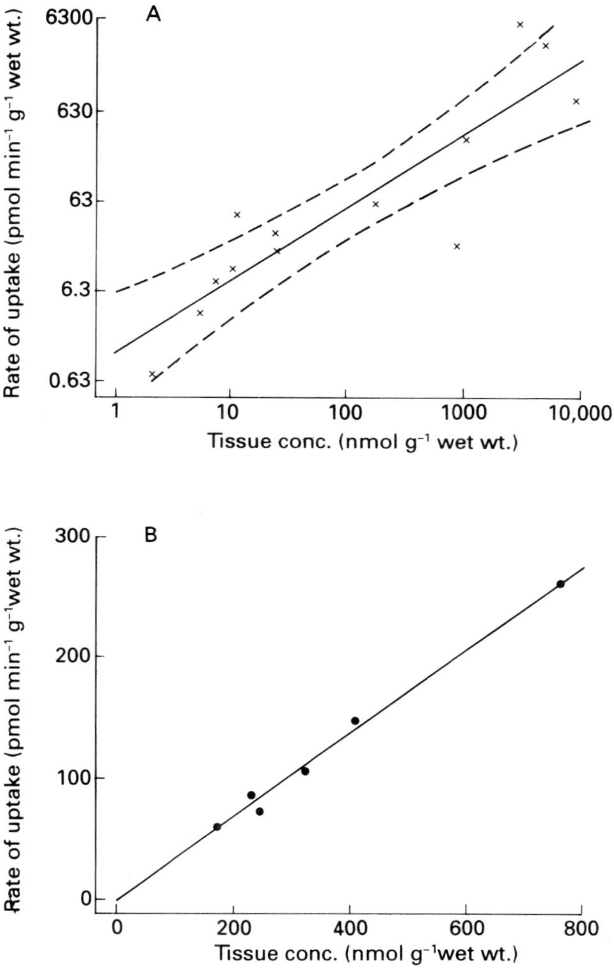

FIG. 1.4 Relationship between rate of uptake and tissue equilibrium concentration for various hydrocarbons bioaccumulated by molluscan species. A: pooled literature data from table 8 of Livingstone (1991a) on uptake of naphthalene, phenanthrene, anthracene, perylene, benzo[a]pyrene, mixed aliphatic/PAHs and heptadecane by variously *Mytilus edulis, Crassostrea virginica, Ostrea edulis, Rangia cuneata* and *Macoma balthica* (95% confidence limits for the regression line shown, corr. coeff. = 0.89, $n = 13$); B: calculated from Stegeman and Teal (1973) for the uptake of mixed hydrocarbons by *C. virginica*.

Macromolecular Adduct Formation

The formation of macromolecular adducts from xenobiotics via biotransformation is indicated to be a widespread phenomenon in marine invertebrates, although caution is required with some of the data, as "non-extractable" radioactivity derived from the radiolabelled xenobiotic could be due to very tightly bound, rather than covalently bound, parent compound (Table 1.6). However, adducts are identified bound to protein, DNA and RNA.

TABLE 1.5 *Bioaccumulation of Hydrocarbons by Crustaceans and Molluscs: Rates of Uptake Compared to Rates of Metabolism (both in pmol min^{-1} g^{-1} wet wt) Calculated for a Tissue Hydrocarbon Concentration of 10 nmol g^{-1} wet weight*

Organism	Rate of uptake[a]	Rate of metabolism[a]
Crustacea	5.78 (2.15–15.6)	1.07 (0.83–1.38)
Calanus helgolandicus	6.95	–
Molluscs	8.40 (3.16–22.3)	0.41 (0.09–1.91)
Crassostrea virginica	3.45	–

[a] Mean value, plus range for 2 standard errors given in parenthesis; rate of uptake calculated from regression equations of Figs. 1.3 and 1.4; rate of metabolism from Table 1.5.

TABLE 1.6 *Macromolecular Adduct Formation in Marine Invertebrates*[a]

Organism	Chemical	Adduct[b]	Percent of total metabolism[c]
PORIFERA			
Tethya lycurium	benzo[a]pyrene	DNA, RNA, protein	–[d]
POLYCHAETA			
Nereis virens	benz[a]anthracene	unidentified	44
Neanthes arenaceodentata	naphthalene	unidentified	–
Abarenicola pacifica	aromatic hydrocarbons	unidentified	–
N. virens	4-monochlorobiphenyl	unidentified	76
MOLLUSCA			
Mytilus edulis	chlorinated paraffins	protein	–
Macoma inquinata	naphthalene	unidentified	–
Physa sp.	benzo[a]pyrene	unidentified	18
Physa sp.	vinyl chloride	unidentified	86
M. edulis	2-acetylaminofluorene	protein	–
Physa sp.	benzidine	unidentified	80
Crassostrea virginica	picric acid	unidentified	–
C. virginica	picramic acid	unidentified	–
C. virginica	phthalate esters[e]	unidentified	6 to 17
Cipangopaludina japonica	fenitrothion	unidentified	trace
CRUSTACEA			
Callinectes sapidus	naphthalene	unidentified	–
Hyalela azteca	anthracene	unidentified	42
C. sapidus	benzo[a]pyrene	unidentified	–
Rhepoxynius abronius	benzo[a]pyrene	protein	–
Eohaustorius washingtonianus	benzo[a]pyrene	protein	–
Daphnia magna	vinyl chloride	unidentified	54
Procambrus clarkii	nitroaromatics[f]	unidentified	3 to 20
Sicyonia ingentis	nitroaromatics[f]	unidentified	3 to 5
Macrobrachium rosenbergii	nitroaromatics[f]	unidentified	2

TABLE 1.6 *Continued*

Organism	Chemical	Adduct[b]	Percent of total metabolism[c]
Palaemonetes pugio	pentachlorophenol	unidentified	15
Penaeus aztecus	phthalate esters[g]	unidentified	8 to 18
Palaemon paucidens	fenitrothion	unidentified	–
ECHINODERMATA			
Strongylocentrotus droebachiensis	2,6–dimethyl-naphthalene	unidentified	34
Strongylocentrotus purpuratus	*p*–nitroanisole	unidentified	–
S. purpuratus	*p*–toluidine	unidentified	–

[a] From Livingstone (1991a) to which the reader is referred for original references.
[b] Unidentified refers to non-extractable radioactivity derived from radiolabelled parent compound.
[c] Adduct formation as a % of total budget of (adducts + polar metabolites + conjugates).
[d] No information.
[e] dibutylphthalate and di(2-ethylhexyl)phthalate.
[f] *p*–nitroanisole, 4–nitro–*m*–cresol and methyl parathion.
[g] As for ref. e plus dimethylphthalate.

Benzo[a]pyrene was indicated to be bound to protein and DNA in *M. edulis, A. rubens* and the crab *Carcinus maenas* (Marsh, Chipman, Den Besten, O'Hara and Livingstone, unpublished data), and to protein in microsomal incubations of digestive gland of *M. edulis* (Livingstone *et al.*, 1990a). DNA adducts were formed in incubations of digestive gland homogenates of *Mytilus galloprovincialis* with BaP and aminofluorene (Kurelec *et al.*, 1988).

Depending on the existence and efficiency of excision mechanisms for adduct removal, of which virtually nothing is known in marine invertebrates, their formation will presumably contribute to the persistence of pollutants in these organisms. Although the database is very limited, there is possibly some indication that hydrocarbons and related compounds are more likely to form adducts than some functional group compounds such as esters and nitroaromatics (Table 1.6). However, species-dependent differences in biotransformation pathways will also obviously be important, e.g. "non-extractable" radioactivity derived from radiolabelled nitroaromatics was more evident for echinoderm than for crustacean species (Livingstone, 1991a).

Metabolite Versus Parent Compound Release

The paradigm of mammalian toxicology essentially states that a major function of biotransformation is to produce water-soluble end products, so expediting removal of the xenobiotic from an organism. To an extent this also appears true for marine invertebrates, given that metabolites and conjugates are detected in the seawater in which the animals are exposed (Livingstone, 1991a). However, in most cases examined, the rate of loss of polar end products is slower than that of the parent compound, this being particularly evident if exposed animals are subsequently transferred to clean seawater. The examples include the metabolism of BaA (McElroy, 1990), PCBs (Ernst *et al.*, 1977) and lindane (Goerke and Ernst, 1980) by *N. virens*, 1-naphthol by *M. edulis* (Ernst, 1979b), 2,4-dichlorophenoxyacetic acid, 2,4,5-trichlorophenoxyacetic acid, *bis*(4-chlorophenyl)acetic acid and phenylacetic acid by *P. argus* (James, 1982), sulfadimethoxine by *H. americanus* (James and Barron, 1988), naphthalene and various other hydrocarbons by *C. sapidus* (Lee *et al.*, 1976), *P. playtceros* (Sanborn and Malins, 1977) and *H. nudus* (Laurén and Rice, 1985), *p*–nitroanisole by *S. purpuratus* (Landrum and Crosby, 1981b), and biphenyl by *Ophiocomina nigra* (Meyer and Bakke, 1977). Biotransformation usually continues during depuration, resulting, in many cases, in an increased build-up of metabolites in the tissues during this period, e.g. *H. nudus* exposed to naphthalene. In this situation, therefore, biotransformation leads to the persistence of derivatives of the xenobiotic. The difference with higher organisms in this respect presumably relates to the situation that polar molecules can only pass through lipophilic membranes with difficulty, and carrier/transport and excretory systems, less developed in marine invertebrates, are required to effect their elimination. Information on these aspects in marine invertebrates is limited. Elimination routes include faeces and "non-particulate" mechanisms in polychaetes, and faeces and urine in crustaceans (Livingstone, 1991a). Binding sites for 2-acetylaminofluorene, and presumably other xenobiotics, have been detected in membrane vesicles from digestive gland, gills and mantle of the freshwater mussel *Anodonta cygnea* (Kurelec and Pivcevic, 1989). "Enterohepatic cycling" of metabolites (phase III metabolism), so delaying their elimination, has also been suggested, e.g. sulfadimethoxine metabolism in *H. americanus* (Barron and James, 1988).

Induction of Biotransformation Enzymes

There is evidence for limited and variable induction of some enzymes by exposure to organic contaminants in marine invertebrates, e.g. MFO system, glutathione *S*–transferase and sulphotransferase (Livingstone, 1991a, b). It is not known whether such increases in enzyme activities have any significant effect on the *in vivo* metabolism of xenobiotics. Increased biotransformation of 2,5,2′,5′-tetrachlorobiphenyl was seen in *Nereis diversicolor* from a contaminated compared to a clean field site, indicating effective induction of the MFO system (Goerke and

Weber, 1990). Increased resistance of the clam *Tapes philippinarum* to pentachlorophenol was correlated with an increase in sulphotransferase activity (Kobayashi, 1985).

Major Pollutant Groups

Polychlorinated biphenyls and Related Compounds

Uptake and depuration of PCBs have been largely studied in polychaetes, molluscs and echinoderms, but their metabolism has only been investigated in polychaetes. Some of these studies, particularly on nereid worms, are described in detail in Chapter 2. Inferences on metabolism have also been made from differences in congener profiles in tissues compared to exposure sources.

Bioaccumulation and depuration are dependent on both degree of chlorination and congener structure. Depuration half-lives (and retention of the compounds) generally increased in bivalves, with increasing hydrophobicity or lipophilicity (chlorination) of the congeners (Table 1.7). In contrast, PCBs with 5 chlorine atoms per biphenyl were accumulated to a greater extent than lower or higher chlorinated congeners (Langston, 1978a), and those with *ortho*-substituted chlorines were retained least (Langston, 1978b), by *M. balthica* and the cockle *Cerastoderma edule*. Similarly, non-*ortho* coplanar congeners were taken up much more slowly than many other ones by the green-lipped mussel *Perna viridis* (Kannan *et al.*, 1989) (see also Chapter 4). Studies on nereid worms revealed the following order of loss from the tissues of mono-, di-, tri-, tetra-, penta- and hexa-chlorobiphenyls (respectively, MCBP, DCBP, TCBP, TeCBP, PCBP and HCBP): *N. virens*: 2,5,4′-TCBP > 2,2′-DCBP > 2,4,6,2′,4′-PCBP (Goerke and Ernst, 1977); 2,3,4,2′,5′-PCBP > 2,4,5,2′,5′-PCBP > 2,4,6,2′,4′-PCBP (Goerke and Ernst, 1986); 4-MCBP was released but 2,4,2′,4′-TeCBP and 2,4,5,2′,4′,5′-HCBP remained high during 8 weeks depuration (McElroy and Means, 1988); *N. diversicolor*: 2,4,5,2′,5′-PCBP > 2,5,2′,5′-TeCBP > 2,3,4,2′,5′-PCBP > 2,3,2′,5′-TeCBP > 2,3,6,2′,5′-PCBP > 2,4,5,2′,4′,5′-HCBP (Goerke and Weber, 1990). Some selective bioaccumulation from contaminated food by *A. rubens* was seen, with possibly a discrimination against certain TeCBP (IUPAC number CB-70) and heptachlorobiphenyls (CB-180) compared to PCBPs and HCBPs (Den Besten *et al.*, 1990). Bioaccumulation by *A. rubens* was seasonally variable (related to lipid levels), and PCBs were present in oocytes and spermatozoa at levels comparable to those of whole gonads.

Several studies have examined metabolism of PCBs in *N. virens*. Metabolism of 4-MCBP occurred but not of 2,4,2′,4′-TeCBP or

TABLE 1.7 *Approximate Depuration Half-Lives of Polychlorinated biphenyls Taken Up by Bivalves*[a]

Species	Degree of chlorination[b]	Half-life (days)	Reference
Cerastoderma edule	3	~7	Langston (1978b)
	4	7–14	
	5	21	
	6	>21	
Perna viridis	2	0.5–2.5	Tanabe *et al.* (1987)
	3	0.5–6.5	
	5	4.9–8.3	
Mytilus edulis	3	16.3	Pruell *et al.* (1986)
	5	27.9	
	6	37–46	
M. edulis	2–4	4.6–9.1	Calambokis *et al.* (1979)
	5–7	20–50	

[a] Half-lives are direct from reference or approximate from body burden data.
[b] Number of chlorine atoms per PCB molecule.

2,4,5,2′,4′,5′-HCBP (McElroy and Means, 1988): rates of metabolism followed the order 2,5,4′-TCBP > 2,2′-DCBP > 2,4,6,2′,4′-PCBP (Ernst *et al.*, 1977). Polar metabolites, conjugates and macromolecular adducts were indicated. Trichlorobiphenylol was the main metabolite of 2,5,4′-TCBP, with probably isomeric hydroxy- and dihydroxycompounds present in small amounts. Calculated rates of metabolism for the different congeners varied from 0.57 to 0.03 pmol min^{-1} g^{-1} for tissue substrate concentrations of 6.3 to 0.2 nmol g^{-1} (Livingstone, 1991a). The rates are thus not markedly lower than for the metabolism of BaA by *N. virens*, viz. 0.8 pmol min^{-1} g^{-1} at tissue concentration of 10 nmol g^{-1} (Table 1.3). Slow elimination of some PCBPs, as well as low occurrence of HCBPs and hepta-CBPs highly chlorinated in one ring only, indicate biotransformation pathways in *N. diversicolor* not yet described for any other organism (Goerke and Weber, 1990). Metabolism of particular congeners (vicinal H atoms in *meta*- and *para*-positions) was indicated to be more marked in *N. diversicolor* than in *M. balthica* (Boon *et al.*, 1989). In contrast to the errant nereids, no metabolism of DCBP, TCBP and PCBP was observed in the sedentary polychaete *Lanice conchilega* (Ernst, 1977). Metabolism of 2,5,3′,4′-TeCBP (CB-70), but not of other potentially metabolizable PCBs (CB-52, 87 and 101; Boon *et al.*, 1989), was indicated in *A. rubens*, particularly in the pyloric caeca (Den Besten *et al.*, 1990).

Congener profiles for PCDFs and PCDDs were very different in *M. edulis* from the likely main source of contamination, municipal fly ash; the former containing mainly tetra-PCDFs and tetra-PCDDS compared to predominantly higher chlorinated congeners in the latter (Miyata *et al.*, 1989). Nothing is known of the metabolism of such compounds in marine invertebrates, but differences in congener water solubilities could not account for the mussel profiles. Other aspects of dioxin bioaccumulation are described in Chapter 3.

Polynuclear Aromatic and Other Hydrocarbons

Marked differences are seen in hydrocarbon metabolism in molluscs compared to crustaceans and polychaetes (Livingstone, 1991a; Table 1.5). Whereas metabolism has been detected in all studies so far of the latter, metabolites or conjugates were not detected in the bivalves *M. edulis, Macoma inquinata, Macoma nasuta* and *Modiolus modiolous* exposed to variously naphthalene, phenanthrene, BaP and other hydrocarbons (these data were not included in Fig. 1.2B and the calculations of Table 1.4). In contrast, 1- and 2-naphthols were formed from naphthalene in the oyster *Ostrea edulis*, and unidentified metabolites and conjugates were detected in the freshwater snail *Physa* sp. exposed to BaP; in the latter case, metabolism was reduced by the presence of the cytochrome P450 inhibitor, piperonyl butoxide. The amount of data available to calculate rates of metabolism in molluscs (Table 1.4) is, therefore, somewhat limited, and included hexachlorobenzene metabolism which fell well below the regression line for functional group compounds (Livingstone, 1991a). In addition to lower rates of metabolism, qualitative differences are also indicated in molluscs compared to crustaceans and polychaetes: the major products of microsomal BaP metabolism in the former are quinones (see below), compared to phenols and dihydrodiols for the latter.

The lower levels of MFO system components and activities in molluscs compared to crustaceans (Table 1.2) do not alone account for the quantitative differences in hydrocarbon metabolism. Marked seasonal variability in MFO (BaP hydroxylase) activity occurs, but is observed for both groups of organisms (Livingstone, 1990, 1991a). An additional factor which could contribute to both the quantitative and qualitative differences is the possibility of cytochrome P450–catalysed one-electron oxidation occurring *in vivo* in molluscs, as has been suggested based on the unusual properties of their MFO system and the predominance of quinones in the *in vitro* metabolism of BaP (Livingstone *et al.*, 1989a, 1990a). Depending on the ionization potential of the PAH, a major fate of the cation radical formed by one-electron oxidation could be covalent binding to protein and other macromolecules: the same mechanism is also thought to be prominent in the formation of quinones from BaP (Fig. 1.5). Adducts were indicated for *Physa* sp. and *M. inquinata* exposed to, respectively, BaP and naphthalene, but were not looked for in the other bivalve studies. Chlorinated paraffins were incorporated into the protein fraction of *M. edulis* (Renberg *et al.*,

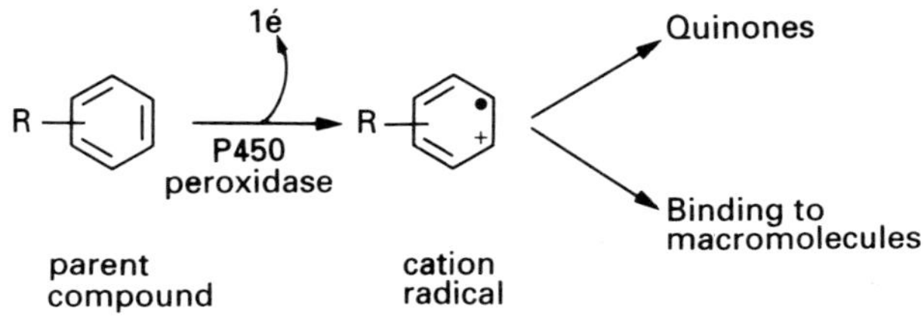

FIG. 1.5 Formation of cation radicals of polynuclear aromatic hydrocarbons, by cytochrome P450 or peroxidase catalysed one-electron oxidation, and their fate in biological systems; R, aromatic groups (modified from Cavalieri and Rogan, 1985, 1990).

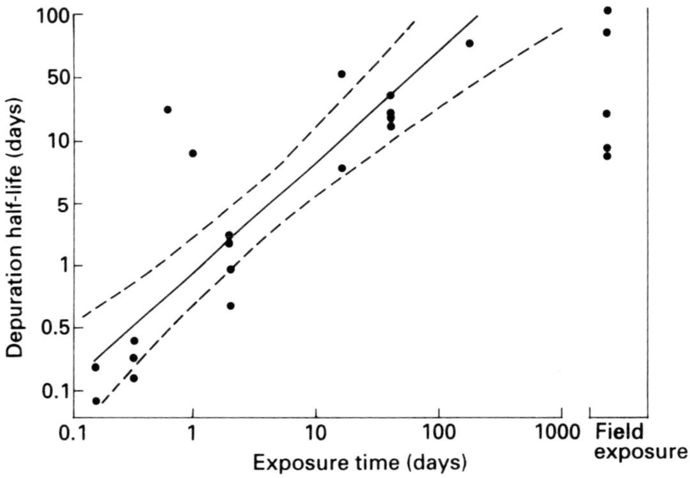

FIG. 1.6 Relationship between exposure time and subsequent depuration half-life for a range of hydrocarbons (paraffins, PAH, alkylated PAH, mixtures, dibenzothiophenes) and molluscan species (*Crassostrea virginica, Macoma balthica, Mercenara mercenaria, Modiolus modiolus, Mytilus edulis, Rangia cuneata*). From Livingstone (1991a) with permission.

1986). The highest rate of metabolism, in relation to tissue hydrocarbon concentration, was recorded when all three metabolite classes (polar metabolites, conjugates and putative adducts) were measured, viz. for *Physa* sp. exposed to BaP.

An increase in depuration half-life following increased periods of previous exposure is a marked phenomenon in molluscs exposed to hydrocarbons (Fig. 1.6).

Short-term exposure is followed by rapid and often complete depuration, in contrast to slow and incomplete elimination for long-term exposure. A tendency to form macromolecular adducts, and presumably eventual excision and release, could contribute to this phenomenon, and to the more stable cellular compartment with a lower rate of hydrocarbon turnover postulated by Stegeman and Teal (1973).

*1,1-*bis*(4-chlorophenyl)–2,2,2–trichloroethane (DDT)*

Information on the fate of DDT in marine invertebrates is limited (Livingstone, 1991a). DDT–dehydrochlorinase activity and metabolism of DDT to 1,1–dichloro–2,2–*bis*(*p*–chlorophenyl)ethane (DDD) and 1,1–dichloro–2,2–*bis*(*p*–chlorophenyl)ethylene (DDE) are observed in lower invertebrates such as free-living planarians. DDT was metabolized to DDD (85% of metabolites), DDE (9%) and 2,2–*bis*(*p*–chlorophenyl) acetic acid (6%) in hepatopancreas of *H. americanus*. The rate of metabolism was 1.3 pmol min^{-1} g^{-1} for a tissue DDT concentration of 4.2 nmol g^{-1}, compared to a calculated 0.6 pmol min^{-1} g^{-1} at 10 nmol g^{-1} for several freshwater crustacean species, including *Daphnia magna*. Traces of DDD and DDE were detected in *Physa* sp. exposed to DDT.

Bis*(tributyl)tin Oxide (TBT)*

The accumulation of TBT by *M. edulis* is ×10 greater than that predicted by octanol–water partition coefficient (Laughlin *et al.*, 1986; Laughlin and French, 1988). Metabolism of TBT was lower in *C. virginica* (traces of dibutyltin detected) than in hepatopancreas of *C. sapidus* and whole-body *P. aztecus*: rates of metabolism of the latter two were about 0.05 pmol min^{-1} g^{-1} for a TBT tissue concentration of 0.014 nmol g^{-1} (Livingstone, 1991a). The composition of metabolites in the crustacean tissues was 67–73% dibutyltin, 17–18% monobutyltin and 9–16% polar metabolites. TBT was taken up from contaminated prey (grass shrimp) by *C. sapidus* and dealkylated to di– and mono-butyltin: with longer-term exposure (8–16 days) the body burden of TBT decreased but the total burden of butyltins remained constant due to retention of the metabolites (Rice *et al.*, 1989).

REFERENCES

Barron, M. G. 1990. Bioconcentration. *Environ. Sci. Technol.* **24**: 1612–1618.

Barron, M. G. and M. O. James. 1988. Fate of sulfadimethoxine in the lobster, *Homarus americanus*. *Mar. Env. Res.* **24**: 85–88.

Bauer, I., K. Weber, W. Ernst and V. Weigelt. 1989. Metabolism of octachlorostyrene in the blue mussel (*Mytilus edulis*). *Chemosphere* **18**: 1573—1579.

Boehm, P. D., J. E. Barak, D. L. Fiest and A. A. Elskus. 1982. A chemical investigation of the transport and fate of petroleum hydrocarbons in littoral and benthic environments: the Tsesis oil spill. *Mar. Env. Res.* **6**: 157–188.

Bjørseth, A., J. Knutzen and J. Skei. 1979. Determinations of polycyclic hydrocarbons in sediments and mussels from Saudafjord, W. Norway, by glass capillary gas chromatography. *Sci. Total Environ.* **13**: 71–86.

Boon, J. P., F. Eijgenraam, J. M. Everaarts and J. C. Duinker. 1989. A structure–activity relationship (SAR) approach towards metabolism of PCBs in marine animals from different trophic levels. *Mar. Env. Res.* **27**: 159–176.

Burns, K. A. 1976. Hydrocarbon metabolism in the intertidal fiddler crab *Uca pugnax. Mar. Biol.* **36**: 5–11.

Burns, K. A. and J. L. Smith. 1981. Biological monitoring of ambient water quality: the case for using bivalves as sentinel organisms for monitoring petroleum pollution in coastal waters. *Estuar. Coast. Shelf Sci.* **13**: 433–443.

Burton, D. T., K. R. Cooper, W. L. Goodfellow Jr. and D. H. Rosenblatt. 1984. Uptake, elimination, and metabolism of ^{14}C–picric acid and ^{14}C–picramic acid in the American oyster (*Crassostrea virginica*). *Arch. Environ. Contam. Toxicol.* **13**: 653–663.

Calambokidis, J., J. Mowrer, M. W. Beug and S. G. Herman. 1979. Selective retention of polychlorinated biphenyl components in the mussel, *Mytilus edulis. Arch. Environ. Contam. Toxicol.* **8**: 299–308.

Cavalieri, E. L. and E. G. Rogan. 1985. One-electron oxidation in aromatic hydrocarbon carcinogenesis. In R. G. Harvey (ed.), *Polycyclic Hydrocarbons and Carcinogenesis.* American Chemical Society, Washington DC, pp. 289–305.

Cavalieri, E. L. and E. G. Rogan. 1990. Radical cations in aromatic hydrocarbon carcinogenesis. *Free Rad. Comms.* **11**: 77–87.

Claisse, D. 1990. Chemical contamination of French coasts. The results of a ten years mussel watch. *Mar. Pollut. Bull.* **20**: 523—528.

Clement, L. E., M. S. Stekoll and D. G. Shaw. 1980. Accumulation, fractionation and release of oil by the intertidal clam *Macoma balthica. Mar. Biol.* **57**: 41–50.

Corner, E. D. S., R. P. Harris, C. C. Kilvington and S. C. M. O'Hara. 1976. Petroleum hydrocarbons in the marine food web: short-term experiments on the fate of naphthalene in *Calanus. J. Mar. Biol. Ass. U.K.* **56**: 121–133.

Crosby, D. G., P. F. Landrum and C. C. Fischer. 1979. In M. A. Q. Khan, J. J. Lech and J. J. Menn (eds.), *Pesticide and Xenobiotic Metabolism in Aquatic Organisms.* Amer. Chem. Soc. Ser. 99, pp. 217–231.

De Leon, I. R., J. B. Ferrario and C. J. Byrne. 1988. Bioaccumulation of polynuclear aromatic hydrocarbons by the clam, *Rangia cuneata*, in the vicinity of a creosote spill. *Bull. Environ. Contam. Toxicol.* **41**: 872–879.

Den Besten, P. J., H. J. Herwig, A. C. Smaal, D. I. Zandee and P. A. Voogt. 1990. Interference of polychlorinated biphenyls (Clophen A50) with gametogenesis in the sea star, *Asterias rubens. Aquat. Toxicol.* **18**: 231–246.

Den Besten, P. J., H. J. Herwig, E. G. van Donselaar and D. R. Livingstone. 1990. Cytochrome *P*–450 monooxygenase system and benzo(a)pyrene metabolism in echinoderms. *Mar. Biol.* **107**: 171–177.

Di Giulio, R. T., P. C. Washburn, R. J. Wenning, G. W. Winston and C. S. Jewell. 1989. Biochemical responses in aquatic animals: a review of determinants of oxidative stress. *Environ. Toxicol. Chem.* **8**: 1103–1123.

Donkin, P. and J. Widdows. 1990. Quantitative structure-activity relationships in aquatic invertebrate toxicology. *Rev. Aquat. Sci.* **2**: 375–398.

Ernst, W. 1977. Determination of the bioconcentration potential of marine organisms. A steady state approach. I. Bioconcentration data for seven chlorinated pesticides in mussels (*Mytilus edulis*) and their relation to solubility data. *Chemosphere.* **6**: 731–740.

Ernst, W. 1979a. Factors affecting the evaluation of chemicals in laboratory experiments using marine organisms. *Ecotoxicol. Environ. Safety.* **3**: 90–98.

Ernst, W. 1979b. Metabolic transformation of 1–(1–^{14}C)naphthol in bioconcentration studies with the common mussel *Mytilus edulis. Veroff. Inst. Meeresforsch. Bremerh.* **17**: 233–240.

Ernst, W., H. Goerke and K. Weber. 1977. Fate of ^{14}C-labelled di-, tri- and penta-chlorobiphenyl in the marine anneli *Nereis virens. Chemosphere* **9**: 559–568.

Farrington, J. W., E. D. Goldberg, R. W. Risebrough, J. H. Martin and V. T. Bowen. 1983. US "mussel watch" 1976–1978: an overview of the trace-metal, DDE, PCB, hydrocarbon, and artificial radionuclide data. *Environ. Sci. Technol.* **17**: 490–496.

Ferraro, S. P., H. Lee II, R. J. Ozretich and D. T. Specht. 1990. Predicting bioaccumulation potential: a test of a fugacity-based model. *Arch. Environ. Contam. Toxicol.* **19**: 386–394.

Feroz, M., A. A. Podowski and M. A. Q. Khan. 1990. Oxidative dehydrochlorination of heptachlor by *Daphnia magna. Pest. Biochem. Physiol.* **36**: 101–105.

Foster, G. D. and D. G. Crosby. 1986. Xenobiotic metabolism of *p*–nitrophenol derivatives by the rice field crayfish (*Procambarus clarkii*). *Environ. Toxicol. Chem.* **5**: 1059–1070.

Foster, G. D. and D. G. Crosby. 1987. Comparative metabolism of nitroaromatic compounds in freshwater, brackish water and marine decapod crustaceans. *Xenobiotica* **17**: 1393–1404.

Foster, G. D. and D. A. Wright. 1988. Unsubstituted polynuclear aromatic hydrocarbons in sediments, clams, and clam worms from Chesapeake Bay. *Mar. Pollut. Bull.* **19**: 459–465.

Foureman, G. L., Z. Ben-Zvi, L. Dostal, J. R. Fouts and J. R. Bend. 1978. Distribution of ^{14}C–benzo(a)pyrene in the lobster, *Homarus americanus*, at various times after a single injection into the pericardial sinus. *Bull. Mt. Desert Isl. Biol. Lab.* **18**: 93–95.

Geyer, H., P, Sheehan, D. Kotzias, D. Freitag and F. Korte. 1982. Prediction of ecotoxicological behaviour of chemicals: relationship between physicochemical properties and bioaccumulation of organic chemicals in the mussel *Mytilus edulis. Chemosphere* **11**: 1121–1134.

Geyer, H., D. Freitag and F. Korte. 1984. Polychlorinated biphenyls (PCBs) in the marine environment, particularly in the Mediterranean. *Ecotoxicol. Environ. Saf.* **8**: 129–151.

Gibson, G. G. and P. Skett. 1986. *Introduction to Drug Metabolism*. Chapman Hall, London.

Gobias, F. A. P. C., A. Opperhuizen and O. Hutzinger. 1986. Bioconcentration of hydrophobic chemicals in fish: relationship with membrane permeation *Environ. Toxicol. Chem.* **5**: 637–646.

Goerke, H. and W. Ernst. 1977. Fate of ^{14}C–labelled di–, tri– and pentachlorobiphenyl in the marine annelid *Nereis virens*. I. Accumulation and elimination after oral administration. *Chemosphere* **9**: 551–558.

Goerke, H. and W. Ernst. 1986. Elimination of pentachlorobiphenyls by *Nereis virens* (Polychaeta) in the laboratory and the marine environment. In *Chemistry in Ecology*, Vol. 2, Gordon and Breach Sci. Pub., Inc., UK, pp. 263–285.

Goerke, H. and W. Ernst. 1980. Accumulation and elimination of ^{14}C–γ–HCH (lindane) in *Nereis virens* (Polychaeta) with consideration of metabolites. *Helgol. Meeresunters* **33**: 313–326.

Goerke, H. and K. Weber. 1990. Population-dependent elimination of various polychlorinated biphenyls in *Nereis diversicolor* (Polychaeta). *Mar. Env. Res.* **29**: 205–226.

Grahl-Nielsen, O., J. T. Staveland and S. Wilhelmsen. 1978. Aromatic hydrocarbons in benthic organisms from coastal areas polluted by Iranian crude oil. *J. Fish. Res. Board Can.* **35**: 615–623.

Guarino, A. M., J. B. Pritchard, J. B. Anderson and D. P. Rall. 1974. Tissue distribution of [^{14}C]DDT in the lobster after administration via intravascular or oral routes or after exposure from ambient sea water. *Toxicol. Appl. Pharmacol.* **29**: 277–288.

Hawker, D. W. and D. W. Connell. 1985. Relationships between partition coefficient, uptake rate constant, clearance rate constant and time to equilibrium for bioaccumulation. *Chemosphere* **9**: 1205–1219.

Hawker, D. W. and D. W. Connell. 1986. Bioconcentration of lipophilic compounds by some aquatic organisms. *Ecotoxicol. Environ. Saf.* **11**: 184–197.

Hummel, H., R. H. Bogaards, J. Nieuwenhuize, L. de Wolf and J. M. van Liere. 1990. Spatial and seasonal differences in the PCB content of the mussel *Mytilus edulis. Sci. Total Environ.* **92**: 155–163.

James, M. O. 1982. Disposition and taurine conjugation of 2,4–dichlorophenoxyacetic acid, 2,4,5–trichlorophenoxyacetic acid, *bis*(4-chlorophenyl)acetic acid, and phenylacetic acid in the spiny lobster, *Panulirus argus. Drug Metab. Disp.* **10**: 516–522.

James, M. O. and M. G. Barron. 1988. Disposition of sulfadimethoxine in the lobster (*Homarus americanus*). *Vet. Human Toxicol.* (*Suppl.*) **30**: 36–40.

Jovanovich, M. C. and K. R. Marion. 1987. Seasonal variation in uptake and depuration of anthracene by the brackish water clam *Rangia cuneata. Mar. Biol.* **95**: 395–403.

Kannan, N., S. Tanabe, R. Tatsukawa and D. J. H. Phillips. 1989. Persistency of highly toxic coplanar PCBs in aquatic ecosystems: uptake and release of coplanar PCBs in green-lipped mussels (*Perna viridis* Linnaeus). *Environ. Pollut.* **56**: 65–76.

Kawano, M., T. Inoue, T. Wada, H. Hikada and R. Tatsukawa. 1988. Bioconcentration and residue patterns of chlordane compounds in marine animals: invertebrates, fish, mammals, and seabirds. *Environ. Sci. Technol.* **22**: 792–797.

Klungsøyr, J., S. Wilhelmsen, K. Westrheim, E. Saetvedt and K. H. Palmork. 1988. The GEEP workshop: organic chemical analyses. *Mar. Ecol. Prog. Ser.* **46**: 19–26.

Knezovich, J. P. and D. G. Crosby. 1985. Fate and metabolism of *o*–toluidine in the marine bivalve molluscs *Mytilus edulis* and *Crassostrea gigas*. *Environ. Toxicol. Chem.* **4**: 435–446.

Knickmeyer, R. and H. Steinhart. 1988. Cyclic organochlorines in the hermit crabs *Pagarus bernhardus* and *P. pubescens* from the North Sea. A comparison between winter and early summer situation. *Neth. J. Sea Res.* **22**: 237–251.

Knutzen, J. and B. Sortland. 1982. Polycyclic aromatic hydrocarbons (PAH) in some algae and invertebrates from moderately polluted parts of the coast of Norway. *Water Res.* **16**: 421–428.

Kobayashi, K. 1985. The effect of the herbicide PCP on short-necked clams and the process of detoxification. In T. OkuTani, T. Tomiyama and T. Hibiya (eds.), *Fisheries in Japan: Bivalves*. Jpn. Mar. Product Photo Material Assoc., pp. 161–165.

Kurelec, B., M. Chacko and R. C. Gupta. 1988. Postlabelling analysis of carcinogen–DNA adducts in mussel, *Mytilus galloprovincialis*. *Mar. Env. Res.* **24**: 317–320.

Kurelec, B. and B. Pivcevic. 1989. Distinct glutathione-dependent enzyme activities and a verapamil–sensitive binding of xenobiotics in a fresh-water mussel *Anodonta cygnea*. *Biochem. Biophys. Res. Commun.* **164**: 934–940.

Landrum, P. F. 1988. Toxicokinetics of organic xenobiotics in the amphipod *Pontoporeia hoyi*: role of physiological and environmental variables. *Aquat. Toxicol.* **12**: 245–271.

Landrum, P. F. and D. G. Crosby. 1981a. Comparison of the disposition of several nitrogen–containing compounds in the sea urchin and other marine invertebrates. *Xenobiotica* **11**: 351–361.

Landrum, P. F. and D. G. Crosby. 1981b. The disposition of *p*–nitroanisole by the sea urchin, *Strongylocentrotus purpuratus*. II. Bioconcentration and biotransformation. *Ecotoxicol. Environ. Safety* **5**: 240–254.

Langston, W. J. 1978a. Accumulation of polychlorinated biphenyls in the cockle *Cerastoderma edule* and the tellin *Macoma balthica*. *Mar. Biol.* **45**: 265–272.

Langston, W. J. 1978b. Persistence of polychlorinated biphenyls in marine bivalves. *Mar. Biol.* **46**: 35–40.

Laughlin, R. B. Jr. and W. French. 1988. Concentration dependence of *bis*(tributyl)tin oxide accumulation in the mussel, *Mytilus edulis*. *Environ. Contam. Toxicol.* **7**: 1021–1026.

Laughlin, R. B. Jr., W. French and H. E. Guard. 1986. Accumulation of *bis*(tributyl)tin oxide by the marine mussel *Mytilus edulis*. *Environ. Sci. Technol.* **20**: 884–890.

Laurén, D. J. and S. Rice. 1985. Significance of active and passive depuration in the clearance of naphthalene from the tissue of *Hemigraspus nudus* (Crustacea: Decapoda). *Mar. Biol.* **88**: 135–142.

Lee, R. F. 1986. Metabolism of *bis*(tributyltin)oxide by estuarine animals. *Mar. Technol. Soc., Washington D.C., Organ. Symp.* Vol. 4, pp. 1182–1188.

Lee, R. F., J. W. Conner, D. Page, L. E. Ray and C. S. Giam. 1982. Cytochrome *P*–450 dependent mixed-function oxygenase systems in marsh crabs. In W. B. Vernberg, A. Calabrese, F. P. Thurberg and F. J. Vernberg (eds.), *Physiological Mechanisms of Marine Pollutant Toxicity*. Academic Press, New York, pp. 147–159.

Lee, R. R., C. Ryan and M. L. Neuhauser. 1976. Fate of petroleum hydrocarbons taken up from food and water by the blue crab *Callinectes sapidus*. *Mar. Biol.* **37**: 363–370.

Little, P. J., M. O. James, J. B. Pritchard and J. R. Bend. 1985. Temperature-dependent disposition of [^{14}C]benzo(a)pyrene in the spiny lobster, *Panulirus argus*. *Toxicol. Appl. Pharmacol.* **77**: 325–333.

Livingstone, D. R. 1990. Cytochrome *P*–450 and oxidative metabolism in invertebrates. *Biochem. Soc. Trans.* **18**: 15–19.

Livingstone, D. R. 1991a. Organic xenobiotic metabolism in marine invertebrates. *Adv. Comp. Environ. Physiol.* **7**: 45–185.

Livingstone, D. R. 1991b. Towards a specific index of impact by organic pollution for marine invertebrates. *Comp. Biochem. Physiol. Part C*: **100**. 151–155.

Livingstone, D. R., R. Arnold, K. Chipman, M. A. Kirchin and J. Marsh. 1990a. The mixed-function oxygenase system in molluscs: metabolism, responses to xenobiotics, and toxicity. *Oceanis* **5**: 331–347.

Livingstone, D. R., P. Garcia Martinez, X. Michel, J. F. Narbonne, S. O'Hara, D. Ribera and G. W. Winston. 1990b. Oxyradical production as a pollution-mediated mechanism of toxicity in the common mussel, *Mytilus edulis* and other molluscs. *Funct. Ecol.* **4**: 415–424.

Malins, D. C., M. S. Myers and W. T. Roubal. 1983. Organic free radicals associated with idiopathic liver lesions of English sole (*Parophrys vetulus*) from polluted marine environments. *Environ. Sci. Technol.* **17**: 679–685.

Malins, D. C. and W. T. Roubal. 1982. Aryl sulphate formation in sea urchins (*Strongylocentrotus droebachiensis*) ingesting marine algae (*Fucus distichus*) containing 2,6–dimethylnaphthalene. *Env. Res.* **27**: 290–297.

Marcus, J. M. and T. D. Mathews. 1987. Polychlorinated biphenyls in blue crabs from South Carolina. *Bull. Environ. Contam. Toxicol.* **39**: 857–862.

Marcus, J. M. and R. T. Renfrow. 1990. Pesticides and PCBs in South Carolina estuaries. *Mar. Pollut. Bull.* **21**: 96–99.

McElroy, A. E. 1990. Polycyclic aromatic hydrocarbon metabolism in the polychaete *Nereis virens*. *Aquat. Toxicol.* **18**: 35–49.

McElroy, A. E., J. W. Farrington and J. M. Teal. 1989. Bioavailability of polycyclic hydrocarbons in the aquatic environment. In U. Varanasi (ed.), *Metabolism of Polycyclic Aromatic Hydrocarbons in the Aquatic Environment*, CRC Press, Boca Raton, Florida, pp. 1–39.

McElroy, A. E. and J. C. Means. 1988. Uptake, metabolism, and depuration of PCBs by the polychaete *Nereis virens*. *Aquat. Toxicol.* **11**: 416–417.

Meyer, T. and T. Bakke. 1977. The metabolism of biphenyl.V. Phenolic metabolites in some marine organisms. *Acta Pharmacol. Toxicol.* **40**: 201–208.

Mix, M. C. 1984. Polycyclic aromatic hydrocarbons in the aquatic environment: occurrence and biological monitoring. In E. Hodgson (ed.), *Reviews in Environmental Toxicology*, Elsevier, Amsterdam, pp. 51–102.

Miyata, H., K. Takayama, J. Ogaki, T. Kashimoto and S. Fukushima. 1987. Polychlorinated dibenzo–*p*–dioxins in blue mussel from marine coastal water in Japan. *Bull. Environ. Contam. Toxicol.* **39**: 877–883.

Miyata, H., K. Takayama, M. Mimura, T. Kashimoto and S. Fukushima. 1989. Specific congener profiles of polychlorinated dibenzo–*p*–dioxins and dibenzofurans in blue mussel in Osaka Bay in Japan: aqueous solubilities of PCDDs and PCDFs. *Bull. Environ. Contam. Toxicol.* **43**: 342–349.

Moore, M. N., D. R. Livingstone and J. Widdows. 1989. Hydrocarbons in marine molluscs: biological effects and ecological consequences. In U. Varanasi (ed.), *Metabolism of Polycyclic Aromatic Hydrocarbons in the Marine Environment,* CRC Press, Boca Raton, Florida, pp. 291–328.

Neff, J. M. 1979. *Polycyclic Aromatic Hydrocarbons in the Aquatic Environment. Sources, Fates and Biological Effects*. Appl. Sci. Publ., London.

Page, D. S. 1989. An analytical method for butyltin species in shellfish. *Mar. Pollut. Bull.* **20**: 129–133.

Phillips, D. J. H. and R. B. Spies. 1988. Chlorinated hydrocarbons in the San Francisco estuarine ecosystem. *Mar. Pollut. Bull.* **19**: 445–453.

Porte, C., D. Barceló, T. M. Tavares, V. C. Rocha and J. Albaigés. 1990. The use of the mussel watch and molecular marker concepts in studies of hydrocarbons in a tropical bay (Todos os Santos, Bahlia, Brazil). *Arch. Environ. Contam. Toxicol.* **19**: 263–274.

Pruell, R. J., J. L. Lake, W. R. Davis and J. G. Quinn. 1986. Uptake and depuration of organic contaminants by blue mussels (*Mytilus edulis*) exposed to environmentally contaminated sediment. *Mar. Biol.* **91**: 497–507.

Ranga Rao, K. R., F. R. Fox, P. J. Conklin and A. C. Cantelmo. 1981. Comparative toxicology and pharmacology of chlorophenols: studies on the grass shrimp, *Palaemontes pugio*. In F. J. Vernberg, A. Calabrese, F. P. Thurberg and W. B. Vernberg (eds), Academic Press, New York, pp. 37–72.

Reichart, W., B. T. Le Eberhart and U. Varanasi. 1985. Exposure of two species of

deposit-feeding amphipods to sediment-associated [^{3}H]benzo[a]pyrene: uptake, metabolism and covalent binding to tissue macromolecules. *Aquat. Toxicol.* **6**: 45–56.
Renberg, L., M. Tarkpea and G. Sundstrom. 1986. The use of the bivalve *Mytilus edulis* as a test organism for bioconcentration studies. II. The bioconcentration of two ^{14}C–labelled chlorinated paraffins. *Ecotoxicol. Environ. Safety* **11**: 361–372.
Rice, C. P. and D. S. White. 1987. PCB availability assessment of river dredging using caged clams and fish. *Environ. Toxicol. Chem.* **6**: 259–274.
Rice, S. D., J. W. Short and W. B. Stickle. 1989. Uptake and catabolism of tributyltin by blue crabs fed TBT contaminated prey. *Mar. Env. Res.* **27**: 137–145.
Sanborn, H. R. and D. C. Malins. 1977. Toxicity and metabolism of naphthalene: a study with marine larval invertebrates. *Proc. Soc. Exp. Biol. Med.* **154**: 151–155.
Sanborn, H. R. and D. C. Malins. 1980. The disposition of aromatic hydrocarbons in adult spot shrimp (*Pandalus platyceros*) and the formation of metabolites of naphthalene in adult and larval spot shrimp. *Xenobiotica* **10**: 193–200.
Satsmadjis, J. and F. Voutsinou-Taliadouri. 1983. *Mytilus galloprovincialis* and *Parapenaeus longirostris* as bioindicators of heavy metal and organochlorine pollution. *Mar. Biol.* **76**: 115–124.
Sericano, J. L., E. L. Atlas, T. L. Wade and J. M. Brooks. 1990. NOAA's status and trends mussel watch program: chlorinated pesticides and PCBs in oysters (*Crassostrea virginica*) and sediments from the Gulf of Mexico, 1986–1987. *Mar. Env. Res.* **29**: 161–203.
Schell, J. D. Jr. and M. O. James. 1989. Glucose and sulphate conjugation of phenolic compounds by the spiny lobster (*Panulirus argus*). *J. Biochem. Toxicol.* **4**:
Sieben, B., N. Bihari, B. Kurelec and R. K. Zahn. 1983. Polycyclic aromatic hydrocarbon levels in waters and sponges of the Northern Adriatic. *Rapp. Comm. Int. Mer Medit.* **28**: 133–134.
Stegeman, J. J. 1981. Polynuclear aromatic hydrocarbons and their metabolism in the marine environment. In H. Gelboin and T.Śo Pop (eds.), *Polycyclic Hydrocarbons and Cancer*, Vol. 3, Academic Press, New York, pp. 1–60.
Stegeman, J. J. and J. M. Teal. 1973. Accumulation, release and retention of petroleum hydrocarbons by the oyster *Crassostrea virginica. Mar. Biol.* **22**: 37–44.
Tanabe, S., R. Tatsukawa and D. J. H. Phillips. 1987. Mussels as bioindicators of PCB pollution: a case study on uptake and release of PCB isomers and congeners in green-lipped mussels (*Perna viridis*) in Hong Kong waters. *Environ. Pollut.* **47**: 41–62.
Uthe, J. F. and C. J. Musial. 1986. Polycyclic aromatic hydrocarbon contamination of American lobster, *Homarus americanus*, in the proximity of a coal-coking plant. *Bull. Environ. Contam. Toxicol.* **37**: 730–738.
Young, D. R., T. C. Heesen and D. J. McDermott. 1976. An offshore biomonitoring system for chlorinated hydrocarbons. *Mar. Pollut. Bull.* **7**: 156–159.
Van der Oost, A. J., H. Heida and A. Opperhuizen. 1988. Polychlorinated biphenyl congeners in sediments, plankton, molluscs, crustaceans, and eel in a freshwater lake: implications of using reference chemicals and indicator organisms in bioaccumulation studies. *Arch. Environ. Contam. Toxicol.* **17**: 721–729.
Varanasi, U., W. L. Reichert, J. E. Stein, D. W. Brown and H. R. Sanborn. 1985. Bioavailability and biotransformation of aromatic hydrocarbons in benthic organisms exposed to sediment from an urban estuary. *Environ. Sci. Technol.* **19**: 836–841.
Webber, E. C., D. R. Bayne and W. C. Seesock. 1989. DDT contamination of benthic macroinvertebrates and sediments from tributaries of Wheeler Reservoir, Alabama. *Arch. Environ. Contam. Toxicol.* **18**: 728–733.
Widdows, J., S. L. Moore, K. R. Clarke and P. Donkin. 1983. Uptake, tissue distribution and elimination of [1–^{14}C]naphthalene in the mussel *Mytilus edulis. Mar. Biol.* **76**: 109–114.
Widdows, J., P. Donkin, P. N. Salkeld and S. V. Evans. 1987. Measurement of scope for growth and tissue hydrocarbon concentrations of mussels (*Mytilus edulis*) at sites in the vicinity of the Sullom Voe oil terminal: a case study. In J. Kuiper and W. J. Van den Brink (eds.), *Fate and Effects of Oil in Marine Ecosystems,* Nijhoff, Dordrecht, pp. 269–277.
Wofford, H. W., C. D. Wilsey, G. S. Neff, C. S. Giam and J. M. Neff. 1981. Bioaccumulation and metabolism of phthalate esters by oysters, brown shrimp, and sheepshead minnows. *Ecotoxicol. Environ. Safety* **5**: 202–210.
Zaroogian, G. E., J. F. Heltsche and M. Johnson. 1985. Estimation of bioconcentration in marine species using structure–activity models. *Environ. Toxicol. Chem.* **4**: 3–12.

CHAPTER 2

Occurrence and Fate of Persistent Pollutants in Marine Invertebrates: Studies with Polychaetes and the Common Mussel, *Mytilus edulis**

W. ERNST

Institute for Polar and Marine Research, Chemistry Section, 2850 Bremerhaven, Germany

INTRODUCTION

Owing to their chemical structures and inherent physicochemical properties, a number of anthropogenically released organic chemicals are bioaccumulated by marine organisms via various routes of uptake. Pollutants considered persistent and toxicologically relevant, such as chlorinated pesticides, polychlorinated biphenyls (PCBs) and other synthetic products, have therefore been determined in numerous monitoring programmes, seeking to reflect pollution more generally by analysing them in a few selected species. On the other hand, surprisingly little is known about the occurrence of metabolites of these organic pollutants and their ecotoxicological properties, although it is well known that marine organisms are capable of metabolizing xenobiotics (Adamson, 1967; Ernst, 1984; James, 1987).

Due to their abundance, experimental suitability and ecological significance, marine invertebrates have been frequently used to explore metabolic pathways and transfer processes of xenobiotics. There is ample evidence for the occurrence in these organisms of enzymes responsible for the biotransformation of organic pollutants, such as the

* This is Contribution No. 424 of the Alfred Wegener Institute for Polar and Marine Research, Bremerhaven.

cytochrome P-450-containing mixed function oxygenase system as proven by a number of authors (Bend *et al.*, 1977; Moore *et al.*, 1980; Lee, 1981, 1981a; Livingstone and Farrar, 1984; Livingstone, 1985; see also Chapter 1).

Investigation of the metabolism of organic pollutants at environmentally relevant concentrations is complicated by the presence of "background" levels of the compounds under study. In order to identify metabolites, the use of radiolabelled compounds is necessary in most cases. But even then the results can frequently be interpreted only by stating that metabolism takes place to a measurable degree. Also, due to the small amounts of metabolites formed, their structures often cannot be elucidated unequivocally.

The chapter describes experiments on the elimination and biotransformation of some halogenated "persistent compounds" by various polychaete species and the common mussel *Mytilus edulis*. Experiments are included which test the reliability of laboratory experiments by comparison with those under natural conditions.

BIOTRANSFORMATION AND ELIMINATION

Polychlorinated Biphenyls (PCBs)

Individual PCB congeners, when applied orally to marine polychaetes such as *Nereis* sp., are biodegraded substantially (Ernst *et al.*, 1977; Goerke and Ernst, 1977). In *Nereis virens*, ^{14}C-labelled di-, tri- and pentachlorobiphenyl were degraded to different polar metabolites during two weeks of oral administration (Table 2.1).

TABLE 2.1 *Formation of Metabolites of ^{14}C-polychlorobiphenyls in* Nereis virens *(as % of total ^{14}C-chlorobiphenyl equivalents accumulated) (Ernst* et al., *1977)*

PCB	unchanged	Metabolites extractable	unextractable
2,2′-Cl_2	73.8	3.0	23.0
2,5,4′-Cl_3	21.6	15.2	63.2
2,4,6,2′-Cl_5	54.1	9.8	36.1

In a subsequent elimination period of 10 weeks, it was found that faecal excretion was a main route for elimination of both parent compounds and metabolites (Table 2.2). Metabolites from 2,5,4′-trichlorobiphenyl, which were more readily produced, could partly be identified as trichlorobiphenylols.

TABLE 2.2 *Composition of Polar Metabolites from Orally Administered ^{14}C-labelled di-, tri- and penta-chlorobiphenyl in Faeces of* Nereis virens *(Ernst* et al., *1977)*

PCB	Total amount[a] (μg)	Unchanged parent compound (%)	Medium polar metabolites (%)	Highly polar metabolites (%)
2,2′-dichlorobiphenyl	4.6	22	17	61
2,5,4′-trichlorobiphenyl	36	3	18	79
2,4,6,2,′4′-pentachloro-biphenyl	36	24	3	73

[a] Unchanged PCB and metabolites.

γ-Hexachlorocyclohexane (Lindane)

Metabolic transformations of lindane (γ-HCH) are well known to occur in mammals, plant and soil (Engst *et al.*, 1979); however, little is known of its fate in aquatic organisms. The common mussel, *M. edulis*, exposed to ^{14}C-γ-HCH at concentrations of up to 4 μg L^{-1}, produced two metabolites of higher polarity, representing about 15% of the accumulated γ-HCH as ^{14}C-γ-HCH equivalents, which could be detected only in the water; elimination of γ-HCH occurred with a half-life of 0.7 day (Ernst, 1975). In the polychaetes, *N. virens*, ^{14}C-γ-HCH was eliminated after uptake from water at similar rates to those found in mussels with half-lives of 2 to 3 days at 6°C and 16°C, respectively (Goerke and Ernst, 1980). Metabolism was substantial, accounting for 20–30% of an accumulated level of 0.4–0.5 ppm, expressed as ^{14}C-γ-HCH equivalents. Several unidentified metabolites were traced in the exposure water including acid-labile compounds, whereas in faeces, which did not represent the major route of elimination for γ-HCH, only unextractable radioactivity was present (Goerke and Ernst, 1980).

Hexachlorobenzene and Octachlorostyrene

Sanborn *et al.* (1977) studied the metabolism of hexachlorobenzene (HCB) in a freshwater fish, the green sunfish, and detected some polar metabolites. In the marine fish, cod and flounder, Ingebrigtsen and Solbakken (1985) studied the distribution of ^{14}C-HCB by whole body autoradiography. From these studies it was concluded that water-soluble metabolites were present in the bile.

More detailed studies with *M. edulis*, exposed to ^{14}C-HCB in seawater, resulted in the detection of six metabolites, three of which were identified in the animals as pentachlorothioanisole, 1.4-*bis*-(methylthio)-2,3,5,6-tetrachlorobenzene and S-(pentachlorophenyl)-thioglycolate (Bauer *et al.*, 1989a). Other metabolites were exclusively

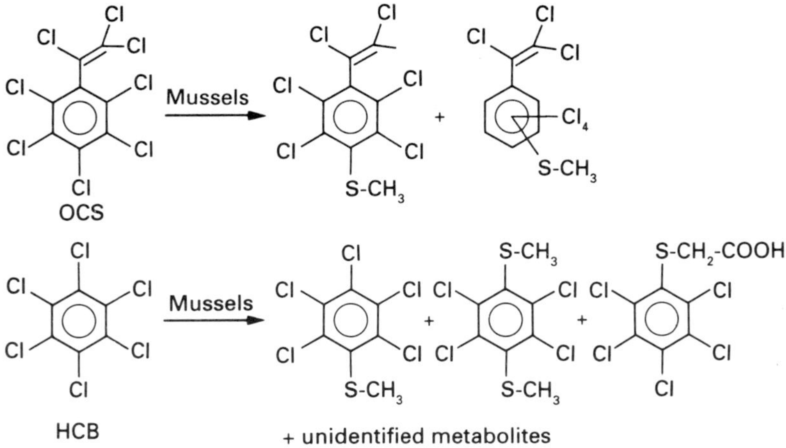

FIG. 2.1 Degradation of octachlorostyrene (OCS) and hexachlorobenzene (HCB) in mussel, *Mytilus edulis*.

detected in the exposure water. After twenty days of exposure at 0.8 μg^{-1} ^{14}C-HCB, the mussels contained as the major metabolite about 200 ng g^{-1} (f.w.) pentachlorothioanisole compared to approx. 800 ng g^{-1} unchanged HCB (Bauer *et al.*, 1989a). Pentachlorophenol, which had been identified as a major degradation product in the green sunfish, *Lepomis cyanellus*, by Sanborn *et al.* (1977), could not be detected in mussels and the exposure water.

When *M. edulis* was exposed to octachlorostyrene (OCS) at 1.5 $\mu g\ L^{-1}$, two isomeric methylthioheptachlorostyrenes were formed in the animals, one of which was identified as 4-methylthioheptachlorostyrene. During an exposure of 240 hours the parent compound reached levels of 300 ng g^{-1} (wet weight) and the sum of the two metabolites approx. 400 ng g^{-1} in the mussels (Bauer *et al.*, 1989b). The pathways of formation of metabolites from HCB and OCS in mussels are summarized in Fig. 2.1.

Laboratory Versus Natural Conditions

Polychlorinated Biphenyls

In order to investigate whether results of experiments obtained in the laboratory can be translated to natural conditions, *N. virens* was used (Goerke and Ernst, 1986). After oral application of a mixture of three isomeric pentachlorobiphenyls, the animals were kept to study elimination under laboratory and field conditions. Under laboratory conditions elimination tanks were supplied with recirculated,

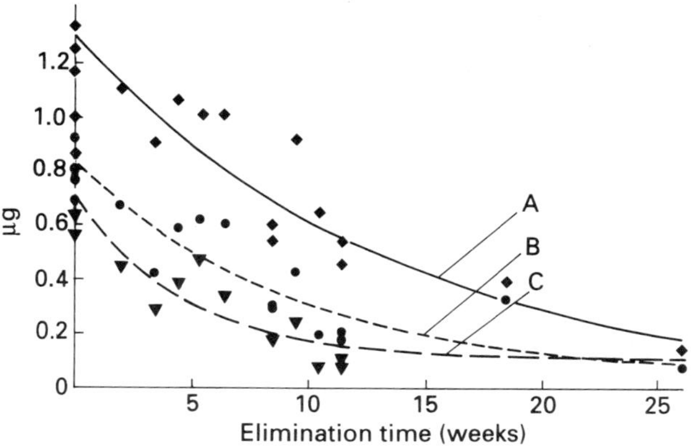

FIG. 2.2 Elimination of 2,4,6,2′,4′-pentachloro-(U-^{14}C)biphenyl equivalents in *Nereis virens* in µg/5 g animal. A: Parent compound + total metabolites. B: Parent compound + metabolites extractable with n-hexane/acetone, 2:1 ^{v}/v. C: Parent compound.

charcoal-filtered seawater. In the field experiment the animals were transferred to a cage, which was moored in the Weser estuary at the polyhaline zone, with Aandera current meter attached to the cage for recording temperature and conductivity. The temperature during the field experiment was 4° higher than in the laboratory experiment carried out at 12°C, but dropped within 15 weeks to 4°C. An estimate of the average temperature was 12°C, as in the laboratory experiment.

As shown in Table 2.3, measured elimination half-lives indicate that PCB elimination kinetics determined in the laboratory can be considered valid for natural conditions, although the variation of temperature during the field experiment might have had an influence on the elimination rate.

Since one of the three pentachlorobiphenyls, the 2,4,6,2′,4′-pentachloroisomer, was ^{14}C-labelled, detailed information could be obtained on its metabolism (Fig. 2.2). Already after termination of uptake, when 10 daily oral doses had been administered resulting in a level of 0.2–0.3 ppm, more than 50% of the accumulated radioactivity could be ascribed to compounds different from the parent compound.

TABLE 2.3 *Elimination Half-lives (weeks) for Three Different Pentachlorobiphenyls in the Marine Polychaete* Nereis virens *in Laboratory and Field Experiments (Goerke and Ernst, 1986)*

PCB	Lab experiment	Field experiment
2,3,4,2′,5′-Cl_5	1.9	1.9
2,4,5,2′,5′-Cl_5	2.8	2.9
2,4,6,2′,4′-Cl_5	4.5	4.5

Species Dependence

It is well known that metabolic capabilities in different aquatic species can be very different. Experimental results obtained with one species can, therefore, not generally be extrapolated to other marine animals. In the following, some experiments with different chemicals might provide an interesting illustration of species differences in this context.

Pentachlorophenol

Pentachlorophenol (PCP) is known to bioaccumulate with bioconcentration factors (BCF) of approximately 300 in fish and mussels: conjugation of PCP with sulfuric acid and glucuronic acid is described in mussels and fish by Kobayashi *et al.* (1970), Ernst (1979) and Glickman *et al.* (1977). In the polychaete *Lanice conchilega*, however, PCP is bioconcentrated with BCFs one order of magnitude higher than in *M. edulis* and fish (Ernst and Weber, 1978). *L. conchilega* did not depurate PCP during a 16-day elimination period and application of ^{14}C-PCP did not lead to the formation of detectable metabolites. For the species *L. conchilega*, PCP can therefore be regarded as persistent and highly bioaccumulating, in contrast to other species from the same location. Similarly, γ-HCH was also bioaccumulated by one order of magnitude higher in *L. conchilega* than in *M. edulis* (Ernst, 1979).

4,5,6-Trichloroguajacol

The fate of 4,5,6-trichloroguajacol (4,5,6-TCG) (Fig. 2.3), one of the products formed by wood pulp bleaching processes, was investigated by Neilson *et al.* (1989). In brackish water model ecosystems, three concentrations of 1, 10 and 50 $\mu g\ L^{-1}$ were applied and the accumulation of 4,5,6-TCG and its metabolites were determined in sediments, macroalgae, invertebrates and fish after an exposure of 16 months. Among the eight different taxa of invertebrates analysed, the polychaete *Nereis diversicolor* produced the widest spectrum of metabolites, whereas fish, the stickleback (*Gasterosteus aculeatus*), accumulated higher levels of 4,5,6-TCG.

Polychlorinated Biphenyls

Elimination kinetics for different PCB compounds were determined in different species by Weber and Goerke (in prep.). The experiments were carried out with shrimps, polychaetes and fish at a salinity of 10‰ simulating estuarine brackish water conditions. Elimination half-lives were significantly different in the different species (Table 2.4) indicating

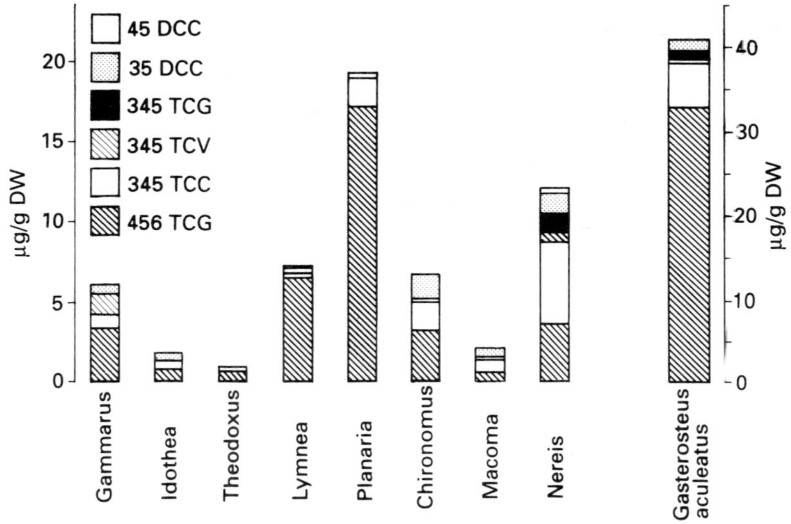

FIG. 2.3 Levels of 4,5,6-trichloroguajacol (4,5,6-TCG) in various invertebrate taxa and fish, stickleback (*Gasterosteus aculeatus*) from Baltic Sea littoral model ecosystems exposed to 50 μg L^{-1} 4,5,6-TCG for 16 months (after Neilson *et al.*, 1989). DCC, dichlorocatechol; TCV, trichloroveratrole; TCC, trichlorocatechol.

TABLE 2.4 *Elimination of PCB Compounds in Different Marine Species (half-lives in weeks) (Weber and Goerke, in prep.)*

PCB	Shrimp *Palaemon longirostris*	Polychaete *Nereis diversicolor*	Fish *Platichthys flesus*
2,5,2′,5′-tetrachlorobiphenyl	0.35	0.66	1.9
2,3,6,2′,5′-pentachlorobiphenyl	0.34	6.9	2.6
2,4,5,2′,5′-pentachlorobiphenyl	0.58	0.56	5.3
2,4,5,2′,4′,5′-hexachlorobiphenyl	6.0	---[a]	9.9

[a] Half-life was not attained after 21 weeks.

that the selection of species plays an important role in obtaining ecotoxicologically relevant data.

Population and Location Dependence

The fate of xenobiotics may also be influenced by the location inhabited by the organism in question. This may be demonstrated by comparison of elimination and metabolic activities in *N. diversicolor*, sampled at two different sites in the Weser estuary (Goerke and Weber, 1990). With polychaetes sampled near Bremerhaven (Site I, mesohaline), a higher degree of metabolism and elimination of PCB compounds was observed compared to animals from a location about

TABLE 2.5 *Elimination Rates of Different Chlorobiphenyls in* Nereis diversicolor *from Different Locations (Goerke and Weber, 1990)*

PCB	Half-life (weeks) site I[a]	site II[a]
2,3,2'5'-tetrachlorobiphenyl	0.92	5.2
2,5,2',5'-tetrachlorobiphenyl	0.66	3.3
2,3,4,2',5'-pentachlorobiphenyl	0.89	4.1
2,4,5,2',5'-pentachlorobiphenyl	0.56	2.6

[a] See text.

TABLE 2.6 *Metabolites of* ^{14}*C-2,5,2',5'-tetrachlorobiphenyl in* Nereis diversicolor *from Different Locations (as % of total* ^{14}*C-tetrachlorobiphenyl equivalents accumulated) (Goerke and Weber, 1990)*

Location[a]	Metabolites extractable[b]	unextractable[b]	Parent compound
Site I	11.8	37.3	50.9
Site II	6.5	14.5	79.0

[a] See text.
[b] n-hexane/acetone 2:1 v/v.

30 km north of Bremerhaven (Site II, polyhaline), as shown in Tables 2.5 and 2.6. Obviously, the mesohaline population had developed an elevated degradation potential in the more polluted area.

SUMMARY AND CONCLUSION

From a number of experiments on the metabolism and elimination of persistent organic compounds, some conclusions can be drawn concerning the suitability of test species, the applicability of test results for ecotoxicological assessments and the quality of environmental analyses, used in, for example, monitoring.

If invertebrates to be used as test organisms, such as polychaetes or mussels, are carefully selected and exposed and maintained under strictly controlled laboratory conditions, results obtained on elimination and metabolism can be translated to natural conditions, at least in the case of a small group of PCB compounds. Furthermore, it is shown that significant amounts of metabolites are formed, which are not as readily eliminated as expected, and may therefore have ecotoxicological consequences.

Metabolic transformations of organic pollutants in invertebrates also influence the transfer of pollutants in the food chain, since substantial degradation and high elimination rates of organic pollutants in lower trophic levels will result in a decrease of biomagnification of the parent compounds (Ernst, 1984). There is little information existing, however, on the occurrence of metabolites of persistent organic compounds in marine species in natural environments. In a survey of the contamination of Dutch coastal waters, Quirijns *et al.* (1979) detected methylthiochlorobenzenes in *M. edulis* at levels of 0.005–0.4 mg/kg wet weight. They stated that these compounds could be attributed to the uptake of pollutants from the River Rhine. The source of the methylthiochlorobenzenes was discussed as being metabolites of quintozene (pentachloronitrobenzene) or polychlorobenzenes, and whether metabolism occurs in mussels or other components of the aquatic environment. Fish, invertebrates and sediments from various stations in the coastal region of Los Angeles (Santa Monica Bay and Palos Verdes) contained substantial amounts of oxygenated forms of DDTs and PCBs (Brown *et al.*, 1986, 1987).

These studies, as well as results reported in this paper, indicate that more consideration should be given to metabolites of persistent compounds as identified in metabolic studies when doing environmental analyses. The lack of data on the occurrence of metabolites of organic pollutants, and the fact that high amounts of total organic halogen were determined in marine species from Norway and Denmark but could not be accounted for as known compounds (Lunde and Gether, 1976; Folke and Birklund, 1986), show that our present knowledge of the body burdens of marine species with halogenated organic compounds is still incomplete.

Elimination rates and metabolic capabilities are principal descriptors for the persistence of recalcitrant organic pollutants in marine biota. Since substantial amounts of metabolites can be produced, even of supposedly persistent chemicals, it can be argued that the term "persistent" should be more specifically defined by indicating species, location and environmental factors. The observed variations of biokinetic parameters should thus be taken into account when test animals are selected for ecotoxicological studies.

REFERENCES

Adamson, R. H. 1967. Drug metabolism in marine invertebrates. *Fedn. Proc. Fedn. Am. Socs. exp. Biol.* **26**: 1047–1055.

Bauer, I., S. Weigelt and W. Ernst. 1989a. Biotransformation of hexa-chlorobenzene in the blue mussel (*Mytilus edulis*). *Chemosphere* **19**: 1701–1707.

Bauer, I., K. Weber and W. Ernst. 1989b. Metabolism of octachlorostyrene in the blue mussel (*Mytilus edulis*). *Chemosphere* **18**: 1573–1579.

Bend, J. R., M. O. James and P. M. Dansette. 1977. *In vitro* metabolism of xenobiotics in some marine animals. *Ann. N.Y. Acad. Sci.* **298**: 505–521.

Brown, D. A., R. W. Gosset, G. P. Hershelman, C. F. Ward, A. M. Westcott and J. N. Cross. 1986. Municipal wastewater contamination in the Southern California Bight: Part I. Metal and organic contaminants in sediments and organisms. *Mar. Environ. Res.* **18**: 291–310.

Brown, D. A., R. W. Gosset and S. R. McHugh. 1987. Oxygenated metabolites of DDT and PCBs in marine sediments and organisms. In J. M. Capuzzo and D. R. Kester (eds.),

Oceanic Progresses in Marine Pollution, Vol. I. Krieger Publ. Company, Malabor, Florida.

Engst, R., R. M. Macholz and M. Kujawa. 1979. Recent state of lindane metabolism. Part II. *Residue Rev.* **72**: 71–95.

Ernst, W. 1975. Aufnahme, Ausscheidung und Umwandlung von Lindan-^{14}C durch *Mytilus edulis. Chemosphere* **4**: 375–380.

Ernst, W. 1979. Factors affecting the evaluation of chemicals in laboratory experiments using marine organisms. *Ecotoxicol. Environ. Saf.* **3**: 90–98.

Ernst, W. 1984. Pesticides and technical organic chemicals. In O. Kinne, (ed.), *Marine Ecology*, Vol. V, *Ocean Management*, Part IV. *Pollution and protection of the seas: Pesticides, domestic wastes, and thermal deformations*, pp. 1627–1709. Wiley and Sons, Chichester, New York, Brisbane, Toronto, Singapore.

Ernst, W., H. Goerke and K. Weber. 1977. Fate of ^{14}C-labelled di-, tri- and pentachlorobiphenyl in the marine annelid *Nereis virens*. II. Degradation and faecal elimination. *Chemosphere* **9**: 559–568.

Ernst, W. and K. Weber. 1978. Chlorinated phenols in selected estuarine bottom fauna. *Chemosphere* **11**: 867–872.

Folke, J. and J. Birklund. 1986. Danish coastal water levels of 2,3,4,6-tetrachlorophenol, pentachlorophenol, and total organohalogens in blue mussels (*Mytilus edulis*). *Chemosphere* **15**: 895–900.

Glickman, A. H., C. N. Statham, A. Wu and J. L. Lech. 1977. Studies on the uptake, metabolism, and disposition of pentachlorophenol and pentachloroanisole in rainbow trout. *Toxicol. appl. Pharmac.* **41**: 649–658.

Goerke, H. and W. Ernst. 1977. Fate of ^{14}C-labelled di-, tri- and pentachlorobiphenyl in the marine annelid *Nereis virens*. I. Accumulation and elimination after oral administration. *Chemosphere* **9**: 551–558.

Goerke, H. and W. Ernst. 1980. Accumulation and elimination of ^{14}C-γ-HCH (lindane) in *Nereis virens* (Polychaeta) with consideration of metabolites. *Helgoländer Meeresunters.* **33**: 313–326.

Goerke, H. and W. Ernst. 1986. Elimination of pentachlorobiphenyls by *Nereis virens* (Polychaeta) in the laboratory and the marine environment. *Chemistry in Ecology* **2**: 263–285.

Goerke, H. and K. Weber. 1990. Population dependent elimination of various polychlorinated biphenyls in *Nereis diversicolor* (Polychaeta). *Mar. Environ. Res.* **29**: 205–226.

Ingebrigtsen, K. and J. E. Solbakken. 1985. Distribution and elimination of ^{14}C-hexachlorobenzene after single oral exposure in cod (*Gadus morhua*) and flounder (*Platichthys flesus*). *J. Toxicol. environ. Health* **16**: 197–205.

James, M. O. 1987. Conjugation of organic pollutants in aquatic species. *Environ. Health Perspect.* **71**: 97–103.

Kobayashi, K., H. Akitake and T. Tomiyama. 1970. Studies on the metabolism of pentachlorophenate, a herbicide, in aquatic organisms. III. Isolation and identification of a conjugated PCP yielded by a shellfish, *Tapes phillippinarum. Bull. Jap. Soc. scient. Fish.* **36**: 103–108.

Lee, R. F. 1981. Mixed function oxygenases (MFO) in marine invertebrates. *Mar. biol. Lett.* **2**: 87–105.

Lee, R. F., S. C. Singer and D. S. Page. 1981. Responses of cytochrome P-450 systems in marine crab and polychaetes to organic pollutants. *Aquatic Toxicology* **1**: 355–365.

Livingstone, D. R. and S. V. Farrar. 1984. Tissue and subcellular distribution of enzyme activities of mixed-function oxygenase and benzo[a]pyrene metabolism in the common mussel *Mytilus edulis. L. Sci. Total Environ.* **39**: 209–235.

Livingstone, D. R. 1985. Responses of the detoxication/toxication enzyme systems of molluscs to organic pollutants and xenobiotics. *Mar. Pollut. Bull.* **16**: 158–164.

Lunde, G. and J. Gether. 1976. Determination of volatility and chemical persistence of lipid-soluble halogenated organic substances in marine organisms. *Ambio* **5**: 180–182.

Moore, M. N., D. R. Livingstone, P. Donkin, B. L. Bayne, J. Widdows and D. M. Lowe. 1980. Mixed function oxygenases and xenobiotic detoxication/toxication systems in bivalve molluscs. *Helgoländer Meeresunters.* **33**: 278–291.

Neilson, A. H., H. Blanck, L. Förlin, L. Landner, P. Pärt, A. Rosemarin and M. Söderström. 1989. Advanced hazard assessment of 4,5,6-trichloroguajacol in the Swedish environment.

In L. Landner (ed.), *Chemicals in the Aquatic Environment*, pp. 329–374. Springer, Berlin, Heidelberg, New York, London, Paris, Tokyo, Hong Kong.

Quirijns, J. K., V. C. G. van der Paauw, M. C. ten noever de Brauw and R. H. de Vos. 1979. Survey of the contamination of Dutch coastal waters by chlorinated hydrocarbons, including the occurrence of methylthiopentachlorobenzene and di(methylthio)tetrachlorobenzene. *Sci. Total Environ.* **13**: 225–233.

Sanborn, J. R., W. F. Childers and L. G. Hansen. 1977. Uptake and elimination of (^{14}C)hexachlorobenzene (HCB) by the green sunfish, *Lepomis cyanellus* Raf. after feeding contaminated food. *J. agric. Food Chem.* **25**: 551–553.

Weber, K. and H. Goerke (in prep.). Species-dependent elimination of various polychlorinated biphenyls in estuarine animals.

Part II

Invertebrates of Higher Trophic Levels

CHAPTER 3

The Distribution of 2,3,7,8-tetrachlorodibenzo-*p*-dioxin in Juvenile Blue Crabs, *Callinectes sapidus*, and the Physiological Effects of Consumption of Food from a Polluted Environment on This Species

ANGELA CRISTINI* and KEITH COOPER†

* *Ramapo College of New Jersey, Mahwah, New Jersey, USA*

† *Rutgers University, Piscataway, New Jersey, USA*

INTRODUCTION

Polychlorinated dibenzo-*p*-dioxins (PCDDs) and polychlorinated dibenzofurans (PCDFs) are a family of compounds with a triple ring nucleus consisting of two benzene rings connected to each other by either a pair or a single oxygen atom respectively. There is the potential of any of the eight substitution positions to react with a halogen atom and a form a variety of compounds with different physical properties and biological activity (toxicity). Dioxins have entered the environment because of the use of certain commercial herbicides, cholorophenol-based chemicals, Kraft bleaching processes and by municipal and industrial waste incineration (Baughman and Meselson, 1973; Crosby *et al.*, 1973; Fishbein, 1976; Hutzinger *et al.*, 1985; Rappe *et al.*, 1987). These compounds remain in aquatic ecosystems for long periods of time and have been detected in the tissues of aquatic species and sediments near industrial locations in North America, Europe and Japan (Shapely, 1973; Kuehl, 1981; Harless *et al.*, 1982; Heida, 1983; Belton *et al.*, 1985; Rappe *et al.*, 1987).

A great deal of interest has been generated in the occurrence and fate of the chlorinated isomers, particularly 2,3,7,8, tetrachlorodibenzo-*p*-dioxin (2,3,7,8-TCDD), because of their toxicity and teratogenic effects at very low concentrations (Bage *et al.*, 1973; Bowes *et al.*, 1973; Kimbrough 1974; Hutzinger *et al.*, 1985; Mason *et al.*, 1985; Cooper *et al.*, 1986; Mason *et al.*, 1987; Wisk and Cooper, 1988). Although it is known that 2,3,7,8-TCDD is a potent inducer of cytochrome P-450 1A1 in mammals and fish (Stegeman *et al.*, 1989), no comparable studies have been carried out in crustacea (James, 1989). The Division of Science and Research of the New Jersey Department of Environmental Protection (NJDEP) examined the levels of dioxins and dibenzofurans in the sediments and biota of Newark Bay/Arthur Kill, New Jersey, USA (Belton *et al.*, 1985). These data indicate that the highest levels of 2,3,7,8-TCDD were found in the mid-gut glands (hepatopancreas) of the blue crab, *Callinectes sapidus*, collected in these waters [mean = 496 parts per trillion (ppt)]. The presence of these chemicals in this commercially and ecologically important species suggest the need for studies examining the physiological effects and the tissue distribution of 2,3,7,8-TCDD in sensitive life stages of *C. sapidus*.

Dioxins have a low water solubility (<12 ppt) and an ability to bind to sediments (Lodge and Cook, 1989; Lakind and Rifkin, 1990). Cantelmo (1977) has shown that *C. sapidus* reduce the water permeability of their gills and guts as a response to acclimation to the low salinities found in estuaries. Therefore, the most likely route of exposure to these crabs is via ingestion. Juvenile blue crabs are known to consume a variety of organisms, including the bivalve mollusc, *Mya arenaria*. Our data indicate that clams of this species from Newark Bay/Arthur Kill contain an average of 137 ppt dioxins (Cristini and Cooper, 1988). Therefore, the present study examined the tissue concentrations of blue crabs fed clams containing radiolabeled TCDD as well as the physiological processes in crabs fed *M. arenaria* from Newark Bay/Arthur Kill.

Maintenance of a viable population of *C. sapidus* depends both on the health of the ecosystem and on the ability of individual crabs to withstand the effects of sublethal concentrations of dioxin on their vital physiological processes. For crustaceans, the processes leading toward and occurring at ecdysis, as well as the process of limb regeneration, are critical and have been shown to be sensitive to the effects of toxicants (Peddicord and McFarland, 1976; Weis, 1977; Fingerman and Fingerman, 1978; Weis, 1978; Rao *et al.*, 1979; Cantelmo *et al.*, 1982; Callahan and Weis, 1983; Wang and Stickle, 1987; Weis *et al.*, 1987; Clare and Costlow, 1989). The length of the intermolt cycle, that is the time between one ecdysis and the next, may have important consequences, particularly for juvenile crabs that have a limited season for growth in northern latitudes. Crustaceans regenerate missing limbs prior to molting. In the Brachyura, the limb grows in a folded position within a layer of cuticle and unfolds at the next molt. Overall growth after each molt, as well as the growth rate of regenerating limbs, represent useful measures of an organism's performance, since growth is an integrated response to the environment, both functionally and temporally. It is also ecologically relevant, since it can be related to success of the population in the field (Sastry and Miller, 1981).

The length of the intermolt cycle, growth after each ecdysis, and the regeneration of limbs are all affected by sublethal concentrations of environmental pollutants. Cantelmo *et al.* (1981) have shown that long-term exposure of juvenile *C. sapidus* to sublethal concentrations of benzene and dimethylnaphthalene increased the length of the intermolt cycle. In addition, exposure to benzene decreased the growth increment per molt and retarded the rate of limb regeneration. Wang and Stickle (1987) reported that exposure of juvenile *C. sapidus* to the water-soluble fraction of S. Louisiana crude oil inhibited growth and molting. Weis and Mantel (1976) found that exposure of regenerating fiddler crabs, *Uca pugilator*, to 1,1-bis (4-chlorophenyl)-2,2,2-trichloroethane (DDT) produced an acceleration of regeneration and molting. This was attributed to the stimulatory effects of the organochloride pesticide on the neuroendocrine system. Exposure of grass shrimp, *Palaemonetes pugio*, to sodium pentachlorophenate did not alter the duration of the intermolt cycle, but caused a dose-related inhibition of limb regeneration (Rao *et al.*, 1981). The present study focused on these three measures (length of the intermolt cycle, growth after ecdysis and limb regeneration) as indicators of sensitivity to chemicals present in the environment for *C. sapidus*.

MATERIALS AND METHODS

Maintenance of Animals

Juvenile *C. sapidus* with carapace widths from 1 to 6 cm were collected from Barnegat Bay, New Jersey, USA. The animals were maintained in individual glass containers containing approximately 700 ml of artificial seawater (Instant Ocean Sea Water Mix) with a constant rate of aeration. The osmolality of the seawater was maintained at approximately 280–380 mosmol, which was the same as ambient seawater from Barnegat Bay. Crabs were kept in an environmental chamber at 23°C on a 12-h light/12-h dark cycle. Before any animal was assigned to either a control or test group it was sexed, weighed, and staged according to the methods of Aiken (1973) to determine its position in the intermolt cycle. The width of the carapace was measured and the crab was induced to autotomize the third appendage on the right or left side. All crabs were acclimated to the laboratory conditions for 2 weeks before tests were initiated. Those in the intermolt stage (C-4) of their cycle were divided into the control and experimental groups for the feeding experiment. Animals were assigned so that each group contained crabs in the same size range with equal numbers of both sexes.

Tissue Distribution Experiment

The bodies of 12–15 soft shelled clams, *Mya arenaria*, from a control area containing background levels of dioxins and dibenzofurans (0.3 ppt), were blended with water, gelatin and tritiated 2,3,7,8-TCDD (98% pure, Chemsyn Science Laboratories, Lenexa, KA) (specific activity 150 DPM/pg of 2,3,7,8-TCDD) to a total volume of 500 ml. The resulting "food gel" contained 50.8 pg/g of 2,3,7,8-TCDD. Crabs were fed 1 g of food gel per day and allowed to eat for 1 h. The water was sampled before and after feeding and was changed after feeding. The stomach, gills, body muscle, claw muscle and mid-gut gland (hepatopancreas) were dissected out after animals had been sacrificed at 24, 48, 72 and 96 h. Any fecal material that was present in the water was also collected. A second study was carried out in order to determine the distribution of 2,3,7,8-TCDD in the hemolymph of the crabs. The food was prepared and the crabs were fed as described above, except that the "food gel" contained 370 pg/g of 2,3,7,8-TCDD. Hemolymph samples were drawn from the base of the walking leg with a 1 ml syringe at 2, 24, 48 and 72 h; the crabs were then sacrificed and the tissues dissected out. All of the samples were dissolved in 1 N NaOH overnight, acidified with 100 μl of glacial acetic acid, brought to 10 ml with Liquiscint and counted in a Tracer Mark III (Tracor Analytic, Elk Grove Village, IL) scintillation counter.

Feeding Experiment

The bodies of *M. arenaria* from the control area (0.3 ppt dioxins and dibenzofurans) and collected from the mud flat in Elizabeth NJ in the Newark Bay ecosystem (137 ppt dioxins and dibenzofurans) were used as the food for this experiment. The food was prepared as described above, except that the radiolabeled dioxin was omitted. The crabs were fed 1 g of the control area or Elizabeth "food gel" 5 days per week for 67 days. They were allowed to eat for 1 h, then the food was removed and the water changed.

Growth of the autotomized limb of the control and experimental animals was monitored throughout the entire molt cycle. The length of the limb bud was measured to the nearest 0.1 mm with vernier calipers.

R values (length of limb bud/carapace width ×100) were calculated according to the method of Bliss and Boyer (1964). The process of setogenesis was also observed on all crabs. *C. sapidus* is a member of the family Portunidae (swimming crabs). The distal segment of the fifth walking leg, therefore, is modified into a flattened "swimming paddle." The entire process of apolysis and setogenesis was observed by examining this segment under a Bausch and Lomb microscope at 100× magnification.

The number of molts and the days to molt were recorded for all animals. The carapace width of all crabs was measured with vernier calipers to the nearest 0.1 mm in the intermolt stage of their molt cycle. The width of the carapace was measured again post molt, when the shell was hard enough. The gain in width was calculated by subtracting the width before from the width after molt. The statistical differences between the control and test animals was determined by ANOVA.

RESULTS AND DISCUSSION

Tissue Distribution Experiment

The fate of radiolabeled 2,3,7,8-TCDD in juvenile blue crabs is shown in Table 3.1. Changing the water 1 h after feeding prevented the accumulation of 2,3,7,8-TCDD in the water from the food. The levels of tritiated 2,3,7,8-TCDD in the water 24 h after feeding were 3–5 DPM above background water and were lower than the levels in the tissues, suggesting that the 2,3,7,8-TCDD accumulated in the tissues as the result of feeding. The 2,3,7,8-TCDD present in the water may have been due to the dioxin adsorbed on to the glass or to the elimination of dioxin from the suspended fecal material.

The mid-gut gland rapidly accumulated the highest concentrations of 2,3,7,8-TCDD and continued to accumulate it throughout the 96-h study (52.5–301.5 pg TCDD equivalents/g wet weight). The gills, stomach and muscle also accumulated 2,3,7,8-TCDD throughout the 96-h study, although the levels were much lower in these tissues. The amount of radiolabeled 2,3,7,8-TCDD recovered in the hemolymph increased from 8.8 pg/ml to 53 pg/ml in 72 h. The levels in the hemolymph compared to the tissue levels at 72 h suggest that the hemolymph could explain the occurrence of 2,3,7,8-TCDD in highly perfused and lipid-poor organs such as the gill and claw muscles. The fecal material contained 2,3,7,8-TCDD, indicating that some excretion occurred or that some material had not been absorbed.

These data demonstrate that juvenile *C. sapidus* can accumulate 2,3,7,8-TCDD present in the food. The natural populations of *M. arenaria* in the Newark Bay/Arthur Kill system contain elevated levels

TABLE 3.1 *The Distribution of 2,3,7,8-TCDD Equivalents (pg equivalents/g wet weight for tissues; pg equivalents/ml for hemolymph) (±SEM) in the Tissues of Juvenile* Callinectes sapidus *Fed Food Containing [H³] 2,3,7,8-TCDD*

Tissue	Time (hours)			
	24	48	72	96
Stomach	5.3 ± 1.5	5.9 ± 1.9	12.2 ± 9.9	30.5 ± 8.9
Midgut gland	52.5 ± 17.4	31.4 ± 14.4	130.5 ± 81.5	301.5 ± 100.1
Gill	2.5 ± 1.1	3.7 ± 2.4	6.2 ± 4.1	15.6 ± 3.6
Body muscle	3.7 ± 0.9	5.6 ± 2.1	7.4 ± 3.8	29.7 ± 9.9
Claw muscle	5.9 ± 1.8	7.8 ± 3.8	12.9 ± 8.8	30.9 ± 9.9
Feces	29.2 ± 15.9	15.5 ± 14.3	70.9 ± 5.8	61.2 ± 25.3

N = five animals sacrificed at each time.
Food = 50.8 pg/g.
Water = 0.02 pg/ml.
Specific activity = 150 DPM/pg of [H³] 2,3,7,8-TCDD.

	2	24	48	72
Hemolymph	8.8 ± 2.8	27.4 ± 3.8	41.5 ± 4.3	53.0 ± 4.7
Stomach	–	–	–	186.1 ± 8.5
Midgut gland	–	–	–	475.3 ± 14.8
Gill	–	–	–	92.5 ± 5.8
Body muscle	–	–	–	135.8 ± 8.9
Claw muscle	–	–	–	62.5 ± 4.4

N = seven animals.
Food = 370 pg/g.
Water = 0.02 pg/ml.
Specific activity = 150 DPM/pg of [H³] 2,3,7,8-TCDD.

of dioxins and dibenzofurans. Miyata *et al.* (1987, 1989) have used the mussel *Mytilus edulis* as a bioindicator for the accumulation of dioxins and have demonstrated correlations between the levels and isomers in fly ash and mussel tissues. Brown *et al.* (1989) have shown that *M. arenaria* concentrated radiolabeled 2,3,7,8-TCDD, present in the food, at the highest levels in the gonad and digestive gland; the gills concentrated 2,3,7,8-TCDD at much lower levels. It is hypothesized that in the field the clams accumulate the dioxins and dibenzofurans when they ingest sediment and food particles as they filter-feed, and possibly also across the surface of the gills. The tissue levels in adult crabs collected in this system are also high, the midgut glands containing the highest levels of any of the animals measured (Belton *et al.*, 1985). *Callinectes sapidus* is known to be a voracious consumer of many benthic invertebrates, including *M. arenaria*. The results of these studies, therefore, could indicate a possible mechanism for uptake of these chemicals through the various trophic levels in the environment.

Feeding Experiment

After 67 days, 50% of the crabs fed clams from Elizabeth molted while 69% of the crabs fed clams from the control area molted. The results of the feeding experiment on molting, time to molt and final stage by size categories are shown in Table 3.2. These data clearly indicate that the animals in the 3–4 cm width size class show the greatest delay in time to molt: when the experiment was terminated, all but one control animal had molted (mean = 45 days) compared to only four treated animals molting (mean = 57 days). The final stage of the one remaining control animal was D, which is the premolt condition. Three out of the seven experimental animals that did not molt were still in the C-4 stage, which is the intermolt condition and four had advanced to the premolt stage. The crabs in the smallest size class (1–3 cm) did not exhibit any significant differences in number of animals that molted or in the average time to molt. However, the two experimental animals that did not molt were still in the C-4 stage of their cycle when the experiment was terminated; in addition, one control animal completed two molt cycles during the study. These data indicate that the chemicals present in the clams from Elizabeth were not affecting the smaller animals until the second or third molt. Although only two of the control animals from the largest size class (4–6 cm) molted, the staging indicates that five were in the premolt condition. None of the treated animals molted and only one was in the premolt condition by the end of the experiment.

TABLE 3.2 *The Results of the Feeding Experiment on Molting [number of animals(%)], Time to Molt (days ± SD) and Stage of the Molt Cycle (number of animals) for Juvenile* Callinectes sapidus *Fed Clams from Newark Bay/Arthur Kill (Elizabeth) and a Control Area*

Carapace width (cm)	Control food	Elizabeth food
	Number of animals molted (%)	
1–3	13 (100%) n=13	15 (88%) n=17
3–4	11 (92%) n=12	4 (36%) n=11
4–6	2 (15%) n=13	0 (0%) n=13
	Days to molt av. (± SD)	
1–3	38 ± 8	43 ± 9
3–4	45 ± 6[a]	57 ± 1[a]
4–6	56 ± 15	no molts
	Final stage in molt cycle (number of animals)	
1–3	All molted	C-4 intermolt (2)
3–4	D Premolt (1)	C-4 Intermolt (3)
		D Premolt (4)
4–6	C-4 Intermolt (6)	C-4 Intermolt (12)
	D Premolt (5)	D Premolt (1)

[a] Sig. at $p < 0.05$.

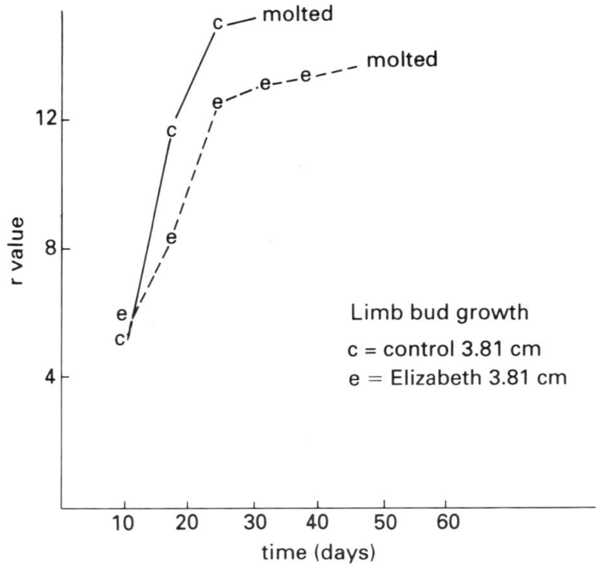

FIG. 3.1 *R* values (length of limb bud/carapace width ×100) of two crabs in the 3–4 cm group that molted. The animal that was fed the clams from Elizabeth exhibited a reduced rate of limb bud regeneration.

These data suggest that there would be a difference between the control and treated crabs if the experiment was continued for a longer period of time.

The progression of *R* values for crabs that molted in the study indicate further effects of the chemicals present in the clams collected in Elizabeth. Figure 3.1. is a plot of *R* values vs. time for two typical animals of similar size that molted. The control showed rapid initial growth followed by a slight plateau before ecdysis took place. The crab that was fed clams from Elizabeth exhibited a much longer plateau in *R* value before molting. Table 3.3 shows the average *R* value at the end of the experiment for the crabs that did not molt. The control animals in the 4–6 cm size class have significantly higher ($p<0.05$) *R* values than the animals fed food from Elizabeth indicating that limb regeneration had progressed further in the control group. The limb buds of the control animals grew continuously, while the animals eating food from Elizabeth appeared to plateau with little or no growth in the limb bud.

These data on molting and limb bud regeneration are similar to the results of the exposure of juvenile blue crabs to benzene, dimethylnaphthalene, and the water-soluble fraction of S. Louisiana crude oil (Cantelmo *et al.*, 1981; Wang and Stickle, 1987). These results also concur with studies on other species of crustacea and suggest that alterations of the molting process and limb regeneration are sensitive

TABLE 3.3 *Results of the Feeding Experiment on Limb Regeneration (Final* R *value) for Juvenile* Callinectes sapidus *Fed Clams from Newark Bay/Arthur Kill (Elizabeth) and from a Control Area*

Carapace width (cm)	Final R value av. ($\pm$SD) R = length of bud/carapace width $\times$100 Control food	Elizabeth food
1–3	All molted n=13	12.51 $\pm$ 0.58 n=2
3–4	16.25 n=1	12.45 $\pm$ 2.92 n=7
4–6	12.41 $\pm$ 2.50 n=11[a]	9.88 $\pm$ 2.92 n=13[a]

[a] Sig. at $p < 0.05$.
R values = length of limb bud/carapace width $\times$100.

and perhaps generalized responses to persistent low-level chemical stress for this group of animals (Weis, 1977; Fingerman and Fingerman, 1978; Rao *et al.*, 1981; Weis *et al.*, 1986; Weis *et al.*, 1987). Benzene and dimethylnaphthalene have been shown to affect the rates of energy production in juveniles of this species at the level of the tissues and the whole animal (Cantelmo *et al.*, 1981).

In addition, both the process of molting and limb regeneration are known to be under neuroendocrine control (Bliss, 1956; Passano, 1960; Kleinholz, 1976). Information on the role of the cytochrome P-450s in steroid metabolism in crustacea is limited (James, 1989), and the effects of xenobiotics on these enzymes have not been been examined. There is evidence in the mammalian literature that 2,3,7,8-TCDD binds to specific forms of P-450 and inhibits catalytic activity (Voorman and Aust, 1987). Singer and Lee (1977) demonstrated that the benzopyrene hydroxylase activity changed during the different stages of the molt cycle of *C. sapidus*. Therefore, if dioxins can alter steroid metabolism, or other enzymes involved in the molt cycle, then delays in molting or complete inhibition of molting might occur. It is also possible that secretion or reception of neuroendocrine factors are altered by chemicals that are bioaccumulated by crustacea. Future studies should focus on how the general metabolism, as well as neuroendocrine secretions and mechanisms, are affected by these chemicals.

The gain in width at molt was measured for control animals and those fed food from Elizabeth during this study. The results are presented in Table 3.4. All animals consuming food from Elizabeth gained significantly less width (p <0.001) than those fed food from the control area. These data indicate that chronic exposure to the chemicals present in the clams living in Newark Bay/Arthur Kill at Elizabeth affect the growth of individual crabs and could possibly affect the size structure of the population in this area. The values are similar to the width gain values reported by Cantelmo *et al.* (1981) for control *C. sapidus* (0.46 cm) and animals exposed to benzene (0.18 cm). Differences

TABLE 3.4 *The Results of the Feeding Experiment on Width Gain at Molt (cm) for Juvenile* Callinectes sapidus *Fed Clams from Newark Bay/Arthur Kill (Elizabeth) and from a Control Area*

Carapace width (cm)	Width gain (cm) (±SEM) Control food	Elizabeth food
1–3	0.530 ± 0.252[a]	0.247 ± 0.102[a]
3–4	0.558 ± 0.116[a]	0.295 ± 0.123[a]

[a] Sig. at $p < 0.001$.

between the two studies could be explained by the size classes that were pooled, exposure routes, and different food and feeding regimes. Wang and Stickle (1987) also report that the percent size increase after molting was significantly reduced in juvenile *C. sapidus* exposed to the water-soluble fraction of oil. The results of these studies suggest that juveniles of this species show a similar response to a variety of organic chemicals present in the environment or in the food.

The present study does not provide data directly linking dioxins and dibenzofurans in the tissues of the clams from Newark Bay/Arthur Kill with the alterations in molting, growth, and limb regeneration. Rather, the experiments show that 2,3,7,8-TCDD can be accumulated in juvenile blue crabs and that physiological processes can be affected by food containing this and many other chemicals present in this severely polluted environment. Future experiments must now focus on the relationships between dioxins and dibenzofurans and the physiological processes associated with molt, and further examine these processes as generalized responses to chemical stressors in the environment.

The data on frequency of molting and growth increment at molt suggest that juvenile animals exposed to chemicals in their environment or in their food could grow less in a season than crabs living in a clean environment. Clearly, data obtained in a laboratory study can only be applied to populations in the field with caution. However, the data on growth are within the range reported for this species in field studies (Darnell, 1959; More, 1969; Perry, 1975). Additionally, this study has shown a reduction in size and alterations in processes associated with molt in animals fed food taken directly from a polluted environment. This suggests that long-term exposure to chemicals in the environment could have detrimental effects on both individuals and populations of *C. sapidus*. There are, however, some studies indicating that populations of animals living in a polluted environment become resistant to the accumulated chemicals. An *in situ* study comparing *M. arenaria* native to the Arthur Kill/Raritan Bay with clams from Long Island Sound caged in Arthur Kill/Raritan Bay indicated the existence of a resistant population (Cristini, 1987). The clams from Long Island Sound caged in the Arthur Kill/Raritan Bay had biochemical energy pools that were significantly lower than the clams native to this area. The data indicated that clams from Long Island were predisposed to be less able to accommodate to the stresses associated with long-term survival in the chemically impacted environment of the Arthur Kill/Raritan Bay, whereas the animals native to this ecosystem appeared to have a predisposition that allowed for their survival. Weis *et al.* (1981, 1982) showed that populations of the killifish, *Fundulus heteroclitus*, from a small tributary to the Arthur Kill, produced eggs and embryos that were more resistant to methyl mercury than did killifish from a control area in Long Island Sound. Laboratory studies indicated that the resistance had a genetic basis and that the animals from the Arthur Kill were near to their tolerance limit. Clearly, future studies must also consider such a possibility for juvenile blue crabs.

Acknowledgments This research was supported by the New Jersey Department of Environmental Protection, Division of Science and Research. The authors are indebted to Susan Bernard and Richard Brown for their many hours of field and laboratory work.

REFERENCES

Aiken, D. E. 1973. Proecdysis setal development, and molt prediction in the American lobster (*Homarus americanus*). *J. Fish. Res. Bd. Canada* **30**: 1337–1344.

Bage, G., E. Cekanova and K. S. Larsson. 1973. Teratogenic and embryotoxic effects of the herbicides di- and trichlorophenoxyacetic acids (2,4-D and 2,4,5-T). *Acta Pharmacol.* **32**: 408–423.

Baughman, R. and M. Meselson. 1973. An analytical method for detecting TCDD in samples from Vietnam. *Environ. Health Perspect.* **5**: 11–46.

Belton, T. J., R. Hazen, B. E. Ruppel, K. Lockwood, R. Muller, E. Sterenson and J. J. Post. 1985. A study of dioxin (2,3,7,8-tetrachlorodibenzo-*p*-dioxin) contamination in select finfish, crustaceans and sediments of New Jersey waterways. Office of Science and Research NJDEP.

Bliss, D. E. and J. R. Boyer. 1964. Environmental regulation of growth in the decapod crustacean *Gecarcinus lateralis*. *Gen. Comp. Endocrinol.* **4**: 15–41.

Bliss, D. E. 1956. Neurosecretion and the control of growth in a decapod crustacean. In K. G. Wingstrand (ed.), *Bertil Hanstrom. Zoological Papers in Honor of his Sixty-fifth Birthday*, Nov. 20, 1956. Zool. Inst., Lund, Sweden, pp. 56–75.

Bowes, G. W., B. R. Simoneit, A. L. Burlingame, B. W. DeLappe and R. W. Risebrough. 1973. The search for chlorinated dibenzofurans and chlorinated dibenzodioxins in wildlife populations showing elevated levels of embryonic death. *Environ. Health Perspect.* **5**: 191–208.

Brown, R. P., A. Cristini and K. Cooper. 1989. The absorption and tissue distribution of 3H-TCDD in the soft-shell clam *Mya arenaria*. SETAC 10th Annual Meeting, Oct. 28–Nov. 2, Ontario, Canada. Abstract #243, p. 250.

Callahan, P. and J. S. Weis. 1983. Methylmercury effects on regeneration and ecdysis in fiddler crabs (*Uca pugilator*, *U. pugnax*) after short-term and chronic pre-exposure. *Arch. Environ. Contam. Toxicol.* **12**: 707–714.

Cantelmo, A. 1977. Water permeability of isolated tissues from decapod crustaceans – 1. Effects of osmotic conditions. *Comp. Biochem. Physiol.* **58A**: 343–348.

Cantelmo, A., L. Mantel, R. Lazell, F. Hospod, E. Flynn, S. Goldberg and M. Katz. 1982. The effects of benzene and dimethylnaphthalene on physiological processes in juveniles of the blue crab, *Callinectes sapidus*. In W. Vernberg, A. Calabrese, F. P. Thurberg and F. J. Vernberg (eds), *Physiological Mechanisms of Marine Pollutant Toxicity*. Academic Press, NY, pp. 349–389.

Cantelmo, A., R. Lazell and L. Mantel. 1981. The effects of benzene on molting and limb regeneration in juvenile *Callinectes sapidus*. *Mar. Biol. Lett.* **2**: 333–343.

Clare, A. S. and J. D. Costlow Jr. 1989. Effects of the insecticide methomyl on development and regeneration in megalop and early postlarval stages of the mud crab, *Rhithropanopeus harrisii* (Gould). In D. Weigmann (ed.), *Pesticides in Terrestrial and Aquatic Environments*. VA Polytech. Inst. Water Resources Inst. Blacksburg, VA, pp. 3–16.

Cooper, K. R., T. H. Umbreit and M. A. Gallo. 1986. The effect of 2,3,7,8-tetrachlorodibenzo-*p*-dioxin on the development and survival of the Japanese medaka (*Oryzias latipes*). *The Toxicologist* **5**: 27–38.

Cristini, A. 1987. An *in situ* study of adenylate energy charge and other biochemical parameters in the bivalve *Mya arenaria* from Raritan Bay and Long Island Sound. In W. B. Vernberg, A. Calabrese, F. Thurberg and F. T. Vernberg (eds), *Pollution Physiology of Estuarine Organisms*. University of South Carolina Press, pp. 231–250.

Cristini, A. and K. Cooper. 1988. Bioavailability and physiological effects of dioxins on the bivalve mollusk *Mya arenaria*. SETAC 9th Annual Meeting, Nov. 3–17, Arlington, VA. Abstract #33, p. 36.

Crosby, D. G., K. W. Moilanen and A. S. Wong. 1973. Environmental generation and degradation of dibenzodioxins and dibenzofurans. *Environ. Health Perspect.* **5**: 259–298.

Darnell, R. M. 1959. Studies of the life history of the blue crab (*Callinectes sapidus* Rathbun) in Louisiana waters. *Trans. Am. Fish. Soc.* **88**: 294–304.

Fingerman, S. and M. Fingerman. 1978. Effects of two polychlorinated biphenyls (aroclor 1242 and 1254) on limb regeneration in the fiddler crab *Uca pugilator*, at different times of the year. *Vie Milieu* **28–29**: 69–75.

Fishbein, L. 1976. Trace organic contaminants in the environment. I. Chlorinated dioxins and dibenzofurans. *Int. J. Ecol. Environ. Sci.* **2**: 69–96.

Harless, R. L., E. O. Oswald, R. G. Lewis, A. E. Dunpuy, D. D. McDaniel and H. Tai. 1982. Determination of 2,3,7,8-tetrachlorodibenzo-*p*-dioxin in freshwater fish. *Chemosphere* **12**: 503–537.

Heida, H. 1983. TCDD in bottom sediments and eels around a refuse dump near Amsterdam, Holland. *Chemosphere* **12**: 503–549.

Hutzinger, O., G. G. Chowdry, B. G. Chitim and L. E. Johnson. 1985. Formation of polychlorinated dibenzofurans and dioxins during combustion, electrical equipment fires and PCB incineration. *Environ. Health Perspect.* **60**: 3–56.

Kimbrough, R. D. 1974. The toxicity of polychlorinated polycyclic chemicals and related compounds. *Crit. Rev. Toxicol.* **2**: 445–486.

Kleinholz, L. H. 1976. Crustacean neurosecretory hormones and physiological specificity. *Am. Zool.* **16**: 151–166.

Kuehl, D. W. 1981. Unusual polyhalogenated chemical residues identified in fish tissue from the environment. *Chemosphere* **10**: 231–268.

Lakind, J. and E. Rifkin. 1990. Current method for setting detection limits in water requires reexamination. *Environ. Sci. Technol.* **24**: 963–965.

Lodge, K. B. and P. M. Cook. 1989. Partitioning studies of dioxin between sediment and water: the measurement of KOC for Lake Ontario sediments. *Chemosphere* **19**: 439–494.

James, M. O. 1989. Cytochrome P-450 monooxygenases in crustaceans. *Xenobiotica* **19**(10): 1063–1076.

Mason, G., T. Sawyer, B. Keys, S. Bandiera, M. Romkes, J. Piskorska-Pliszczynska, B. Zmudzka and S. Safe. 1985. Polychlorinated dibenzofurans (PCDFs): correlation between *in vivo* and *in vitro* structure-activity relationships. *Toxicology* **37**: 124–138.

Mason, G., T. Zacharewski, M. A. Denomme, L. Safe and S. Safe. 1987. Polybrominated dibenzo-*p*-dioxins and related compounds: quantitative *in vivo* and *in vitro* structure-activity relationships. *Toxicology* **44**: 245–272.

Miyata, H., K. Takayama, J. Ogaki, T. Kashimoto and S. Fukushima. 1987. Polychlorinated dibenzo-*p*-dioxins in blue mussel from marine coastal water in Japan. *Bull. Environ. Contam. Toxicol.* **39**: 872–883.

Miyata, H., K. Takayama, M. Mimura, T. Kashimoto and S. Fukushima. 1989. Specific congener profiles of polychlorinated dibenzo-*p*-dioxins and dibenzofurans in blue mussel in Osaka Bay in Japan: aqueous solubilities of PCDDs and PCDFs. *Bull. Environ. Contam. Toxicol.* **43**: 342–340.

More, W. R.1969. A contribution to the biology of the blue crab (*Callinectes sapidus* Rathbun) in Texas, with a description of the fishery. *Tex. Parks Wildl. Dept. Tech. Serv.* **1**: 1–31.

Passano, L. M. 1960. Molting and its control. In T. Waterman (ed.), *Physiology of Crustacea*, Vol. 1. *Metabolism and Growth.* Academic Press, NY, pp. 473–536.

Peddicord, R. and V. McFarland. 1976. Effects of suspended dredged material on the commercial crab, *Cancer magister*. In P. Krenkel, J. Harrison and J. Burdick III (eds), *Dredging and Its Environmental Effects.* American Society of Civil Engineers, NY, pp. 633–644.

Perry, H. M. 1975. The blue crab fishery in Mississippi. *Gulf Res. Rep.* **5**: 39–57.

Rao, K. R., F. R. Fox, P. J. Conklin, A. C. Cantelmo and A. C. Brannon. 1979. Physiological and biochemical investigations of the toxicity of pentachlorophenol to crustaceans. In W. B. Vernberg, F. P. Thurberg, A. Calabrese and F. J. Vernberg (eds), *Marine Pollution: Functional Responses.* Academic Press, NY, pp. 307–340.

Rao, K. R., F. R. Fox, P. J. Conklin and A. C. Cantelmo. 1981. Comparative toxicology and pharmacology of chlorophenols: studies on the grass shrimp, *Palaemonetes pugio*. In F. J. Vernberg, A. Calabrese, F. P. Thurberg and W. B. Vernberg (eds), *Biological Monitoring of Marine Pollutants.* Academic Press, NY, pp. 37–72.

Rappe, C., R. Anderson, P. Bergqvist, C. Brohede, M. Hansson, L. Kjeller, G. Lindstrom, S. Marklund, M. Nygren, S. E. Swanson, M. Tysklind and K. Wiberg. 1987. Overview on the environmental fate of cholorinated dioxins and dibenzofurans. Sources, levels and isometric pattern in various matrices. *Chemosphere* **16**: 1603–1618.

Sastry, A. N. and D. C. Miller. 1981. Application of biochemical and physiological responses to water quality monitoring. In F. J. Vernberg, A. Calabrese, F. P. Thurberg and W. B. Vernberg (eds), *Biological Monitoring of Marine Pollutants.* Academic Press, NY, pp. 265–294.

Shapely, D. 1973. Herbicides: AAAS study finds dioxin in Vietnamese fish. *Science* **180**: 285–286.

Singer, S. C. and R. F. Lee. 1977. Mixed function oxygenase activity in blue crab *Callinectes sapidus*: tissue distribution and correlation with changes during molting and development. *Biol. Bull.* **153**: 377–386.

Stegeman, J. J., M. R. Miller and D. E. Hinton. 1989. Cytochrome P-450 1A1 induction and localization in endothelium of vertebrates (Teleost) heart. *Molecular Pharmocology* **36**: 723–729.

Voorman, R. and S. D. Aust. 1987. Specific binding of polyhalogenated aromatic hydrocarbon inducers of cytochrome P-450 to the cytochrome and inhibition of its estradiol 2-hydroxylase activity. *Toxicology and Applied Pharmacology* **90**: 69–78.

Wang, S. Y. and W. B. Stickle. 1987. Bioenergetics, growth and molting of the blue crab, *Callinectes sapidus*, exposed to the water-soluble fraction of South Louisiana crude oil. In W. B. Vernberg, A. Calabrese, F. P. Thurberg and F. J. Vernberg (eds), *Pollution Physiology of Estuarine Organisms.* Univ. of South Carolina Press, Columbia, SC, pp. 107–126.

Weis, J. S. 1978. Interactions of methylmercury, cadmium, and salinity on regeneration in the fiddler crabs *Uca pugulator*, *Uca pugnax*, and *Uca minax*. *Mar. Biol.* **49**: 119–124.

Weis, J. S., J. Gottlieb and J. Kwiatkowski. 1987. Tributyltin retards regeneration and produces deformities of limbs in the fiddler crab *Uca pugulator*. *Arch. Environ. Contam. Toxicol.* **16**: 321–326.

Weis, J. S. 1977. Limb regeneration in fiddler crabs: species differences and effects of methlymercury. *Biol. Bull.* **152**: 263–274.

Weis, J. S. and L. H. Mantel. 1976. DDT as an accelerator of regeneration and molting in fiddler crabs. *Estuarine Coastal Mar. Sci.* **4**: 461–466.

Weis, J. S., N. Donellon and F. Barnwell. 1986. Regeneration of tubercles on the limbs of *Uca pugulator* and effects of mercury and cadmium on their growth. *J. Crust. Biol.* **6**: 648–651.

Weis, J. S., P. Weis, M. Heber and S. Vaidya. 1982. Investigations into mechanisms of heavy metal tolerance in killifish (*Fundulus heteroclitus*) embryos. In W. B. Vernberg, A. Calabrese, F. Thurberg and F. J. Vernberg (eds), *Physiological Mechanisms of Marine Pollutant Toxicity*. Academic Press, NY, pp. 311–330.

Weis, J. S., P. Weis, M. Heber and S. Vaidya. 1981. Methylmercury tolerance of killifish (*Fundulus heteroclitus*) embryos from a polluted vs. non-polluted environment. *Marine Biology* **65**: 283–287.

Wisk, J. and K. R. Cooper. 1988. Lethal and sublethal potency of various dioxin congeners to the Japanese medaka embryo, *Oryzias latipes*. *The Toxicologist* **8**: 54.

CHAPTER 4

The Relationship Between PCBs in Biota and in Water and Sediment from New Bedford Harbor: a Modeling Evaluation

JOHN P. CONNOLLY

HydroQual, Inc., 1 Lethbridge Plaza, Mahwah, New Jersey 07430 and Environmental Engineering and Science Program, Manhattan College, Riverdale, New York 10471, USA

INTRODUCTION

From the 1940s to the late 1970s New Bedford Harbor, a tidal estuary located at the head of Buzzards Bay in southeastern Massachusetts, received PCB containing industrial waste waters from two electronic component manufacturing plants. These discharges resulted in PCB concentrations in sediments that ranged as high as 100,000 ppm and concentrations in biota that exceeded the Food and Drug Administration (FDA) Action Limit of 2 μg/g (previously 5 μg/g). As a result of this PCB contamination, finfish and shellfish fisheries in the harbor, and in bordering areas of Buzzards Bay, have been closed since 1979. Of particular concern in this regard are the lobster and winter flounder whose fisheries are economically important to the area. The region has been designated as a "SuperFund Site" and a Remedial Investigation/Feasibility Study (RI/FS) has recently been completed. As part of this study, a model of PCB homologs 3 through 6 (tri-, tetra-, penta- and hexachlorobiphenyl) in the lobster and winter flounder food chains was developed and applied to the harbor and Buzzards Bay. This model is an extension of a modeling framework previously applied to food chains in Lake Michigan (Thomann and Connolly, 1984), and the James River estuary (Connolly and Tonelli, 1985).

The model has been successfully calibrated to data collected as part of the RI/FS. Calibration was achieved with a single set of bioenergetic parameters and assimilation efficiency and excretion rates which were defined by literature-based relationships to octanol/water partition coefficient (K_{ow}). It is the purpose of this paper to discuss the importance of various processes to the accumulation of PCBs and how these processes control the relationship between PCBs in biota and in water and sediment.

PCB CONTAMINATION IN NEW BEDFORD HARBOR

PCB measurements made prior to 1979 in water, sediment and biota in New Bedford Harbor revealed a significant concentration gradient. Highest concentrations were observed in the upper estuary, and these declined downriver and through an outer harbor area into Buzzards Bay (study area is shown in Fig. 4.1). Consistent with this gradient, the fishery closure established in 1979 by the Massachusetts Department of Public Health is divided into three areas. The area north of the hurricane barrier (i.e. the inner harbor, area 1) is closed to the taking of all finfish, shellfish and lobsters. The outer harbor (area 2), extending from the hurricane barrier to a line from Wilber Point to Ricketsons Point, is closed to the taking of lobster and bottom-feeding finfish including eel, scup, flounder and tautog. Between the outer harbor and a line from the southern tip of West Island to Round Hill Point (area 3) only the lobster fishery is closed. Food chains were considered for each of the closure areas. In addition, a fourth area representative of the Buzzards Bay region south of area 3 was modeled. Area 1 is restricted to the region between the hurricane barrier and the route 6 bridge, the area assumed to be the habitat of the species of interest.

As part of the RI/FS, Battelle Ocean Sciences (1990) conducted cruises in New Bedford Harbor and Buzzards Bay in September–October 1984, November–December 1984 and June–July 1985 to collect water column, sediment and biota samples. These samples were analyzed for PCB congener concentrations and the results were reported on a homolog basis. For the purposes of the food chain model, concentrations were averaged for each of the four areas. Because cruise to cruise variability was not great (Battelle Ocean Sciences, 1990) concentrations were also averaged across cruises to establish a data set for calibration of the food chain model.

Water Column and Sediment Concentrations

Average concentrations of PCBs dissolved in the water column for homologs 2 through 9 from each of the four areas are presented in

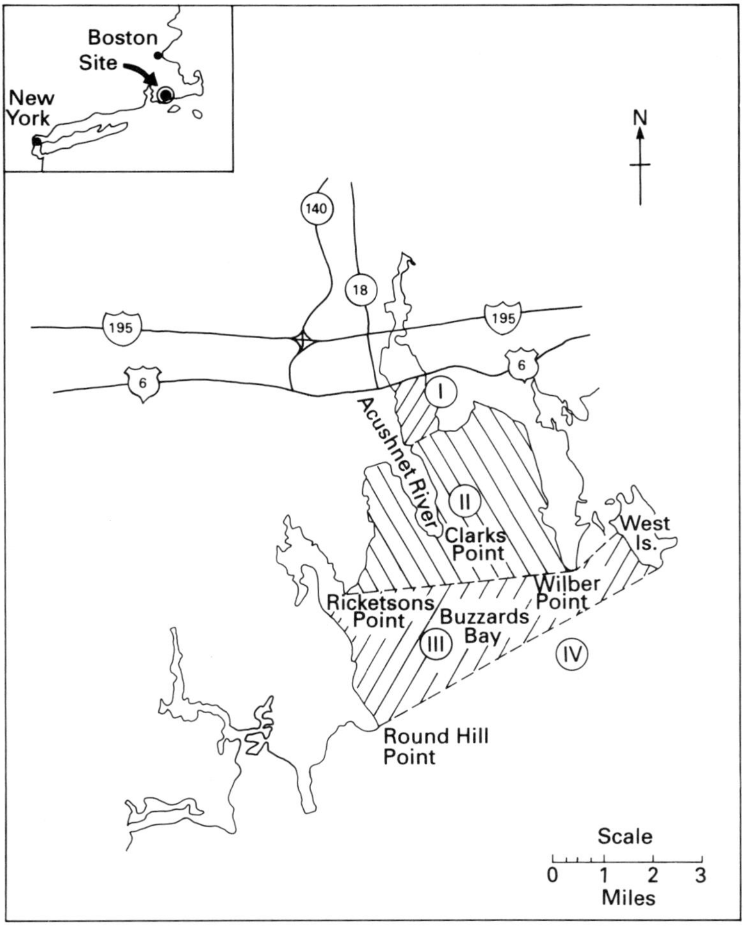

FIG. 4.1 New Bedford Harbor study area indicating the four segments used for analysis of PCBs in the food chain.

Table 4.1 and Fig. 4.2. Comparing the areas, a dramatic decline in PCB concentration is evident proceeding from the lower inner harbor (area 1) to Buzzards Bay (area 3). Over this region concentrations decline by a factor of *ca*. 30; from a total dissolved PCB concentration of about 70 ng/L to about 2 ng/L. Moving out into Buzzards Bay from area 3 to area 4, concentrations remain approximately constant. In all of the areas homologs 3 through 6 comprise greater than 90% of the total PCB. Homolog 2 is significant in areas 1 and 2. The relatively high values for homolog 9 in areas 1 and 2 may be an analytical artifact.

TABLE 4.1 *Average Measured PCB Homolog Concentrations in the Water, Sediment and Biota from the Four Areas*

Homolog	Water ng/L	Sediment μg/g C	Flounder μg/g lipid	Polychaete μg/g lipid	Lobster μg/g lipid	Crab μg/g lipid	Mussel μg/g lipid	Clam μg/g lipid
Area 1								
2	2.8	9.1	2.6			1.33	2.32	3.2
3	23.4	53.3	56.2			29.8	42.67	41.6
4	27	108	162.6			75.5	87.74	82.1
5	12	121.2	224.2			95.7	77.4	65.9
6	2.7	46.8	104			39	23.04	24.1
7	1.1	8.1	8.9			4.62	1.31	2.27
8	0.84	4.3	1.46			0.63	0.07	0.31
9	2.2	6.26	0.21			0.38	0.03	0.16
Area 2								
2	0.87	6.17	1	4.1	0.33	5.07	5.13	0.13
3	6.5	23	15.4	36.8	4.75	6.37	59.76	2.13
4	7.9	42.4	50.4	93.4	17.7	18.3	150.2	7.39
5	4.6	52.1	85.7	213.3	31.7	46.6	158.6	8.82
6	1.3	16.7	46.2	96.3	22.6	30.8	54.6	3.5
7	0.33	4.69	4.6	18.9	2.4	4.34	3.35	0.44
8	0.24	4.12	0.82	5	0.38	0.68	0.21	0.07
9	0.47	6.47	0.14		0.23	0.26		0.1
Area 3								
2	0.08	0.53	0.62	0.72	0.16	1.99	0.24	0.16
3	0.68	2.9	3.39	4.6	1.04	11.59	7.43	2.43
4	1	6.19	16.4	9.7	5.32	30.77	25.5	8
5	0.64	10.45	35.9	19.5	13.88	64.24	40.3	11.67
6	0.2	4.97	22.3	8.8	11.58	38.85	17.9	6.51
7	0.03	0.79	2.45	1.09	1.48	5.63	1.24	0.77
8	0.02	0.21	0.38	0.31	0.28	0.73	0.04	0.21
9	0.02	0.12	0.06	0.11	0.21	0.22	0.02	0.18
Area 4								
2	0.07	0.28	0.48	0.63	0.05	0.1	0.02	0.07
3	0.6	1.25	0.95	2.46	0.25	0.5	0.18	0.52
4	0.74	2.42	4.65	7.67	1.27	1.55	0.7	1.61
5	0.55	3.6	11.2	18.25	4.35	4.39	1.33	2.27
6	0.19	1.8	7.7	11.57	4.52	3.26	0.77	1.39
7	0.04	0.37	0.96	1.97	0.63	0.55	0.08	0.2
8	0.02	0.22	0.16	0.43	0.13	0.11	0.01	0.04
9	0.02	0.27	0.02	0.2	0.09	0.11	0.03	0.02

Surface sediment concentrations decline in similar fashion to the dissolved concentrations in the water column (Table 4.1 and Fig. 4.3), although the rate of decline is less. From area 1 to area 3 total PCB concentrations decline from about 360 μg/gC to about 25 μg/gC, a factor of 14. Unlike those in the water column, sediment concentrations in area 4 are lower than in area 3. The average total PCB concentration of 10 μg/gC is about the same factor below that observed in area 1 as found for dissolved PCB. Nearly all of the PCB in the sediment is in the homolog range of 3 to 6.

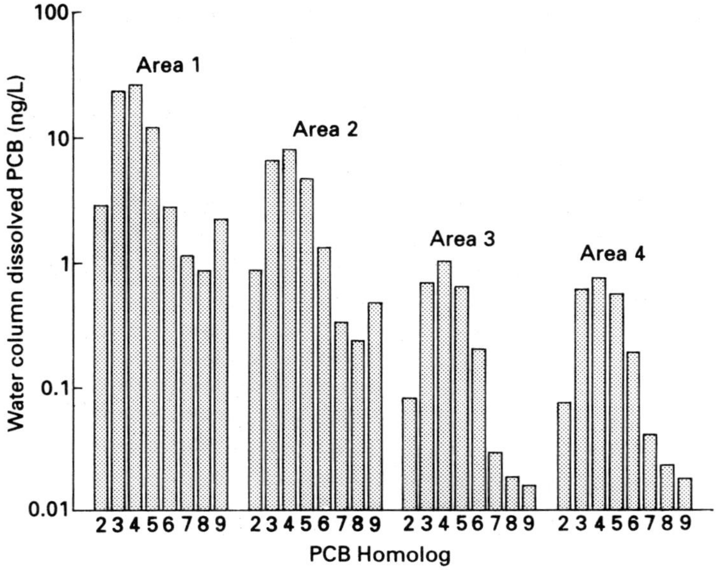

FIG. 4.2 Average water column dissolved PCB homolog concentrations for each of the four areas.

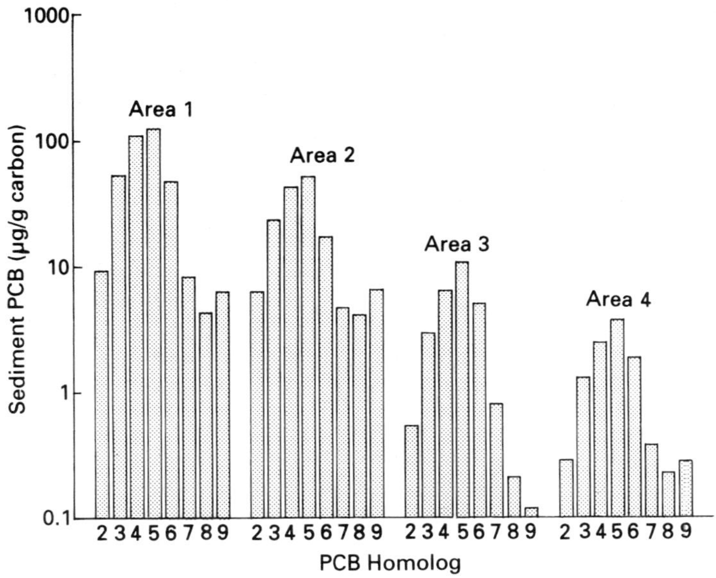

FIG. 4.3 Average sediment PCB homolog concentrations for each of the four areas.

Although limited measurements of water column particulate organic carbon (C) are available to compute carbon normalized water column particulate PCB, the computed averages indicate concentrations significantly less than those observed in the surface sediment. Values of total PCB for the four areas, proceeding from area 1, are about 125, 45, 8 and 2 μg/gC, as compared to surface sediment values of 360, 160, 25 and 10 μg/gC. These differences between the water column and the sediment suggest that the system is depurating PCBs in response to cessation of active loading.

Concentrations in Biota

PCB concentrations were measured in mixed species of polychaetes, hard clams (*Mercenaria mercenaria*), mussels (*Mytilus edulis*), spider crabs (*Chionecetus opilio*), winter flounder (*Pseudopleuronectes*) and lobsters (*Homarus americanus*). The average lipid-normalized concentrations of homologs 2 through 9 are presented for each species and each area in Table 4.1. As with the water and sediment, homologs 3 through 6 account for greater than 90% of the total PCBs in all of the animals sampled. Concentrations decline from area 1 through area 4 in a manner consistent with that observed for water and sediment. Flounder and polychaetes exhibit the highest whole body concentrations, probably reflecting their greater exposure to sediment PCBs. The lowest concentrations are found in the clam. Concentrations decline by factors of about 10 to 20 from area 1 to area 4. The pattern of decline differs somewhat from species to species. Both clams and crabs exhibit similar concentrations in areas 2 and 3, whereas the concentrations in area 3 flounder and lobster are about 2 and 3 less than the values observed in area 2. Similarly, polychaete concentrations are about the same in areas 3 and 4 in contrast to the significantly lower values found in area 4 for all of the other species. These differences in the spatial pattern of PCB contamination probably reflect differences in PCB exposure. In particular, sedentary species collected in locations where the PCB exposure concentrations were very different from the average concentration of the area may not reflect the area to area patterns as well as mobile species whose exposure is integrated over a larger region.

Average bioaccumulation factors, BAFs (i.e. biota concentration in units μg/kg lipid divided by dissolved concentration in units μg/L), are consistent with previous analyses of field data (Connolly and Pedersen, 1987; Thomann, 1989), increasing with chlorine number, and thus K_{ow}, to a maximum value and then decreasing (Fig. 4.4). For all species the maximum BAF occurs for hexachlorobiphenyl. The drop in BAF for the higher chlorinated homologs has been attributed to a decline in transfer efficiency within the gut (Thomann, 1989).

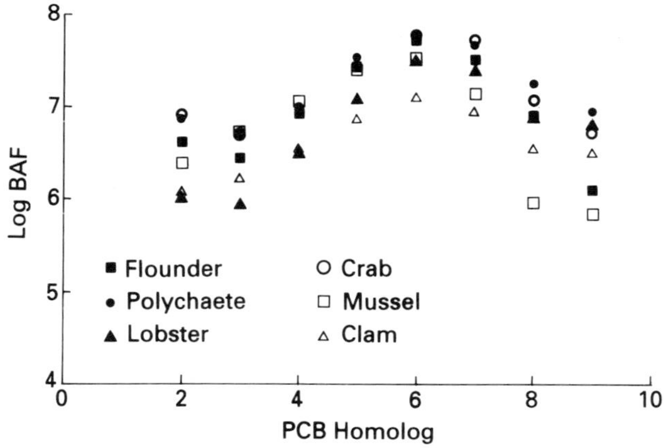

FIG. 4.4 Bioaccumulation factors (BAFs) averaged over the entire study area in relation to homolog for each of the species sampled.

Ratio of Edible to Whole Body PCB

Measurements of PCBs in flounder and lobster muscle tissue and in lobster hepatopancreas indicate a proportionality to whole body concentration that is independent of homolog (Fig. 4.5). Average ratios are 1.35 for lobster muscle-to-whole body, 0.18 for flounder muscle-to-whole body and 92 for lobster hepatopancreas-to-whole body (Battelle Ocean Sciences, 1990). Thus the 10 μg/g wet weight concentration in area 1 flounder is equivalent to about 2 μg/g muscle, indicating that these flounder are at PCB levels about equal to the FDA Action Limit. For lobster both the muscle and hepatopancreas are edible, and conversion of whole body to edible tissue requires knowledge of the weight fractions of each. Limited data from the USEPA Narragansett Laboratory (Richard Pruell, personal communication) indicate that muscle and hepatopancreas constitute about 91% and 9% of edible tissue, respectively. From these data the lobster edible-to-whole body PCB concentration ratio is about 9.2. Thus the maximum lobster concentrations of about 1 μg/g live weight are equivalent to edible concentrations of about 9 μg/g, considerably above the FDA Action Limit.

The tissue-to-whole body ratios are independent of homolog because, presumably, they are essentially a measure of the ratio of lipid content between the tissues and whole body. Lipophilic chemicals such as PCBs are deposited predominantly in lipids. The concentration in any tissue is dependent on the lipid fraction of that tissue and the partitioning characteristics of the chemical. The ratio between two tissues is the ratio of the products of lipid mass and chemical concentration in the

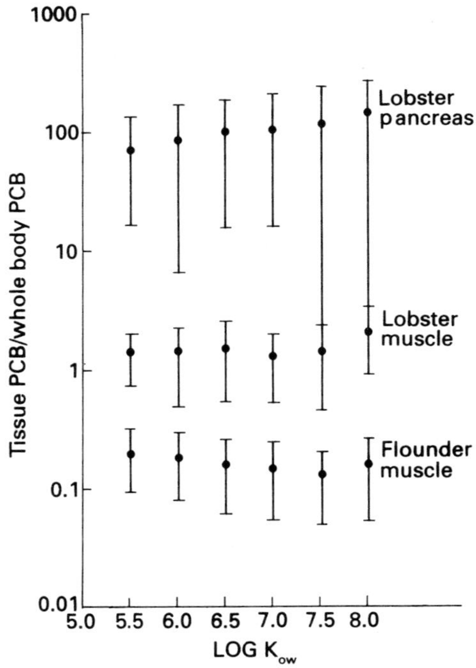

FIG. 4.5 Mean (± standard deviation) ratio of PCB homolog concentration in muscle or pancreas to PCB homolog concentration in whole body of New Bedford Harbor lobsters and flounders.

lipid. Because the chemical concentration in lipid should be equal for both tissues, the ratio reduces to the ratio of lipid masses. Thus the ratio is independent of homolog.

Distribution of PCB Homologs

Although homologs 3 through 6 constitute greater than 90% of the total PCB concentration observed in the water, sediment and biota, the distribution of total PCB among these homolog groupings differs between these phases. As shown in Fig. 4.6, the distributions shift toward the more highly chlorinated homologs as one moves from water to sediment to flounder. Trichloro- and tetrachlorobiphenyl comprise the largest fractions of water column dissolved PCB, whereas tetrachloro- and pentachlorobiphenyl are the dominant fractions in water column particulates, sediment and flounder. The flounder have the highest percentages of pentachloro- and hexachlorobiphenyl. The distributions for water column and sediment solids are essentially identical.

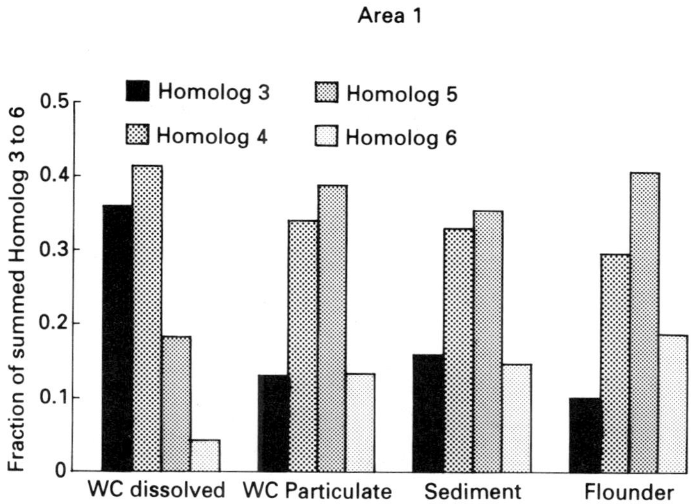

FIG. 4.6 Fraction of PCB homologs 3 through 6 in each homolog for various phases. Data are from area 1.

This shift toward the more highly chlorinated homologs results in the increase in BAF with homolog that is shown in Fig. 4.4. Because Fig. 4.4 indicates that the BAF reaches a maximum and then declines, the homolog shift does not persist to the most highly chlorinated homologs.

As illustrated in Fig. 4.7, the homolog distribution of all the phases changes with distance from the PCB sources (i.e., going from area 1 to area 4). In both the surface sediment and the flounder the fractions of tri- and tetrachlorobiphenyl decrease and the fractions of hexa- and heptachlorobiphenyl increase. Similar behavior is seen for the water column dissolved chemical, except that the changes are not as dramatic. These changes are likely the result of differences in the fate of the homologs as they are transported through the system. Such changes are consistent with a loss of lower chlorinated PCBs to volatilization.

MODELING OF PCB ACCUMULATION

Mass Balance Equation

The accumulation of toxic chemicals by aquatic animals may be described by the following equation (Thomann and Connolly, 1984):

$$\frac{dv_i}{dt} = K_{ui}c + \sum_{j=1}^{n} a_{ij}C_{ij}v_j - (K_i + G_i)v_i \qquad (1)$$

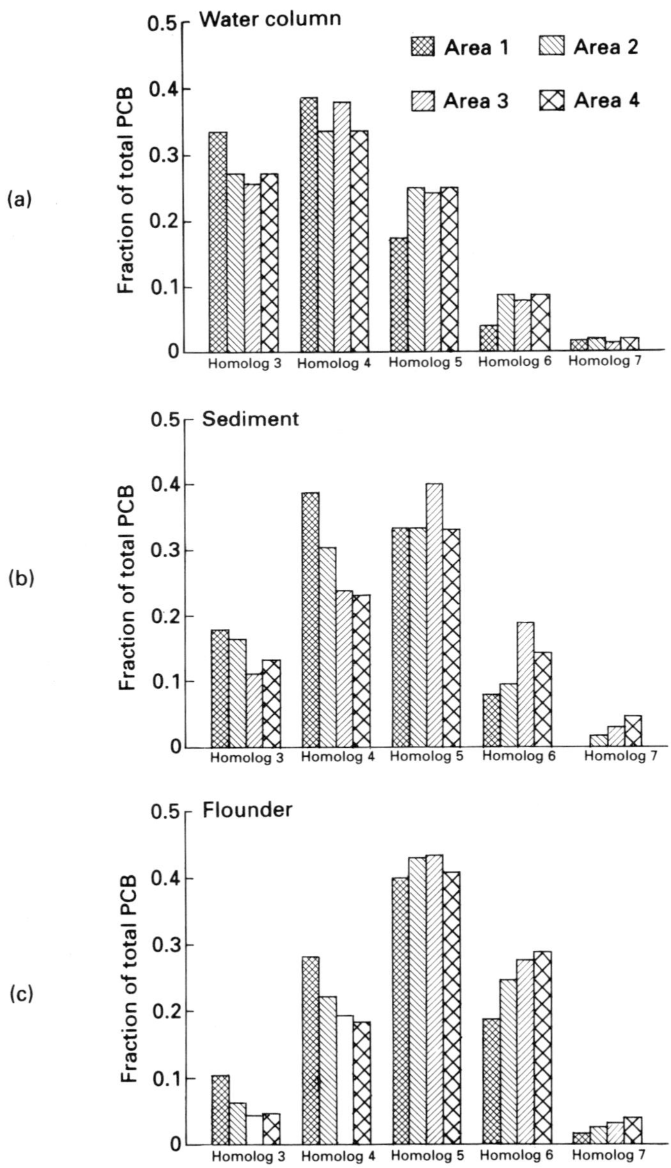

FIG. 4.7 Fraction of total PCB concentration in various homologs in each of the four areas for (a) water column dissolved PCB, (b) sediment PCB, and (c) flounder PCB.

where v_i = concentration of chemical in species i in the food chain [μg/g(w), where g(w) is grams wet weight],
K_{ui} = rate constant for chemical uptake across the gill of species i [L/g(w)-d],
K_i = rate constant for excretion of chemical by species i (l/d),
a_{ij} = efficiency at which ingested chemical from prey j is assimilated by species i,
C_{ij} = ingestion or consumption rate of species i on species j [g(w)prey/g(w)predator/d],
G_i = growth rate of species i [g(w)/g(w)/d],
c = dissolved chemical concentration (μg/L),
n = number of species (including different year classes of a single species) preyed on by species i.

The first term of equation (1) represents the direct uptake of chemical by the animal from water. The second term represents the flux of chemical into the animal through feeding. The third term is the loss of chemical due to desorption and excretion plus the change in concentration due to growth. Equation (1) is applied to each of the animals that comprise the food chain. For the upper levels of the food chain, changes in chemical concentration with age are sometimes significant and each year class of the species at these levels is modeled separately. The parameterizations of each of the coefficients in equation (1) and the values chosen are presented by Connolly (1991).

The calculation is driven by externally specified concentrations of water column dissolved chemical and sediment dissolved and particulate chemical. The sediment particulate organic carbon is the base of the benthic food chain. The base of the pelagic food chain is water column phytoplankton. Chemical concentration in phytoplankton is calculated from the water column dissolved concentration by the use of a partition coefficient.

Food Chain Structure

Determination of the appropriate species to include in the model is based on a review of published stomach content data for lobster and flounder. From these data species are chosen to represent the next lower level of their food chains. Representative species are then chosen for each lower trophic level in succession to the phytoplankton-detritus level. At any level a single species is sometimes sufficient, because the members of that level generally have similar bioenergetic characteristics and chemical concentrations. Additional prey animals are incorporated as necessary to account for differing prey chemical concentrations or multiple vectors of chemical transfer to the predator (i.e., sediment vs. water column).

The lobster is the top predator in a three-level food chain (Battelle Ocean Sciences, 1990). In the model, crabs, mussels and polychaetes are defined as the prey species of the lobster. Phytoplankton and sediment detrital organic material constitute the base of the food chain. Polychaetes are included, even though they are a less important food source than juvenile crabs and mussels, because they provide a different route of chemical transfer to the lobster. The small crabs and mussels transfer chemical from the water column to the lobster through consumption of phytoplankton. The polychaetes transfer chemical from the sediment to the lobster through their ingestion of sediment particles.

The winter flounder is an omnivore. Field studies clearly indicate that benthic animals form the bulk of the diet (Battelle Ocean Sciences, 1990). In the model, young-of-the-year winter flounder are assumed to consume equal percentages of phytoplankton and polychaetes. Older flounder are assumed to consume polychaetes only. In this structure the polychaetes are viewed as representatives of the variety of benthic animals that are prey for the flounder. Mature flounder exhibit local onshore–offshore migration in response to temperature changes. However, migration was judged not to be significant for the less than 6-year-old animals included in the model (Battelle Ocean Sciences, 1990) and was not considered.

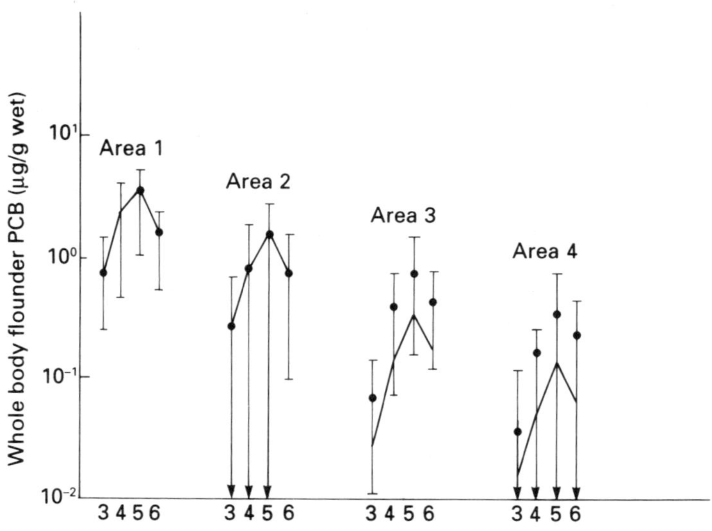

FIG. 4.8 Comparison of observed (mean and range) and computed flounder whole body PCB homolog concentrations for each of the four areas.

Model Calibration

The final calibrations were the results of model runs to determine a consistent set of parameters that were in agreement with literature or laboratory values and reproduced the observed average PCB concentrations reported during the sampling cruises in 1984 and 1985.

The food chain model was calibrated for homologs 3, 4, 5 and 6 and for total PCB. The model successfully reproduced the variation in body burdens across the homologs and over the entire food chain (Connolly, 1991). Comparisons between observed and computed concentrations in flounder are presented in Fig. 4.8 as an example of how well the model fit the data. As seen in this figure, the model also reproduced the spatial concentration gradients evident in the data, although some bias was evident in areas 3 and 4. In these areas the computed values fell within the range of the data, but they were consistently below the mean.

FACTORS CONTROLLING PCBs IN BIOTA

Sources and Sinks of Biota PCB Concentrations

The model indicates that for all of the homologs food constitutes the major route of PCB uptake for animals above the base of the food chain, becoming increasingly important as PCB chlorine content increases (Connolly, 1991). This conclusion is consistent with previously published generic models of organic chemical accumulation (Connolly and Pedersen, 1987; Thomann, 1989). It occurs because as a chemical becomes more hydrophobic and lipophilic its concentration in the plankton and sediment that constitute the base of the food chain becomes increasingly greater than its concentration in the water.

Because food constitutes the major vector of contamination, the phytoplankton bioconcentration factor (BCF) and the assimilation efficiency of ingested chemical are controlling parameters in the model. Successful calibration required that the BCF be constant for all homologs and that assimilation efficiency decline as PCB chlorination increases (Fig. 4.9). A constant BCF is consistent with field data from other systems (Connolly, 1991). The reduction in assimilation efficiency signifies a reduction in the bioavailability of the chemical to which the animal is exposed. I have shown elsewhere (Connolly, 1991) that this reduction is forced by the data and is not the result of the particular parameter set used in the calibration. Assuming the basic validity of the model (i.e., equation 1), other hypotheses cannot explain the observed biota PCB concentrations in relation to the observed water and sediment PCB concentrations.

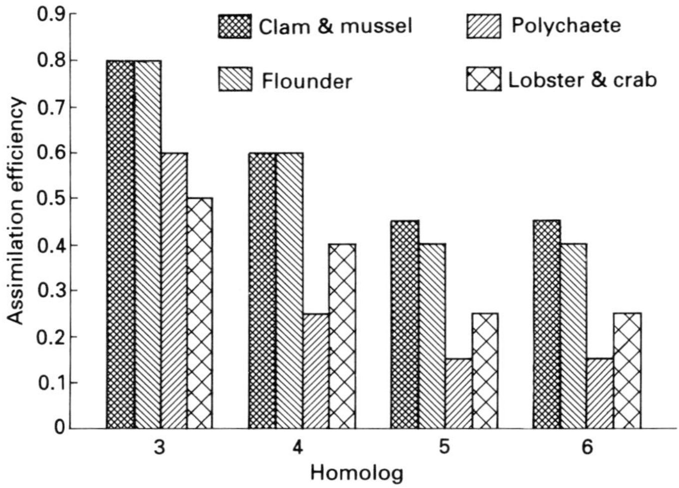

FIG. 4.9 Chemical assimilation efficiencies used in the calibrated model for the various species and each of the homologs.

PCB concentration is reduced through both excretion and growth. While growth rate is a constant loss mechanism for all the homologs, it is well known that excretion rate declines as chlorine content increases because of increasing lipophilicity. For chemicals with log K_{ow} values greater than about 6.5 to 7, the limited available data suggest that the excretion rate tends to become approximately constant (Gobas *et al.*, 1989). The rates at which the biota in the model excrete the lower chlorinated homologs are much larger than the rates at which they grow. For example, the growth rate of 3-year-old flounder in the model is 0.0018/d, whereas the excretion rate of trichlorobiphenyl is about 0.018/d. At the point where excretion rate levels off (i.e., penta- to hexachlorobiphenyl), the rates are of the same order (3-year-old flounder excretion rate is about 0.018/d) and both processes are equally important as sinks of PCB concentration. At this point, the rates of PCB concentration loss become essentially independent of homolog. This functions to limit the increase in PCB accumulation observed as a function of homolog at the lower homologs. However, as will be discussed, the dominant control on accumulation is a decline in uptake rate due to the decline in assimilation efficiency of ingested PCBs.

Metabolism of PCBs was not explicitly considered in defining the rates of excretion. Recent experimental data suggest that some species are capable of metabolizing individual PCB congeners. For example, Goerke and Weber (1990) found that the rates of elimination of 2,5,2′,5′- and 2,3,2′,5′-tetrachlorobiphenyl, and 2,4,5,2′,5′- and 2,3,4,2′,5′-pentachlorobiphenyl by the polychaete *Neries diversicolor* differed between animal populations by factors of from 2 to 5, with one population apparently able to metabolize the congeners. By examining PCB congener ratios between species, Boon *et al.* (1989) concluded that certain congeners were metabolized and that the significance of this metabolism varied between species. Whereas these data suggest that metabolism occurs, its significance with regard to homolog level excretion remains problematical. The high BAF values observed in New Bedford Harbor and elsewhere tend to indicate that it is not likely a major factor in the PCB dynamics of aquatic food chains.

Factors Controlling the PCB Homolog Distribution

Two significant and related features of the data analyses presented earlier are the differences in the contributions of each homolog to total PCB in biota as compared to water and sediment and the BAF–homolog or, equivalently, the BAF–K_{ow} relationship. The model provides a framework within which these observed relationships may be explained. Both are consequences of differences between homologs in the rates of uptake and loss of PCBs by the biota.

For animals with a largely water column based food chain, such as the filter feeding mussel and hard clam, the model indicates that the observed relationship between BAF and K_{ow} in which BAF tends to a maximum at hexachlorobiphenyl, is a direct result of a decrease in the bioavailability of ingested chemical and a lack of increase in phytoplankton BCF with increasing PCB chlorine content. This is illustrated in Fig. 4.10, which shows the average over the four areas of

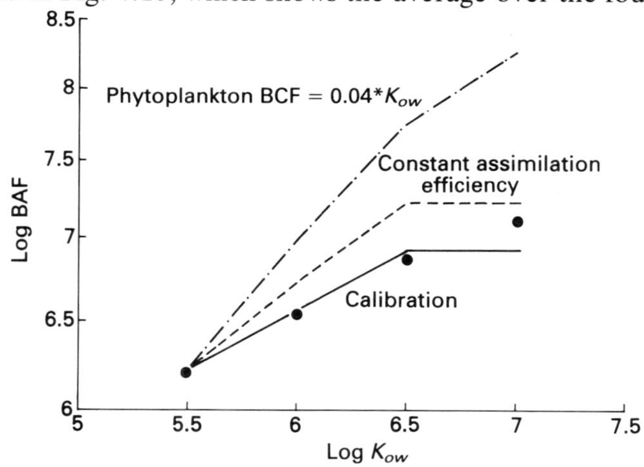

FIG. 4.10 Comparison of the average observed homologs 3 through 6 BAFs in the hard clam with values computed by the calibrated model and alterations of the model that use either a constant assimilation efficiency or a phytoplankton BCF that is proportional to the K_{ow} of the homolog.

the observed hard clam homolog BAFs and the BAFs computed in the calibration and if chemical assimilation efficiency is assumed constant at the value used for trichlorobiphenyl or if the phytoplankton BCF is made proportional to the K_{ow} of the homolog. Although both factors are significant in achieving a reasonable fit to the data, the constant phytoplankton BCF is much more important. Using a constant assimilation efficiency overestimates the BAF by as much as half an order of magnitude. Allowing the phytoplankton BCF to be proportional to K_{ow} results in an over-prediction of as much as one and one-half orders of magnitude.

For animals with a largely benthic based food chain, such as the polychaete and the flounder, phytoplankton partitioning is unimportant. Sediment forms the base of this food chain and its equivalent BCF or partition coefficient (i.e., sediment PCB/water column dissolved PCB), which is fixed by observation, is an important factor in the observed relationship of BAF and K_{ow}. Although it increases with PCB chlorination or K_{ow} it tends to level off between penta- and hexachlorobiphenyl (Fig. 4.11). While this leveling off does influence the flounder BAF, the dominant effect is the decline in assimilation efficiency. As shown in Fig. 4.11, maintaining a constant assimilation efficiency results in an over-prediction of as much as one and one-half orders of magnitude. The effect is greater than that shown for the hard clam, because the flounder are one step higher in the food chain and are effected by the assimilation efficiency of the polychaete as well as their own efficiency.

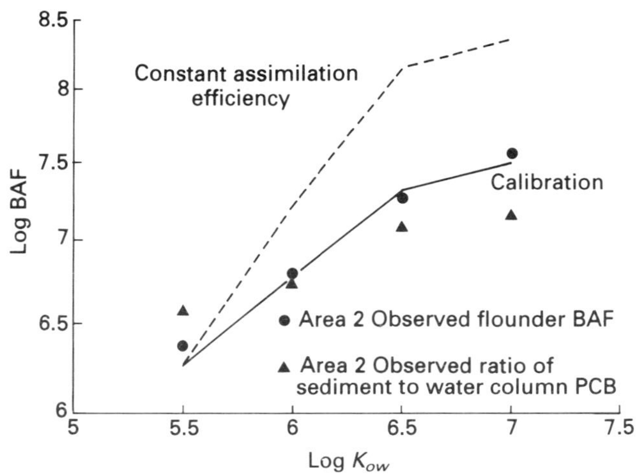

FIG. 4.11 Comparison of the area 2 observed homologs 3 through 6 BAFs in the flounder with values computed by the calibrated model and an alteration of the model that uses a constant assimilation efficiency.

Because the model was not calibrated for homologs above hexachlorobiphenyl, it does not extend into the K_{ow} region in which the BAF begins to decline. However, since the model indicates that the uptake of the highest homologs is almost entirely from food, the only way in which a BAF decline may occur is if the chemical assimilation efficiency continues to decline. Thomann (1989) has demonstrated this using a generic food chain model.

An inverse relationship between chemical assimilation efficiency and K_{ow} has been found in several laboratory studies which examined highly hydrophobic chemicals with log K_{ow} values greater than about 7 (Bruggeman *et al.*, 1984; Muir and Yarechewski, 1988). However, the results of this modelling evaluation of field data indicate that an inverse relationship extends down to log K_{ow} values of about 5.5 to 6.

CONCLUSIONS

1. BAFs of PCB homologs in aquatic animals increase with chlorine number to a maximum at hexachlorobiphenyl. Above this point they tend to decline with chlorine number. This pattern occurs for all of the species examined, including animals exposed only to water column PCBs (i.e., hard clam and mussel), those exposed to a combination of sediment and water column PCBs (i.e., crab, lobster and flounder) or those exposed only to sediment PCBs (i.e., polychaetes).

2. The distribution of PCBs between muscle and whole body in flounder and in lobster and between hepatopancreas and whole body in lobster is independent of the chlorine number of the PCB. This observation suggests that, since PCBs are stored in lipids, the distribution simply reflects the distribution of lipids between the tissues.

3. The homolog distribution of PCBs differs between water, sediment and biota from New Bedford Harbor. Water tends to have a higher proportion of di-, tri- and tetrachlorobiphenyl than either sediment or biota. Biota have the highest proportion of penta- and hexachlorobiphenyl.

4. A food chain model calibrated to PCB homologs 3 through 6 and total PCB indicates that most to nearly all of the PCBs in animals above the base of the food chain comes from ingested prey. Excretion constitutes the major PCB concentration sink for tri- and tetrachlorobiphenyl. Growth becomes an important sink for penta- and hexachlorobiphenyl, being of the same order as the excretion rate.

5. The model indicates that the assimilation efficiency of ingested PCBs declines with increasing PCB chlorination above trichlorobiphenyl.

6. For water column animals the observed BAF–K_{ow} relationship is largely a consequence of the decline in assimilation efficiency and the lack of increase in phytoplankton BCF with K_{ow}.

7. For animals receiving significant PCBs from the sediment, the observed BAF–K_{ow} relationship is largely a consequence of the decline in assimilation efficiency and the relationship of sediment and water column PCBs.

Acknowledgments This work was carried out at HydroQual, Inc. under subcontract to Battelle Ocean Sciences as part of the USEPA New Bedford Harbor RI/FS Modeling Program. John P. St. John of HydroQual directed the project and his contribution to this work is gratefully acknowledged. Maureen Casey and Leslie Sparrow of HydroQual participated in the data analysis and model calibration. The support of the Battelle Project Manager Richard McGrath is also gratefully acknowledged.

LITERATURE CITED

Battelle Ocean Sciences. 1990. *New Bedford Harbor Modeling Program Final Report.* United States Environmental Protection Agency, Boston, MA.

Boon, J. P., F. Eijgenraam and J. M. Everaarts. 1989. A structure–activity relationship (SAR) approach towards metabolism of PCBs in marine animals from different trophic levels. *Mar. Environ. Res.* **27**: 159–176.

Bruggeman, W. A., A. Opperhuizen, A. Wijbenga and O. Hutzinger. 1984. Bioaccumulation of superlipophilic chemicals in fish. *Environ. Toxicol. Chem.* **7**: 173–189.

Connolly, J. P. 1991. Application of a food chain model to PCB contamination of the lobster and winter flounder food chains in New Bedford Harbor. *Environ. Sci. Technol.* **25**.

Connolly, J. P. and C. J. Pedersen. 1987. A thermodynamic-based evaluation of organic chemical accumulation in aquatic organisms. *Environ. Sci. Technol.* **22**: 99–103.

Connolly, J. P. and R. Tonelli. 1985. Modeling kepone in the striped bass food chain of the James River Estuary. *Est. Coast. Shelf Sci.* **20**: 349–366.

Gobas, F. A. P. C., K. E. Clark, W. Y. Shiu and D. Mackay. 1989. Bioconcentration of polybrominated benzenes and biphenyls and related superhydrophobic chemicals in fish: Role of bioavailability and elimination into the feces. *Environ. Toxicol. Chem.* **8**: 231–245.

Goerke, H. and K. Weber. 1990. Population-department elimination of various polychorinated biphenyls in *Neries diversicolor* (Polychaeta). *Mar. Environ. Res.* **29**: 205–226.

Muir, D. C. G. and A. L. Yarechewski. 1988. Dietary accumulation of four chlorinated dioxin congeners by rainbow trout and fathead minnows. *Environ. Toxicol. Chem.* **7**: 227–236.

Thomann, R. V. 1989. Bioaccumulation model of organic chemical distribution in aquatic food chains. *Environ. Sci. Technol.* **23**: 699–707.

Thomann, R. V. and J. P. Connolly. 1984. Model of PCB in the Lake Michigan lake trout food chain. *Environ. Sci. Technol.* **18**: 65–71.

Part III

Fish

CHAPTER 5

Chlorinated and Aromatic Hydrocarbons in Bottom Sediments, Fish and Marine Mammals in US Coastal Waters: Laboratory and Field Studies of Metabolism and Accumulation

USHA VARANASI, JOHN E. STEIN, WILLIAM L. REICHERT, KAREN L. TILBURY, MARGARET M. KRAHN and SIN-LAM CHAN

Environmental Conservation Division, National Marine Fisheries Service, Seattle, USA

INTRODUCTION

Organochlorines, such as polychlorinated biphenyls (PCBs), are among the most widespread and persistent chemical contaminants in near coastal environments. Commercial PCBs (e.g., Aroclors[1]) are Mixtures of chlorinated biphenyl isomers and congeners, which differ both in the degree and ring positions of chlorination, resulting in mixtures of compounds of differing toxicity. The chemical stability, toxicity and widespread use of these contaminants prior to 1975 raise considerable environmental concern. For example, even after almost two decades since the manufacture of these compounds was banned in the USA, substantial concentrations of these chemicals are frequently detected in biota, ranging from benthic invertebrates to marine mammals [1–6].

A variety of adverse effects of PCBs in terrestrial and aquatic organisms has been reported (see references cited in [5,7,8]). However, relatively few studies with marine species have focused on effects in

[1] Mention of trade names is for information only and does not constitute endorsement by the US Department of Commerce.

animals exposed to sediment-sorbed PCBs, although it has been demonstrated in laboratory studies that PCBs are bioavailable from the sediments to benthic marine species [8–11]. In the absence of such data on biological effects of sediment-sorbed PCBs, direct consequences of chronic exposure and long-term accumulations of these persistent pollutants in marine organisms residing in urban environments are difficult to assess. Additionally, indirect effects, such as infections from microorganisms leading to mortality from suspected immunomodulatory effects of organochlorines, are even more difficult to ascertain.

Besides PCBs, other chlorinated hydrocarbons (CHs) found in near coastal environments include several persistent insecticides: the chlorinated ethane derivatives such as DDT (1,1,1-trichloro-2, 2-bis[*p*-chlorophenyl]ethane); the chlorinated cyclodienes such as chlordane, aldrin, dieldrin, endrin and toxaphenes; and the hexachlorocyclohexanes such as lindane. However, several of these compounds are detected in sediment and biota, primarily DDT and its degradation products appear to occur in US coastal waters at concentrations of potential toxicological concern [12]. It is well recognized that DDT and its metabolites (DDD and DDE) bioaccumulate in natural food chains and that high tissue concentrations of DDTs (DDT plus DDD and DDE) are associated with reproductive failure in several species of birds [13]. Although the acutely toxic nature of waterborne DDT to a variety of aquatic organisms is well documented [14], there is extremely limited information on the toxicity of sediment-sorbed DDTs.

The term "persistent pollutants" is often applied to organochlorines because of their chemical stability in the environment due to minimal degradation occurring either via microbial metabolism or via metabolism by marine macrofauna. Therefore, compounds such as aromatic hydrocarbons (AHs) do not fit within this traditional definition of persistent pollutants, although many members of this class of pollutants are highly resistant to microbial degradation and, hence, persist in the aquatic environment, particularly in sediments. The AHs are not commonly considered as persistent pollutants because of their extensive metabolism and excretion by fish and mammals [15–20]. Nevertheless concentrations of these ubiquitous pollutants may be increasing in coastal environments because of increasing urbanization of the coastal areas. Moreover, large numbers of organisms, such as many invertebrates, in the aquatic environment possess limited or no capability to metabolize these compounds [21,22], hence accumulation of these contaminants in invertebrates, which are important food organisms for fish and mammals, is also of concern. Added to this concern are the findings that certain AHs exert their toxicity only after being metabolized, and some of these metabolites can bind covalently to cellular macromolecules and hence may be relatively more persistent in biota than the parent AHs. Considering these properties of AHs, we have included them in our discussion of persistent pollutants in marine organisms.

The majority of AHs associated with coastal sediments originate from petroleum and combustion products (e.g., from fossil fuel combustion and burning of wood), with surface water runoff from urban areas being an important source of AHs to coastal waters. The suite of AHs detected in aquatic systems is composed of alkylated and unsubstituted compounds of fused benzene rings and can be generally divided into two classes: the lower molecular weight compounds with 1–3 rings (LAHs), and the higher molecular weight compounds with 4–6 rings (HAHs). Relative to the LAHs, the HAHs are hydrophobic and tend to remain more tightly sorbed to sediment. Although considerable information exists on the toxicity of the "water-soluble fractions" of mixtures of AHs (e.g., crude oils and fuel oils) in several marine species [23], there is little information on sediment levels of either LAHs or HAHs that could be considered a threshold for adverse biological effects. In mammals and fish certain HAHs, such as benzo[a]pyrene (BaP), have been shown to be carcinogenic [24, 25]. Additionally, some AHs are known to impair reproductive processes in mammals and fish [26–28], and recent studies indicate that flatfish residing in selected areas where sediments are contaminated with high concentrations of AHs exhibit inhibited ovarian maturation and reduced spawning success [29–31]. Hence, AHs, specifically HAHs, are clearly a toxicologically important class of environmental contaminants.

Information on the bioavailability and disposition (e.g., uptake, metabolism, tissue accumulation and distribution, and excretion) of chemical contaminants in marine biota is essential to establish cause-and-effect relationships between chemicals in the marine environment and observed biological abnormalities. For example, because AHs are biotransformed in fish and mammals, certain biological effects such as liver tumors in fish or reproductive dysfunction in fish or mammals, which may be due to AHs, have been attributed at times to organochlorines that persist in the body unchanged and, hence, are more easily measured, rather than being attributed to AHs, which are rapidly metabolized to toxic metabolites that are difficult to measure, and thus not analyzed routinely. It is important, therefore, to understand the interrelationships between these co-occurring toxicologically significant environmental contaminants.

Accordingly, in this overview consisting mostly of our studies, we will focus on the uptake and metabolism of PCBs and AHs with respect to their bioaccumulation and tissue distribution in benthic invertebrates and fish exposed to an urban sediment in the laboratory, as well as in organisms sampled from urban waterways. A few examples are selected from a large database originating from our long-term monitoring of chemical contaminants in benthic fish and sediments from over 70 US coastal sites to elucidate relationships among various parameters of exposure of marine fish to PCBs and AHs in urban environments [6, 32]. Furthermore, we have used laboratory studies [10, 11, 33] on uptake of sediment-sorbed contaminants by invertebrates and fish to improve our understanding of the role of metabolism and food chain transport in determining both body burden and profiles of these contaminants in marine fish sampled in our monitoring studies. The original works (as cited) from which these data are selected for discussion herein include various statistical treatments to evaluate correlations among parameters of exposure and to determine the statistical significance of the findings. In this review we will cite some of these findings, but will not include entire data sets that were used for statistical analyses. Finally, some of our recent findings on both the levels and relative proportions (profiles) of organochlorines in marine mammals—pinnipeds and cetaceans—will be described in relation to our results on profiles of these compounds in fish and bottom sediments where these marine mammals may have fed during their seasonal migration.

LABORATORY STUDIES

Sediment from a contaminated estuary in Puget Sound, WA (Duwamish Waterway) was used to examine the uptake of AHs and PCBs by benthic invertebrates [clams (*Macoma nasuta*); amphipods (*Rhepoxynius abronius* and *Eohaustorius washingtonianus*); and fish (English sole, *Parophrys vetulus*)]. The PCBs in the test sediment (Table 5.1) from the Duwamish Waterway consist primarily of tri- to nona-chlorobiphenyls (2.2 μg/g dry weight); the other CHs (e.g, DDTs) were present at less than 10 ng/g dry weight. Hence, radiolabeled PCBs (^{14}C-Aroclor) were added to assess the bioavailability and bioaccumulation of sediment-associated CHs. Additionally, the test sediment contained a broad spectrum of AHs (32 μg/g dry weight), ranging from 2–6 ring compounds including significant concentrations of a carcinogenic AH (Table 5.1). Accordingly, ^{3}H-BaP was added to the sediment to serve as an indicator compound to obtain information on the metabolism and disposition of sediment-associated HAHs by these organisms [10, 11].

By the use of several different analytical techniques (e.g., radiometric analyses to detect both parent compounds and metabolites and gas chromatographic (GC) analyses to detect individual AHs and CHs), evidence was obtained for the uptake of sediment-associated AHs and PCBs by the benthic invertebrates and fish under comparable conditions

TABLE 5.1 *Concentrations (ng/g dry weight) of Selected Aromatic Hydrocarbons (AHs) and Chlorinated Hydrocarbons (CHs) in Sediment from the Duwamish Waterway*

AHs		CHs	
Naphthalene	400 ± 20[a]	Hexachlorobenzene	1.3 ± 0.2
2-Methylnaphthalene	160 ± 1	Lindane	2.4 ± 0.3
1-Methylnaphthalene	110 ± 1	Heptachlor	ND[b]
Biphenyl	33 ± 2	Aldrin	ND
2,6-Dimethylnaphthalene	67 ± 4	α-Chlordane	2.0 ± 0.2
Acenaphthene	310 ± 20	*Trans*-nonachlor	2.0 ± 0.2
2,3,5-Trimethylnaphthalene	40 ± 5	Mirex	3.5 ± 0.7
Fluorene	390 ± 30	*o,p'*-DDE	ND
Phenanthrene	2400 ± 200	*p,p'*-DDE	8.0 ± 0.2
Anthracene	610 ± 20	*o,p'*-DDD	ND
1-Methylphenanthrene	240 ± 10	*m,p'*-DDD	ND
3,6-Dimethylphenanthrene	88 ± 4	*p,p'*-DDD	5 ± 5
Fluoranthene	3900 ± 200	*o,p'*-DDT	ND
Pyrene	4800 ± 50	*p,p'*-DDT	13 ± 1
Benz(a)anthracene	2000 ± 50	Dichlorobiphenyls	36 ± 5
Chrysene	2900 ± 50	Trichlorobiphenyls	78 ± 6
Benzo(e)pyrene	2500 ± 200	Tetrachlorobiphenyls	370 ± 5
Benzo(a)pyrene	2300 ± 50	Pentachlorobiphenyls	800 ± 20
Perylene	650 ± 40	Hexachlorobiphenyls	580 ± 30
Dibenz(a,h)anthracene	470 ± 5	Heptachlorobiphenyls	200 ± 20
Benzofluoranthenes	4900 ± 200	Octachlorobiphenyls	52 ± 7
Indenopyrene	1800 ± 100	Nonachlorobiphenyls	47 ± 3
Benzo(g,h,i)perylene	1300 ± 50	Trichlorobutadiene	2.3 ± 0.3
		Tetrachlorobutadiene	11 ± 1
		Pentachlorobutadiene	3.2 ± 0.1
		Hexachlorobutadiene	1.4 ± 0.1

[a] The data are shown as mean ± SD.
[b] ND = not detected. Detection limits for individual AHs were 10 to 40 ng/g dry weight, and detection limits for CHs varied from 0.6 to 4.9 ng dry weight (adapted from [11]).

(Table 5.2). The extent of accumulation of these contaminants varied markedly among the different species. Factors related to the structure of the contaminants (e.g., molecular weight, polarity), lipid contents of tissues, feeding strategy of the test organisms and their capacity to metabolize these compounds appeared to all have significantly influenced the extent of accumulation.

Aromatic Hydrocarbons

Data in Table 5.2 show that HAHs were clearly accumulated by the clams and amphipods, whereas accumulation of LAHs was less; the presence of a smaller proportion of LAHs in clams and amphipods relative to sediment may have been due to the rapid excretion of these compounds compared to HAHs. The original paper [10] showed that the naphthalenes were present only in trace amounts in clams and amphipods. Widdows *et al.* [34] have reported that the excretion of

TABLE 5.2 *Concentrations of Aromatic Hydrocarbons (AHs) and Chlorinated Hydrocarbons (CHs) in Fish and Invertebrates Exposed in the Laboratory to a Contaminated Sediment from Puget Sound, WA*[a]

Species	Exposure (days)	Total AHs (ng/g)	% of total AHs LAHs[b]	HAHs	Total CHs (ng/g)	% of total CHs PCBs[c]	other CHs[c]	DDTs[c]
Clams	30	890 ± 90[d]	5	95	230 ± 30	96	0.4	3
Shrimp	30	nd[e]	–	–	430 ± 100	97	1	2
Amphipod[d]								
E. washingtonianus	7	4600 ± 300	2	98	1100 ± 80	94	<1	5
R. abronius	7	1700 ± 400	5	95	1700 ± 600	97	<1	<2
English sole								
liver	30	nd	–	–	1600 ± 300	98	<1	1
muscle	30	nd	–	–	100 ± 50	96	1	3
Sediment		16 000 ± 2500	16	84	990 ± 200	96	1	3

[a] The sediment was collected from the contaminated Duwamish Waterway (adapted from [10, 11]).
[b] LAHs = lower molecular weight AHs; HAHs = higher molecular weight AHs.
[c] PCBs = polychlorinated biphenyls; DDTs = DDTs, DDEs, DDDs; other CHs = see Table 5.1 for list of analytes.
[d] All values (mean ± SD) are based on wet weight tissue or sediment.
[e] nd = not detected.

naphthalene from mussels (*Mytilus edulis*) is biphasic with a half-life of 1.3 h for the "fast" component. Earlier studies [16, 35, 36] in which fish were exposed to radiolabeled AHs in sediment or the diet have demonstrated that one- to five-ring AHs are readily bioavailable. However, once taken up, the decline in tissue concentrations of radioactivity derived from benzenes and naphthalenes is considerably greater than that for BaP-derived radioactivity. It has been suggested that LAHs can be directly released from an organism either as parent AHs through gills and skin [34, 37, 38] or as metabolites through feces (via bile for fish) and urine, whereas metabolism of more lipophilic HAHs is necessary to facilitate excretion. Thus, the preferential accumulation of HAHs in tissues of clams, as shown in Table 5.2, was apparently due to the minimal capacity of clams to metabolize AHs and the relative retention of these compounds is due to their hydrophobicity. As will be described in the latter section on field studies, the preferential retention of HAHs in invertebrates observed in the laboratory exposure is also observed in invertebrates or stomach contents (e.g., molluscs, arthropods, annelids, etc.) of fish sampled from urban areas.

In contrast to clams, fish which efficiently metabolize most AHs (see references cited in [20, 22]) appear to biotransform AHs at a rate equal to the rate of uptake of parent compounds, hence parent AHs were not detected in fish exposed to urban sediment in the laboratory (Table 5.2). For example, the results on the disposition of radiolabeled BaP in fish show high concentrations of BaP metabolites in bile, confirming that BaP taken up from the sediment was metabolized and excreted by fish (Fig. 5.1). Subsequently, Stein *et al.* [11] reported that the bile of fish placed on the same urban sediment contained fluorescent metabolites of 2–5 ring AHs, thus demonstrating uptake, metabolism and excretion of a broad spectrum of sediment-sorbed AHs. It appears, therefore, that the uptake of sediment-associated AHs by fish may be similar to that of clams, even though parent AHs were not detected in tissues of the test fish.

The use of radiolabeled BaP added to the sediment allowed us to assess the concentrations of both the parent AH (e.g., BaP) and metabolites in all species. The results show that the tissue-to-sediment concentration ratio (TSR, based on wet weight of tissue and sediment) of BaP-derived radioactivity was considerably less for fish than that for clams (Fig. 5.1), which suggests that the extent of accumulation of AHs and their metabolites was inversely proportional to the ability of the organism to metabolize AHs. It should be noted that the test sediment contained significant concentrations of PCBs, which are known inducers of mixed function oxidases (MFO); hence the exposure of fish to PCBs may have enhanced their ability to metabolize AHs [39–41]. Connor

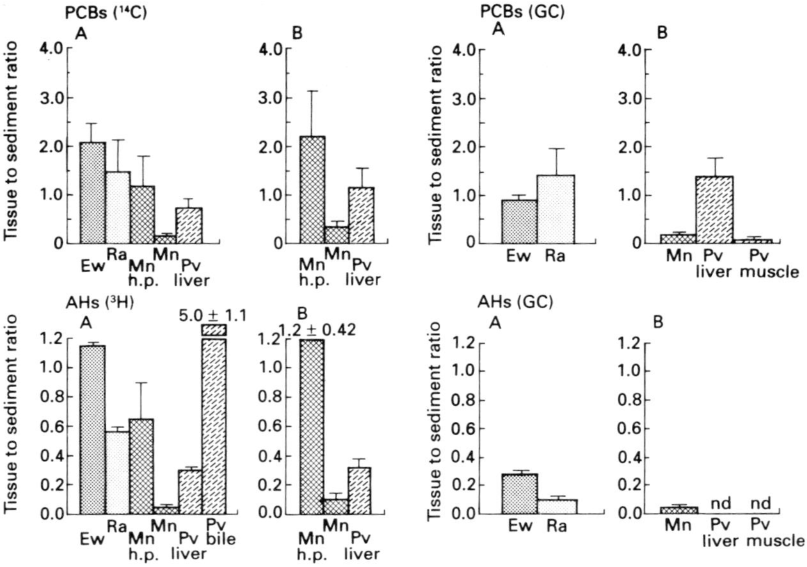

FIG. 5.1 Tissue-to-sediment concentration ratios for polychlorinated biphenyls (PCBs) and aromatic hydrocarbons (AHs) in whole body, (unless otherwise noted h.p.– hepatopancreas), of the amphipods *Eohaustorius washingtonianus* (Ew) and *Rhepoxynius abronius* (Ra), the clam *Macoma nasuta* (Mn) and the benthic fish English sole *Parophrys vetulus* (Pv) sampled after (A) 7 days' (10 days' for Pv) and (B) 30 days' exposure to Duwamish Waterway sediment spiked with trace levels of ^{14}C-PCBs and ^{3}H-BaP. The concentrations of PCBs and AHs were determined by gas chromatographic analysis (GC) (see Table 5.3 for List of Analytes) and by radiometric analyses, which measures both parent compounds and metabolites, of ^{14}C-PCBs or ^{3}H-benzo[a]pyrene (adapted from [10, 11] and Reichert, unpublished results).

[42] has suggested that there is an inverse relationship between MFO activity and TSRs of AHs in marine organisms. However, other factors such as feeding strategy and rate of excretion may also influence TSRs for AHs. For example, in our study [10] both amphipod species metabolized BaP to a greater extent than did clams; however, the tissue concentrations of parent AHs (specifically HAHs) as determined by GC analyses or as extrapolated from radiometric analyses of BaP and metabolites were significantly higher for both species of amphipods after 1 week of exposure than those for clams, even when clams were exposed to the test sediment for 30 days (Fig. 5.1). It is apparent, therefore, that MFO activity alone cannot adequately predict TSRs for AHs. Nevertheless, when TSRs for the two species of amphipods were compared, *R. abronius*, which has a greater ability to metabolize BaP [33], did have significantly lower concentrations of AHs as measured by GC analyses and by radiometric analyses, which account

for both the parent AH and metabolites. Thus, an inverse relationship between ability to metabolize xenobiotics and TSRs may hold well for organisms within a suborder.

These results (Fig. 5.1), showing lower TSRs for AHs in *R. abronius* compared to *E. washingtonianus*, have significant implications with respect to potential differential susceptibilities of these crustaceans to sediment-sorbed toxicants [43]. Further investigation [33] of metabolism of ^{3}H-BaP by amphipods showed that in *R. abronius* a significantly higher level of BaP metabolites were covalently bound to macromolecules (DNA, RNA and protein) (Fig. 5.2). Because HAHs exert their toxicity via metabolism and binding of reactive metabolites to DNA and protein, the higher levels of BaP metabolites bound to macromolecules in *R. abronius* after exposure to sediment-associated BaP may explain, in part, the reported sensitivity of *R. abronius* to urban sediments containing toxicants such as AHs [44, 45]. Thus, relationships among contaminant levels in sediment, body burden of such toxicants in resident organisms and resultant toxic effects can be best evaluated by understanding the fate of the contaminants, including their activation and detoxication. Moreover, methodologies that allow measurements of metabolites of contaminants are needed to expand our definition of persistent pollutants, because in some cases toxic metabolites may be more persistent than the parent compounds in the exposed organism, and measurements of parent compounds alone may lead to erroneous conclusions with regard to observed toxicity in the exposed organism.

Polychlorinated Biphenyls

The same experiments [10, 11] discussed above also provided information on the bioavailability and bioaccumulation of PCBs in molluscs, crustaceans and fish. The data obtained by either the measurement of radioactivity from the added ^{14}C-PCBs (Aroclor 1254) or GC analyses of PCBs in sediment and organisms show that all organisms accumulated PCBs from sediment (Table 5.2). The results of GC analyses of whole body homogenates showed that clams bioaccumulated substantially lower concentrations of PCBs after 30 days' exposure than both amphipod species after 7 days (Fig. 5.1). One factor that may contribute to this difference is lipid levels in tissues of the test organisms (see references cited in [4, 8, 41]). While lipid contents were not measured for all organisms, the ratios of both the lipid contents and PCB concentrations (Fig. 5.1) in liver and muscle of English sole were similar.

Interestingly, the results depicted in Fig. 5.1 show that *R. abronius* and *E. washingtonianus* accumulated comparable concentrations of PCBs when exposed to the same test sediment. These results demonstrate that, unlike AHs, the PCBs were probably not significantly

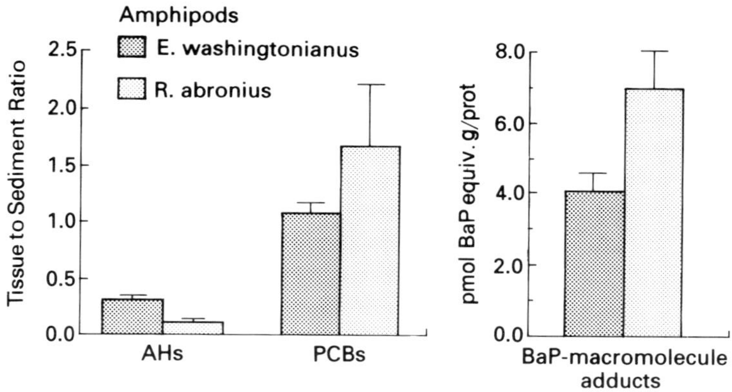

FIG. 5.2 Tissue-to-sediment concentration ratios for aromatic hydrocarbons (AHs) and polychlorinated biphenyls (PCBs) as measured by GC analysis, and levels of benzo[a]pyrene (BaP)-macromolecule (i.e., protein, RNA, DNA) adducts in the amphipods *Eohaustorieus washingtonianus* and *Rhepoxynius abronius* exposed to Duwamish Waterway which was spiked with trace levels of ^{3}H-BaP (adapted from [10, 33]).

metabolized by either amphipod species and hence bioaccumulation of these compounds was comparable for both species. Moreover, based on whole body concentrations measured either radiometrically or via GC, the two amphipod species accumulated substantially higher concentrations of PCBs than did clams. This finding, showing higher accumulation of PCBs and AHs by the amphipod species combined with their ability to activate AHs to potentially toxic metabolites that bind to cellular macromolecules (Fig. 5.2), may explain, at least in part, the relatively high susceptibility of amphipods compared to other invertebrate species (e.g., molluscs) to environmental contamination and the extensive use of certain amphipods as test species in sediment toxicity bioassays [46].

Data in Table 5.2 show that substantial bioaccumulation of PCBs, as compared to AHs, occurred in fish after 30 days' exposure to the urban sediment. These results suggest that fish do not metabolize PCBs appreciably. It should be noted, however, that Stein *et al.* [9, 11] showed that as much as 50% of PCB-derived radioactivity in bile of English sole exposed to ^{14}C-PCBs added to either the urban or the relatively uncontaminated sediments was due to metabolites of PCBs; nevertheless, the amount of PCB-derived radioactivity was substantially less than the amount of BaP-derived radioactivity released into bile. It appears that fish, and presumably many crustaceans, can metabolize certain PCB congeners, albeit to a considerably lesser extent than AHs. Hence, metabolism may not play a significant role in determining the body burden of PCBs as it does for AHs in crustaceans and fish. For example, the TSR for radiolabeled PCBs in English sole liver was

TABLE 5.3 *List of Analytes*[a]

Aromatic hydrocarbons (AHs)	Chlorinated hydrocarbons (CHs)
2- and 3-ring aromatic hydrocarbons (LAHs)	PCBs
Naphthalene	Trichlorobiphenyls
2-Methylnaphthalene	Tetrachlorobiphenyls
1-Methylnaphthalene	Pentachlorobiphenyls
Biphenyl	Hexachlorobiphenyls
2,6-Dimethylnaphthalene	Heptachlorobiphenyls
Acenaphthylene	Octachlorobiphenyls
Acenaphthene	Nonachlolrobiphenyls
2,3,5-Trimethylnaphthalene	Decachlorobiphenyl
Fluorene	
Phenanthrene	
Anthracene	DDTs
1-Methylphenanthrene	
	o,p'-DDD
4- through 6- ring aromatic hydrocarbons (HAHs)	*p,p'*-DDD
	o,p'-DDE
Fluoranthene	*p,p'*-DDE
Pyrene	*o,p'*-DDT
Benz[a]anthracene	*p,p'*-DDT
Chrysene	
Benzo[b]fluoranthene	other CHs
Benzo[k]fluoranthene	
Benzo[e]pyrene	Hexachlorobenzene
Benzo[a]pyrene	Lindane (*gamma*-BHC)
Perylene	Heptachlor
Indeno[1,2,3-cd]pyrene	Aldrin
Dibenz[a,h]anthracene	Heptachlorepoxide
Benzo[ghi]perylene	*Alpha*-chlordane
	Trans-nonachlor
	Dieldrin
	Mirex

[a] Compounds analyzed and quantified in samples from laboratory and field studies.

comparable to that of hepatopancreas in clams which, as discussed above, accumulated significantly higher concentrations of parent AHs compared to fish after 30 days' exposure.

Based on these laboratory studies, which compared uptake, metabolism and bioaccumulation of PCBs and AHs, it is tempting to propose that the body burden of PCBs in fish and crustaceans may be useful in estimating exposure to AHs, in areas where AHs co-occur with PCBs. The importance of such a relationship between co-occurring contaminants is obvious; however, a number of issues need to be considered first. For example, PCB concentrations in environmentally exposed organisms generally would be at a steady state, that is, the rates of uptake and excretion will be equal. Thus, even for highly persistent pollutants, such as PCBs, the body burden at any specified time would underestimate the total amount processed by an organism under chronic

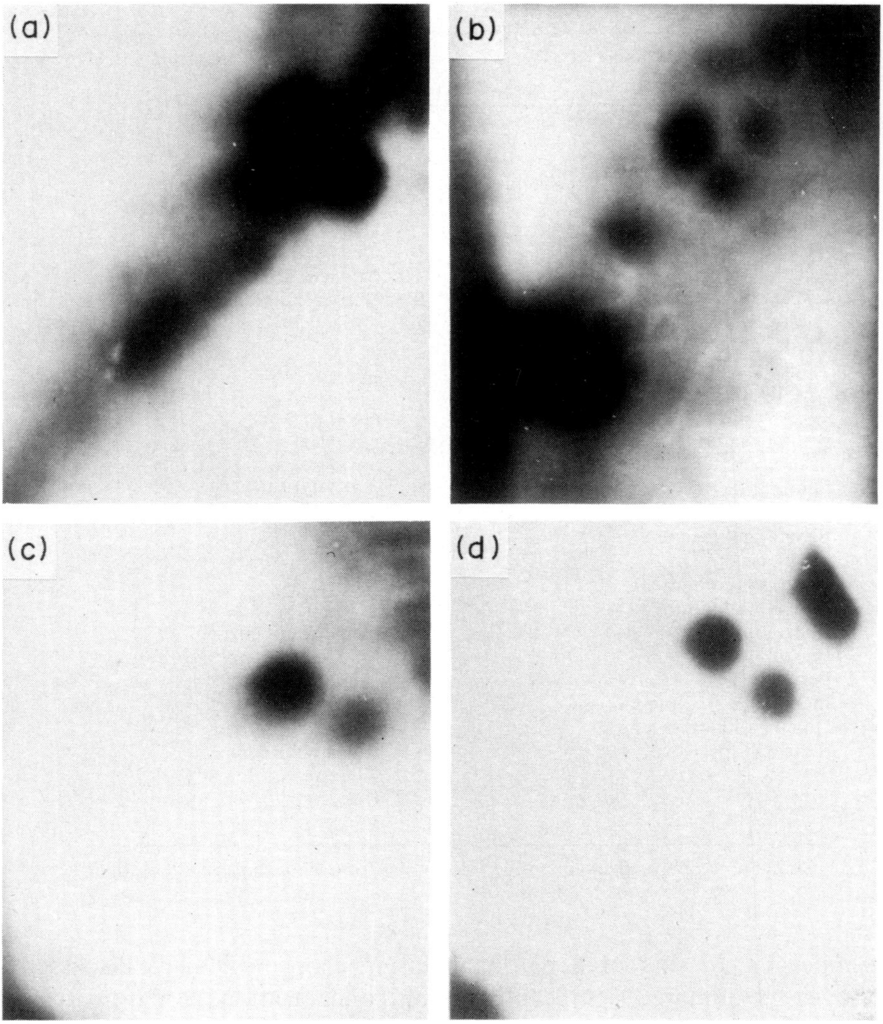

FIG. 5.3 Representative autoradiograms of thin-layer chromatograms of ^{32}P-labeled hepatic DNA adducts in winter flounder from Boston Harbor, MA (a) and Long Island Sound [New Haven, CT (b); Norwalk, CT (c), and Niantic, CT (d)] (adapted from [74, 75]).

exposure. Furthermore, interactive influences of AHs and PCBs on uptake and metabolism (discussed later) of individual classes of pollutants must be considered. Moreover, because the manufacture of PCBs and DDTs has been banned in the USA, the levels of these pollutants in the marine environment may decrease over time, whereas AH levels may actually increase. It is apparent, therefore, that the methodologies (Fig. 5.3) that allow more direct measures of chronic exposure to AHs are needed. At present, a powerful combination of methodologies would appear to be the analyses of CHs, especially congener-specific analyses, and the analyses of metabolites of AHs bound to tissue macromolecules (e.g., DNA, protein).

FIELD STUDIES

Contaminants in Sediment, Invertebrates and Fish

Figure 5.4 shows the sites from which sediment, invertebrates and benthic fish species are collected for the National Benthic Surveillance Project (NBSP) of NOAA's National Status and Trends Program as well as for other regional surveys to measure concentrations of chemical contaminants (listed in Table 5.3) and associated biochemical and biological effects. A summary of the concentrations of organic pollutants in sediments from the NBSP sites in the West Coast of USA sampled regularly over a 5-year period (1984–8) is shown in Fig. 5.5 to illustrate the types of data collected. The results show that more than half the

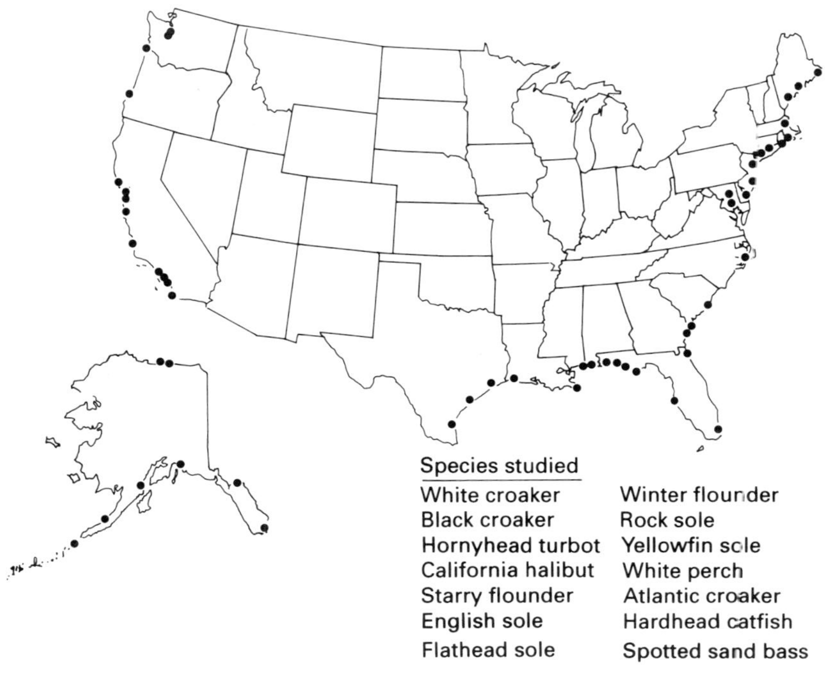

FIG. 5.4 The map shows sites where sediments and fish species (list given) are sampled regularly under NOAA's National Benthic Surveillance Project and other regional surveys to assess levels of chemical pollutants (see list of analytes in Table 5.3) and associated biochemical and biological effects in fish.

TABLE 5.4 *West Coast Sites Sampled at Least Once Between 1984 and 1988 as Part of the National Benthic Surveillance Project (NBSP)*

State	Site	Site abbrev.
CA	S. San Diego Bay	sSDBy
CA	National City	NtCy
CA	West Harbor Island	WHIs
CA	Shelter Island	ShIs
CA	N. San Diego Bay	nSDBy
CA	Outside San Diego Bay	oSDBy
CA	Dana Point	DnPt
CA	San Pedro Outer Harbor	SPOHb
CA	Long Beach	LnBh
CA	Seal Beach	SlBh
CA	San Pedro Canyon	SPCn
CA	E. Santa Monica Bay	eSMBy
CA	W. Santa Monica Bay	wSMBy
CA	Santa Monica Basin	SMBn
CA	Monterey Bay	MtBy
CA	San Pablo Bay	SPbBy
CA	Castro Creek	CsCk
CA	Southampton Shoal	SoSh
CA	Oakland	Oakl
CA	Hunters Point	HnPt
CA	Oakland Estuary	OkEs
CA	Redwood City	RdCy
CA	Bodega Bay	BdBy
OR	Coos Bay	CsBy
OR	Columbia River Estuary	ClREs
WA	Nisqually Reach	NsRh
WA	Commencement Bay	CmBy
WA	Elliott Bay	ElBy
AK	Lutak Inlet	LtIn
AK	Nakhu Bay	NkBy
AK	Skagway	Skg
AK	Kamishak Bay	KmBy
AK	Port Valdez	PtVz
AK	Boca de Quadra	BdQd
AK	Dutch Harbor	DuHb
AK	Port Moller	PtMl
AK	Chukchi Sea	ChkS
AK	Endicott Field	EnFl
AK	Oliktok Point	OlkPt

Adapted from [6].

sites had significantly higher levels of AHs and more than two-thirds of the sites contained significantly higher levels of CHs in sediments compared to the reference sites where the levels of these pollutants are less than 10 ppb (sites shown in Fig. 5.5 are listed in Table 5.4). In these surveys, lipid contents of tissues were not measured; however, age of fish was determined and generally the same age group of a particular test species is sampled from all sites. For details of the statistical treatment of the data and for discussion of

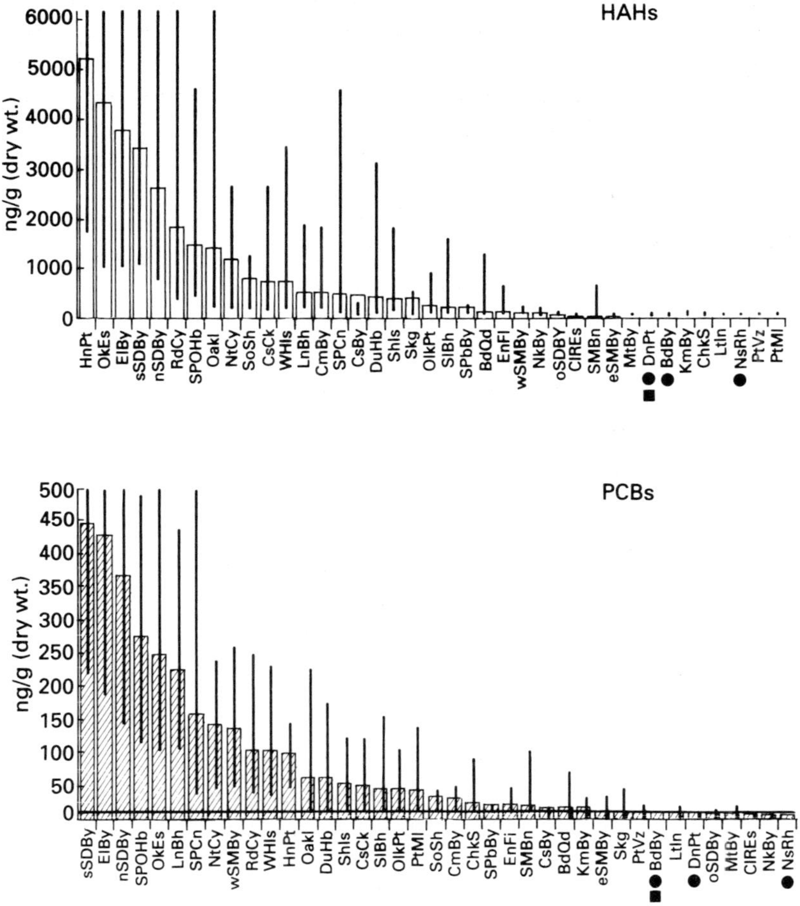

FIG. 5.5 Concentrations of (4–6 rings) aromatic hydrocarbons (HAHs) and polychlorinated biphenyls (PCBs) in sediment from West Coast sites sampled for NOAA's National Benthic Surveillance Project. Samples collected from 1984 to 1988. Site names are given in Table 5.4. ● comparison, reference sites. ■ reference threshold, for statistical comparisons (GT method) HAHs = 9.2 ng/g dry weight, PCBs = 10 ng/g dry weight (adapted from [6]).

individual sites and fish species, the reader is referred to our reports [6, 32].

The data in Table 5.5, on the concentrations of AHs and CHs in sediment and invertebrates (sampled directly or as stomach contents of fish) and fish from selected sites on both the Atlantic and Pacific coasts of the USA, are used to assess the validity of the hypotheses formulated from the laboratory studies about uptake and biological fate of sediment-sorbed contaminants. For example, the laboratory studies discussed earlier indicate that sediment-sorbed AHs and CHs are bioavailable to benthic fish and invertebrates; however, the extensive metabolism of AHs compared to CHs by fish and certain crustaceans limits biomagnification of the former compounds in higher trophic level organisms. Moreover, laboratory studies showing accumulation of sediment-sorbed AHs and CHs by a number of invertebrates which are important food organisms for many fish species, suggest that fish in urban environments can be exposed to these pollutants through the food chain. At present, however, it is not known if metabolites of AHs bound to macromolecules in food organisms can be assimilated by fish or mammals.

A summary of the results of statistical analyses, given in Table 5.6, shows strong statistical correlations among the levels of contaminants in sediment and the measures of exposure in fish, indicating that these sediment-sorbed compounds were bioavailable to fish. For example, the levels of AHs in sediment are significantly positively correlated with the levels of fluorescent aromatic compounds in bile, a measure of recent exposure of fish to AHs [47]. Furthermore, Table 5.6 reveals statistically significant correlations between the concentrations of PCBs, DDTs and other CHs in (a) sediment and liver tissue, (b) sediment and stomach contents, and (c) stomach contents and liver tissue, indicating bioavailability of sediment-sorbed CHs.

Bioavailability and Biomagnification

An obvious reason for measuring the concentrations of pollutants in stomach contents of fish was to assess exposure of fish to these contaminants through the diet. The concentrations of contaminants in stomach contents also provide information on the level of contamination in benthic invertebrates and, thus, provides an additional measure of chemical contamination at a particular site. The data in Table 5.5, which are a part of the larger data set for the NBSP [6, 32, 48], show a consistent trend of biomagnification of CHs from sediment to stomach contents and to liver of fish. The biomagnification for CHs from the sediments to stomach contents, which for many fish species were largely composed of molluscs, annelids and arthropods, indicates that the invertebrates were bioaccumulating these sediment-associated pollutants. Additionally, polychaetes sampled from three of the sites, Hunters Point, CA, Long Island Sound and New Haven, CT, show concentrations of CHs that were closely comparable to the concentrations of these pollutants in stomach contents of the fish species collected from the respective sites (Table 5.5). The biomagnification of CHs from the food organisms to liver of test fishes ranged from 2 to 10 fold. The lipid contents of benthic invertebrates and fish were not measured in these surveys, and hence the term biomagnification is used in a general context here; however, the consistent trend of increasing concentrations of CHs from sediment to stomach contents to fish liver strongly suggests biomagnification via food chain transport of these sediment-sorbed contaminants. This finding is not surprising, as considerable literature exists on biomagnification of PCBs via the food chain in terrestrial and aquatic organisms, based on both laboratory and field studies (see references cited in [4, 5, 8, 49–51]). In contrast to the findings showing biomagnification of CHs from sediment to stomach contents (i.e., food organism) of benthic fish, the concentrations of AHs in stomach contents of fish varied considerably, but in most cases was similar to or lower than the concentrations in sediment. The finding of generally lower concentrations of AHs in stomach contents relative to sediment is not surprising. A partial explanation could be that food organisms may have already metabolized AHs, because many invertebrates and fish found in the stomach contents are capable of metabolizing AHs [20–22]. In addition, once taken up, the AHs in stomach contents may be further metabolized by the test fish species as it is known that the MFOs in the gastrointestinal tract of fish can mediate metabolism of AHs [52]. However, it should be noted that the polychaetes sampled from the three sites shown in Table 5.5 also contained significantly lower concentrations of AHs compared to the concentrations in sediment. While metabolism is still a viable factor in explaining the lower concentrations of AHs observed in polychaetes [53], another possible explanation for the lower concentrations of AHs in biota compared to sediments is that only a portion of total AHs present in urban sediments may be readily bioavailable for uptake. Our laboratory study [10] also showed a differential uptake of added radiolabeled BaP relative to BaP already present in the urban sediment by invertebrates. These results support the hypothesis that only a portion of sediment-sorbed AHs in urban environments is readily bioavailable, the rest being tightly bound to particles, thus limiting uptake by organisms, but that can be chemically extracted from sediments [10, 54, 55]. Farrington *et al.* [54] have proposed that AHs released into marine ecosystems by oil spills or chronic oil inputs would be less tightly incorporated into particulate matter compared to AHs from pyrogenic sources and that the former will be more readily bioavailable, but AHs from both sources would be extracted from sediment by solvents. This explanation also supports the observations of the biomagnification of CHs from sediment to invertebrates and fish if it is assumed that the bioavailability of CHs deposited into the marine environment is comparable to the more readily bioavailable portion of AHs. However, further laboratory studies on bioavailability of sediment-sorbed contaminants are needed to confirm these observations.

Because sediment is an intimate part of the stomach contents of many

TABLE 5.5 *Concentrations of Aromatic Hydrocarbons (AHs) and Chlorinated Hydrocarbons (CHs) in Sediment, Fish and Invertebrates*[a]

	Total AHs (ng/g)	% of total AHs		Total CHs (ng/g)	% of total CHs		
		LAHs[b]	HAHs		PCBs[c]	DDTs	other CHs
Alaska							
Dutch Harbor—Cycle III							
Sediment	710 ± 260[d]	17	83	61 ± 30	98	1	1
Flathead Sole Stom. Contents	nd (1)	–	–	530 (1)	97	2	1
Flathead Sole Liver	e	–	–	2300 ± 560	95	3	2
West Coast							
Elliott Bay Cycle—Cycle III							
Sediment	7700 ± 3700	17	83	480 ± 160	96	3	1
English Sole Stom. Contents	210 000 (1)	10	90	1500 (1)	74	19	7
English Sole Liver	e	–	–	12 000 ± 5000	92	6	2
Hunters Point—Cycle IV							
Sediment	14 000 ± 9400	7	93	130 ± 22	91	6	3
Polychaetes	3 500 (1)	9	91	730 (1)	89	8	3
Starry Flounder Stom. Contents	1700 (1)	6	94	690 (1)	89	7	4
Starry Flounder Liver	e	–	–	8700 ± 3300	75	18	7
Hunters Point—Cycle V							
Sediment	12 000 ± 6700	17	83	170 ± 97	93	6	1
White Croaker Stom. Contents	180 (1)	11	89	1100 (1)	88	8	4
White Croaker Liver	e	–	–	2800 ± 310	75	20	5
San Pedro Outer Harbor—Cycle IV							
Sediment	1900 ± 800	10	90	450 ± 310	40	59	1
White Croaker Stom. Contents	2100 (1)	9	91	3500 (1)	37	62	1
White Croaker Liver	e	–	–	17 000 ± 11000	29	70	1
N. San Diego-Summer—Cycle V							
Sediment	3900 ± 1800	8	92	170 ± 25	94	5	1
White Croaker Stom. Contents	920 (1)	2	98	2900 (1)	90	7	3
White Croaker Liver	e	–	–	9300 ± 4100	85	12	3

S. San Diego-Summer—Cycle V							
Sediment	4700 ± 2600	9	91	240 ± 110	93	4	3
Barred Sand Bass Stom. Contents	600 (1)	2	98	1400 (1)	93	5	2
Barred Sand Bass Liver	e	–	–	12 000 ± 2700	90	7	3
Northeast Coast							
W. Long Island Sound—Cycle IV							
Sediment	6000 ± 4700	20	80	260 ± 110	90	6	4
Nephtys sp	300 ± 51	4	96	1200 ± 19	91	6	3
Winter Flounder Stom. Contents	2000 (1)	25	75	1300 (1)	93	4	4
Winter Flounder Liver	e	–	–	6500 ± 3400	86	9	5
New Haven—Cycle IV							
Sediment	3900 ± 660	15	85	260 ± 75	93	3	4
Glycera sp	940 ± 360 (2)	1	99	2400 ± 1000 (2)	91	5	4
Winter Flounder Stom. Contents	960 ± 49 (2)	14	86	2400 ± 92 (2)	92	4	4
Winter Flounder Liver	e	–	–	12 000 ± 6400	88	8	4
Raritan Bay—Cycle V							
Sediment	8000 ± 480	12	88	600 ± 110	90	7	3
Winter Flounder Stom. Contents	3000 (1)	7	93	3500 (1)	88	9	3
Winter Flounder Liver	e	–	–	12 000 ± 1100	84	11	5
New Bedford Harbor—Cycle V							
Sediment	2300 ± 1800	13	87	3400 ± 3200	99	0.6	0.4
Winter Flounder Stom. Contents	150 (1)	7	93	14 000 (1)	99	0.5	0.5
Winter Flounder Liver	e	–	–	37 000 ± 2300	99	0.6	0.4
Boston Harbor—Cycle V							
Sediment	7200 ± 2800	14	86	330 ± 72	97	1	2
Winter Flounder Stom. Contents	1300 (1)	28	72	2000 (1)	92	4	4
Winter Flounder Liver	e	–	–	12 000 ± 830	87	6	7

[a] Selected values from a larger database collected for NBSP (adapted from [6,32])
[b] LAHs—low molecular weight AHs; HAHs—high molecular weight AHs.
[c] PCBs—polychlorinated biphenyls. Other CHs—see Table 5.3. DDTs—1,1,1-trichloro-2,2-bis(*p*-chlorophenyl)ethane.
[d] Values (mean ± SD are ng/g dry weight). n = 3 unless otherwise noted.
[e] The liver samples were not analyzed for AHs.

benthic fish and invertebrates, some portion of the body burden (e.g., liver) of pollutants in fish may be derived from direct desorption of contaminants from sediment [35, 56, 57] ingested by the fish or that present in the gut of the food organisms ingested by the fish. In a recent preliminary study in which English sole were force-fed an urban sediment containing AHs and PCBs as well as trace levels of radiolabeled BaP and PCBs, the fish appeared to take up both classes of contaminants to a greater extent when the sediment was fed together with clam homogenates (Reichert *et al.*, unpublished observations). Thus, the presence of food organisms may facilitate desorption of sediment-sorbed contaminants, which adds another dimension to our overall consideration of uptake of contaminants by fish when they consume benthic food organisms from urban environments. These latter findings serve to emphasize the complexity of relating contaminant levels in sediments to levels in the tissues of fish and their food organisms, while demonstrating the importance of such measurements in environmental monitoring of contaminant uptake and associated biological effects.

TABLE 5.6 *Correlation Coefficients Between Concentrations of Chemicals in Sediment and Liver or Bile Using Spearman's Rank Correlation Method*

Chemical class[a]	Spearman's rank correlation Sediment vs. liver or bile[b]	
LAHs	0.572**	(Bile FACs-L)[c]
HAHs	0.619**	(Bile FACs-H)[c]
AHs	0.639**	(Bile FACs-L+H)
PCBs	0.656**	
DDTs	0.623**	
Non-DDT pesticides	0.586*	

[a] See Table 5.3 for list of analytes and Table 5.4 for list of sites (adapted from [6]).
[b] All concentration values were site means; $n = 66$ for AHs and 74 for chlorinated compounds. Species: four-horn sculpin, flathead sole, English sole, starry flounder, white croaker, hornyhead turbot, and barred sand bass. Based on analyses of samples collected for NBSP from 1984 to 1988.
[c] FACs-L = fluorescent aromatic compounds at naphthalene wavelengths to represent LAHs; FACs-H = fluorescent aromatic hydrocarbons at BaP wavelengths to represent HAHs.

* = $p < 0.001$; ** = $p < 0.0001$.

Further evaluation of the data in Table 5.5 shows that the relative proportion of PCBs to total CHs decreases from sediment to fish, regardless of whether PCBs constituted as much as 98% or as low as 40% of the total CHs in sediment. The magnitude of decline in the proportion of PCBs in fish liver compared to that in sediment varied, but the trend was highly consistent. The explanation for this trend is not clear, but the data suggest that PCBs compared to other CHs, especially DDTs, are either less bioavailable or more rapidly excreted from the liver or both. Whether metabolism plays a significant role in determining the relative proportion of PCBs to DDTs in fish tissues is not clear. For example, some PCB congeners are metabolized to more polar compounds which are not included in the measurements, whereas certain metabolites of DDTs, such as DDD and DDE, are included in the analyses.

Metabolism and Body Burden

Many factors influence the body burden of lipophilic contaminants. Notable among them is the ability of the organism to convert these compounds into more water-soluble products. As discussed earlier, Stein *et al.* [11] reported that bile of English sole placed on an urban sediment with added radiolabeled PCBs and BaP contained about 50% and 99% of PCB- and BaP-derived radioactivity, respectively; as metabolites, however, the amount of PCB metabolites was substantially less than BaP metabolites. Thus, fish are apparently able to metabolize PCBs, but to a considerably lesser extent. Moreover, PCBs were largely converted into glucuronide conjugates, whereas BaP was converted to a greater extent into putative thioether conjugates, such as glutathione conjugates. Studies with rodents have shown that glucuronide conjugates are more rapidly released from the liver into gall bladder, whereas glutathione conjugates are preferentially retained [58]. Thus, PCB metabolites may not be as persistent in the liver as metabolites of AHs. This is further supported by the finding that the liver of English sole did not contain appreciable levels of PCB metabolites either unbound or bound to macromolecules (e.g., protein) compared to low but significant levels of BaP metabolites either bound to hepatic macromolecules or present as thioether conjugates. This difference in the binding of BaP and PCBs to macromolecules in English sole may be due in part to different mechanisms of metabolism. Preston *et al.* [59] reported that 2,2′,5,5′-tetrachlorobiphenyl, a component of Aroclor 1254, is predominantly (>90%) metabolized by a non-arene oxide pathway in the rat. In contrast, BaP is primarily metabolized via arene oxide pathways. Metabolism by a non-arene oxide pathway leading to hydroxy derivatives may result in less conjugation with glutathione and less binding to hepatic macromolecules. It is possible, however, that certain specific congeners of PCBs may be significantly activated to reactive intermediates that bind to protein and DNA [5, 49] and accordingly comparison of binding levels of BaP, a single AH, to that of Aroclor 1254, a mixture of many components, may not be strictly valid. To assess the significance of metabolism in accumulation of PCBs and their derivatives in marine organisms, further studies with toxicologically important congeners are warranted. However, before initiating such studies it is necessary to determine which of these congeners are present in appreciable concentrations in the marine environment.

Contaminants in Marine Mammals

A substantial body of evidence exists, based on studies with a variety of marine mammals, to demonstrate that these top predators in the marine food chain may accumulate high levels of organochlorines in their lipid-rich tissues (see references cited in [8]). It is not the intent of this paper to review this vast literature, but to describe findings of our recent studies. We have examined levels of CHs in liver and blubber of several marine mammal species from both the East and West Coasts of the USA. These animals were collected after stranding or taken under subsistence or other harvest permits. In contrast to our long-term monitoring studies with fish, the quality of tissue samples was often compromised and the number of marine mammals available for analysis was often limited. However, certain patterns of accumulation of CHs demonstrated the usefulness of evaluating these data in light of profiles of CHs in sediment and fish.

The data in Table 5.7 show that several marine mammal species had high concentrations of CHs in liver and blubber. As expected, concentrations of CHs were highly correlated with lipid contents of tissues, with blubber having considerably higher concentrations of both lipids and CHs. Interestingly, in Steller's sealions (Table 5.7) the concentration of CHs observed were substantially greater than the concentrations detected in harbor seals, when both species were sampled from relatively pristine waters in Alaska. Such differences could be explained, in part, on the basis of differences in the migration patterns of these animals. Harbor seals have a limited range of migration and may show levels and profiles of CHs similar to those observed for the Alaskan marine environment, whereas the CH profiles in sealions may reflect sources far from the site of capture. The relative proportions of PCBs, DDTs and other CHs in harbor seal tissues resemble those in fish and sediment sampled in an Alaskan site, whereas the pattern in several of the sealions is similar to that observed for certain sites in southern California where fish, invertebrates and sediment show relatively high proportions of DDTs compared to other sites in our field surveys (Table 5.5). Furthermore, the data for pilot whale tissues collected from the Northeast coast of the USA show that the pattern of CHs in these mammals also closely resemble the proportion of CHs in sediment and fish from that region (Table 5.5).

TABLE 5.7 *Concentrations of Chlorinated Hydrocarbons (CHs) in Marine Mammals*

Species	Tissue (*n*)	PCBs[a]	DDTs	Other CHs
Harbor seal (*Phoca vitulina richardii*)				
	liver (9)	21 ± 5[b]	8 ± 4	4 ± 2
	blubber (7)	340 ± 110	260 ± 99	130 ± 56
Steller's sealion (*Eumetopias jubatus*)				
	liver (8)	290 ± 340	210 ± 280	94 ± 81
	blubber (8)	23 000 ± 37 000	20 000 ± 35 000	5200 ± 6600
Northern fur seal (*Callorhinus ursinus*)				
	liver (2)	77 ± 32	86 ± 36	14 ± 8
	blubber (2)	1300 ± 570	1800 ± 800	630 ± 230
Pilot whale (*Globicephala macrorhynchus*)				
	liver (5)	520 ± 510	250 ± 350	78 ± 89
	blubber (1)	9900	2600	1400
Pilot whale fetus				
	liver (5)	160 ± 130	110 ± 120	23 ± 27
	blubber (1)	2100	610	370
Bottlenose dolphin (*Tursiops truncatus*)				
	liver (20)	3600 ± 5200	2900 ± 7300	600 ± 900
	blubber (19)	31 000 ± 32 000	20 000 ± 25 000	7400 ± 7400

[a] PCBs—polychlorinated biphenyls. Other CHs—see Table 5.3.
[b] Values (mean ± standard deviation deviation are ng/g wet weight).

These comparisons are not meant to imply that these marine mammals accumulated CHs from the sites that show similar profiles of CHs in sediments, but suggest that accurate knowledge of migratory patterns of marine mammals, together with chemical analyses of tissues and stomach contents, and sediments from various sites where they feed may be useful in understanding the types of profiles of persistent pollutants found and in eventually linking sources to the chemical profiles. For example, tissues of bottlenose dolphins from a mass stranding in the Gulf of Mexico in 1990 show very high levels of CHs (Table 5.7), whereas sediments from this region generally show relatively low levels of these contaminants; moreover, relatively high proportions of DDT, the parent compound, in tissues of some of these dolphins emphasizes the need for detailed knowledge of feeding habits and migration patterns of marine mammals. The high concentrations of CHs in lipid-rich tissues of these top predators in the marine food chain not only indicate their exposure to these persistent pollutants, but also suggest potential exposure to co-occurring toxic contaminants such as AHs which are not readily measured in marine mammal tissues due to metabolism.

Because of the concern and awareness that chemical contaminants, such as CHs and AHs, may act directly or indirectly to bring about deleterious consequences on the health and reproduction in these ecologically important animals, it is essential that a scientifically credible and comprehensive database on levels and types of contaminants is generated, using state-of-the-art analytical techniques. Presently, we are assessing exposure to chemical pollutants in a variety of marine mammals using a suite of chemical and biochemical measurements (e.g., CHs in various tissues and stomach contents, metabolites of AHs in bile and hepatic and extrahepatic DNA adducts) as well as quality assurance/quality control procedures similar to those used for our studies with marine fish and invertebrates. Whenever possible, a suite of measurements indicating exposure is used. Such comprehensive information is necessary to improve our understanding of how certain persistent pollutants may affect the health of marine mammals and the quality of their habitats.

Congener-Specific Toxicity of Planar PCBs

Of the possible 209 PCB congeners, only a few—especially the planar congeners with no chlorine substitution at the *ortho* positions of the biphenyl moiety—are demonstrably or potentially toxic [5, 7]. McFarland and Clarke [5] suggest that if potential toxicity, environmental prevalence and relative abundance in animal tissues are used as criteria, only 36 PCB congeners are environmentally relevant. Furthermore, a subgroup of nine planar congeners that occur frequently in environmental samples and that exhibit polycyclic aromatic hydrocarbon-type induction of MFO activity have the greatest toxic potential [5]. These planar congeners may account for most of the toxicity exerted by PCBs in the environment [5, 7, 60].

Several recent studies report the presence of low levels of these planar PCBs in a variety of marine mammals [4, 5, 61] and fish [4, 60]. However, most routine analytical procedures used to separate and quantify PCBs in field surveys do not permit clear resolutions of planar congeners which are often present in very low concentrations in fish. Moreover, the currently available methods for isolation and analysis of the planar PCB congeners [4, 5, 60] are laborious and, hence, not readily amenable for analyzing large numbers of samples from field surveys. Recently, we have obtained promising results for isolation of planar congeners by modifying a simple high-performance liquid chromatographic procedure [62] for use with environmental samples [63]. We first tested this procedure with Aroclor samples and then applied it to tissues of a marine mammal.

Our preliminary results from the analysis of blubber of bottlenose dolphin (*Tursiops truncatus*) stranded in the Gulf of Mexico in 1990 (see previous section) showed the presence of a number of planar PCBs

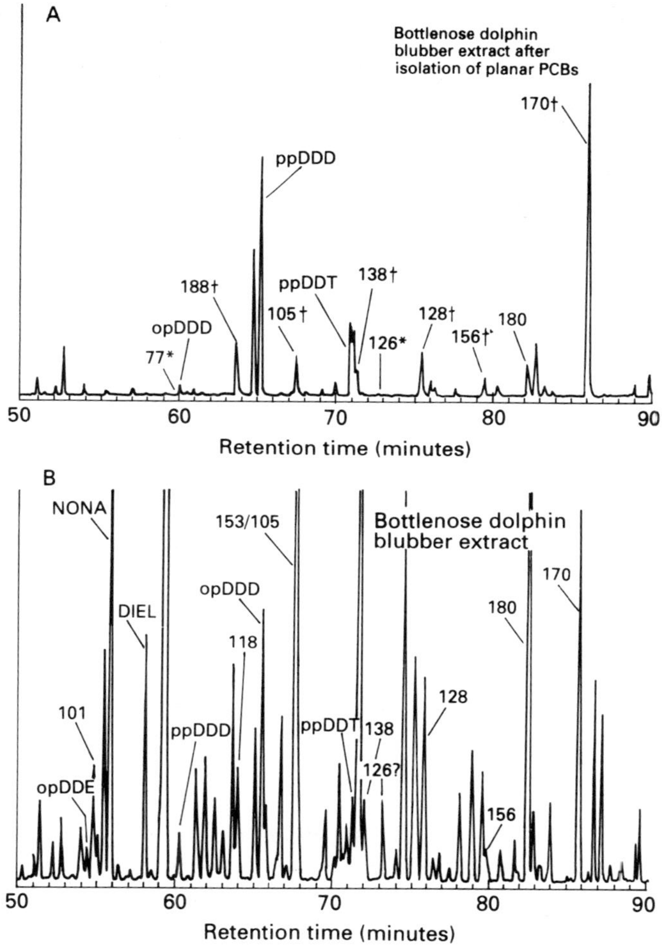

FIG. 5.6 Chromatogram (GC/ECD) of cleaned up extracts (by size-exclusion HPLC) from the blubber of a bottlenose dolphin before (B) and after (A) HPLC chromatography on 2-(1-pyrenyl)ethyldimethylsilylated silica ([63] modified from [62]) to isolate the planar PCBs. The PCBs are identified by IUPAC number and are also identified in (A) by the toxicity groupings of McFarland and Clarke [5]: Group IA (*), the three most potent congeners (pure 3-methylcholanthrene-type inducers; 169 not present); and Group IB (†), mixed-type inducers that are prevalent in the environment. Abbreviations: NONA = cis-nonachlor; DIEL = dieldrin; *op*-, *pp*-DDT = 2,4′- or 4,4′-DDT; *op*-, *pp*-DDD = 2,4′- or 4,4′-DDD; *op*-, *pp*-DDE = 2,4′- or 4,4′-DDE.

which have been cleanly separated from other PCBs and pesticides that interfere with their quantitation (compare Fig. 5.6A with 6B) [63]. Additionally, the accurate quantitation of the planar congeners allowed determination of their toxic 2,3,7,8-tetrachlorodibenzo-*p*-dioxin (TCDD) equivalence (i.e., the concentration factored with toxic potency relative to TCDD [7, 40, 60]) and found it to be comparable to the values reported by Tanabe *et al.* [61] for several marine mammal species. The availability of such a simple and rapid procedure should greatly facilitate the assessment of the environmental occurrence and the potential impact of these congeners in marine organisms sampled in field surveys.

Interactions Among Co-occurring Contaminants

While progress has been made during the last decade to understand the role of various endogenous (e.g., age, sex, lipid contents, reproductive cycle) and environmental (e.g., sediment characteristics, season, salinity) factors on uptake, metabolism and bioaccumulation of environmental pollutants in marine organisms, our understanding of synergistic or antagonistic effects of co-occurring contaminants is more limited. For example, there are indications that the carcinogenic potency of AHs and other agents may be strongly influenced in fish by the exposure to PCBs, and the magnitude and type (synergistic or antagonistic) of effect are dependent on the timing of PCB exposure [64]. It is believed that such interactive effects of PCBs are due, at least in part, to their ability to induce MFOs that mediate the metabolism of carcinogens.

In urban environments, fish are exposed to myriad pollutants; hence, it is important to investigate whether exposure of fish to PCBs and AHs alters the uptake and metabolism of these pollutants in marine organisms. Stein *et al.* [9] studied interactive effects of sediment-associated PCBs and BaP to show that the presence of PCBs increased the uptake of BaP in English sole liver, as indicated by the level of BaP-derived radioactivity which comprised primarily of metabolites. Because certain AHs are biotransformed in the liver to form reactive metabolites that bind to DNA, these results showing that exposure of fish to PCBs increases uptake of AHs may have considerable toxicological significance. Moreover, hepatic PCB concentrations were lower in sole exposed to both PCBs and BaP than those exposed to PCBs alone, suggesting alterations in uptake of PCBs in the presence of BaP. These studies suggesting interactive effects among co-occurring pollutants have obvious toxicological implications. Moreover, it is important to understand such interrelationships between these pollutants if we are to consider using one class of contaminant as an indicator of exposure to another class, as discussed below.

To test the hypothesis put forth earlier that the levels of PCBs in marine fish may be used to approximate exposure to co-occurring AHs, the data on concentrations of AHs and PCBs in sediment and fish collected in the National Benthic Surveillance Project were further evaluated. The relationships in Fig. 5.7 were derived from a data set

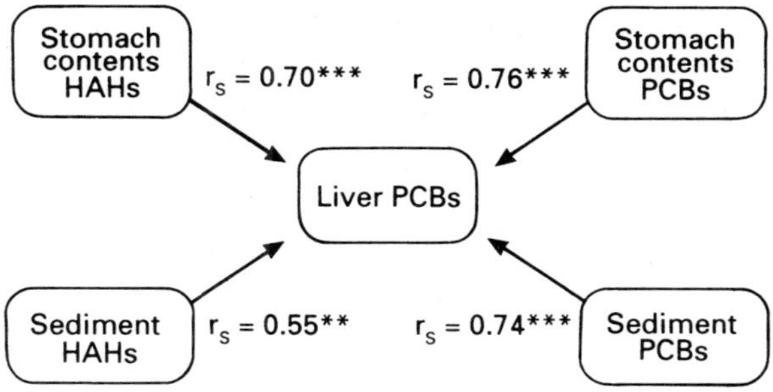

FIG. 5.7 Relationships between liver PCBs in white croaker and contaminants in sediment or stomach contents in samples collected between 1984 and 1988 in NOAA's National Benthic Surveillance Project ($n = 31$; ** = $p \leq 0.01$; *** = $p \leq 0.0001$).

generated over a 5-year period (1984–8) for white croaker from sites on the west coast of the USA (Fig. 5.4) to show that PCB concentrations in liver of benthic fish were statistically significantly correlated with concentrations of both PCBs and co-occurring AHs in sediment and stomach contents. Hence, it appears that liver PCBs could be used, with caution, to approximate levels of AH-exposure in fish and presumably in other organisms (e.g., crustaceans and marine mammals) that metabolize these pollutants.

However, as discussed earlier, potentially more useful parameters to indicate exposure of fish to aromatic contaminants such as AHs would include the measurement of fluorescent aromatic compounds in bile and hepatic DNA-xenobiotic adducts. The levels of fluorescent aromatic compounds in bile are determined using a method [47] that measures fluorescence at wavelength pairs appropriate for LAHs and HAHs and the data show that this method is applicable over a broad range of species [6]. Additionally, the presence of specific metabolites of AHs in bile of fish from urban areas has been confirmed with GC/MS analysis of selected samples [65]. Laboratory studies [56, 66–68] have shown that the concentrations of metabolites of AHs in bile of fish decline substantially within approximately 2 weeks after the contaminant exposure is terminated, whereas DNA-xenobiotic adducts in liver are relatively more persistent and, hence the latter measure may be more useful in integrating cumulative exposure of fish to AHs.

In recent years considerable advances have been made to measure xenobiotics bound to hepatic DNA in fish sampled from urban environments [20, 69]. This is particularly important because the binding of xenobiotics to DNA, a process known as adduct formation, is generally widely accepted from studies with mammalian species (e.g., mice, rats) to be a necessary early step in chemical carcinogenesis [24]. The use of a ^{32}P-postlabeling technique [70, 71] that allows visualization and quantification of hydrophobic xenobiotics, such as AHs, bound to DNA (Fig. 5.3) has increased our ability to assess exposure of fish to pollutants that are biotransformed and thereby allows the assessment of their potential toxicity. We first applied this technique to several benthic fish species exposed in the laboratory to (a) a variety of AHs (e.g., BaP, dibenz[a,h]-anthracene and chrysene) that are found in urban sediment; (b) crude oils; and (c) mixtures of contaminants extracted from urban sediments [20, 66]. These studies allowed us to compare the patterns of hepatic DNA adducts in fish exposed in the laboratory with the adduct profiles in fish sampled from urban areas [72]. Moreover, these studies also show that DNA-xenobiotic adducts are relatively more persistent than parent AHs or unbound metabolites in marine fish [56, 73]. The results in Fig. 5.8 show the concentrations of AHs and PCBs in sediment along with concentrations of hepatic PCBs and DNA-xenobiotic adducts in winter flounder (*Pseudopleuronectes americanus*) sampled from four sites on the northeast coast of the USA; it is evident that the levels of hepatic PCBs and DNA adducts exhibited similar trends [72], and thus hepatic PCBs appear to provide an overall estimate of exposure of fish to these two co-occurring classes of hydrophobic pollutants in urban environments. Additionally, a similar trend was found between concentrations of hepatic PCB and DNA adducts in English sole sampled from five sites in Puget Sound, WA [72]. At present, we are measuring DNA adducts in a variety of benthic fish from both the east and west coasts of the USA and preliminary results suggest that the levels of these adducts are correlated with levels of AH contamination in sediment. Thus, both the concentrations of PCBs, specifically planar congeners, and levels of DNA adducts may provide an overall indication of the relatively long-term exposure history of fish, and possibly mammals, to toxicologically significant environmental pollutants. Use of such state-of-the-art techniques to assess the levels of toxic persistent pollutants in marine organisms increases our ability to link contaminant exposure to biological effects—an essential requirement for understanding whether and how chemical pollution affects the marine ecosystem.

Acknowledgments—Studies discussed here are funded, in part, by the National Marine Mammal Tissue Bank and stranding program of National Marine Fisheries Service (NMFS), NOAA and the National Status and Trends Program, National Ocean Service (NOS), NOAA. While it is not possible to thank individually the staff of the Environmental Conservation Division who have contributed to the work described here, our special thanks are to Donald W. Brown, Douglas Burrows, Dr. Robert Clark, William Gronlund, Dr. John Landahl, Dr. Bruce McCain, Mark Myers, Susan Pierce, Paul Plesha, Herb

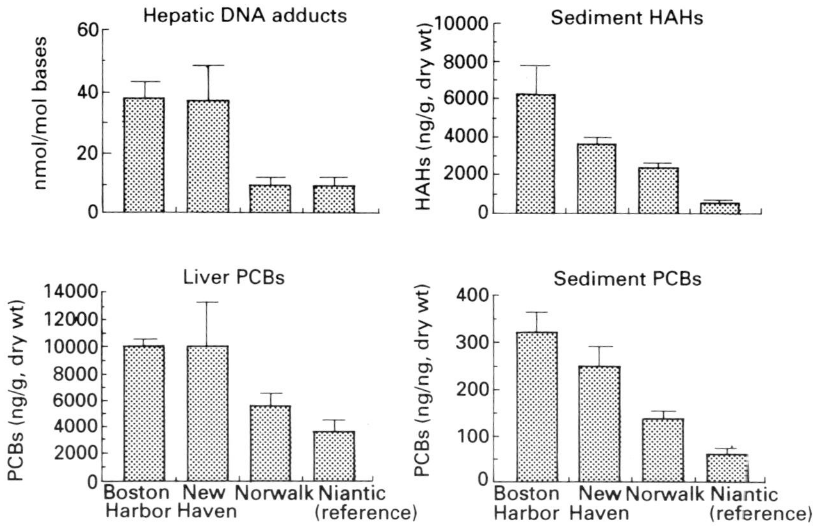

FIG. 5.8 Levels of hepatic DNA-xenobiotic adducts and polychlorinated biphenyls (PCBs) in winter flounder and concentrations of high molecular weight aromatic hydrocarbons (HAHs) and PCBs in sediment from sites in Boston Harbor, MA and Long Island Sound (adapted from [48, 74]).

Sanborn, Carla Stehr, Catherine Wigren and Gina Ylitalo, who have played major roles in coordinating and conducting sampling, analyses and statistical treatment of data from field surveys. Special thanks are also given to Bich-Thuy L. Eberhart and Tom Hom who contributed significantly to the success of the laboratory studies described herein. We are most grateful to the following for providing us valuable marine mammal samples: Dr. Paul Becker, OMA, NOS, Anchorage, AK; Dr. Greg Early, New England Aquarium, Boston, MA; Kathy Frost, Alaska Department of Fish and Game, Anchorage, AK; Larry Hansen, Southeast Fisheries Science Center, NMFS, Miami, FL; Dr. Thomas Loughlin, Alaska Fisheries Science Center, NMFS, Seattle, WA; and Dr. Steven Wise, National Institute of Standards and Technology, Gaithersburg, MD. Finally, we thank Dr. James Meador for promptly and thoughtfully reviewing the manuscript and Ms. Shirley Perry who worked tirelessly to get the manuscript typed.

REFERENCES

1. **Sugiura, K., N. Ito, N. Matsumoto, Y. Mihara, K. Murata, Y. Tsukakoshi and M. Goto.** 1978. Accumulation of polychlorinated biphenyls and polybrominated biphenyls in fish: limitation of "correlation between partition coefficients and accumulation factors". *Chemosphere* **7**: 731–736.
2. **Brown, J. F. Jr, R. E. Wagner, D. L. Bedard, M. J. Brennan, J. C. Carnahan and R. J. May.** 1984. PCB transformations in Upper Hudson sediments. *Northeast. Environ. Sci.* **3**(3/4): 167–179.

3. **Boon, J. P., P. J. H. Reijnders, J. Dols, P. Wensvoort and M. T. J. Hillebrand.** 1987. The kinetics of individual polychlorinated biphenyl congeners in female harbour seals (*Phoca vitulina*), with evidence for structure-related metabolism. *Aquat. Toxicol.* **10**: 307–324.
4. **Tanabe, S., N. Kannan, A. Subramanian, S. Watanabe and R. Tatsukawa.** 1987. Highly toxic coplanar PCBs: Occurrence, source, persistency and toxic implications to wildlife and humans. *Environ. Pollut.* **47**: 147–163.
5. **McFarland, V. A. and J. U. Clarke.** 1989. Environmental occurrence, abundance, and potential toxicity of polychlorinated biphenyl congeners: Considerations for a congener-specific analysis. *Environ. Health Perspect.* **81**: 225–239.
6. **Varanasi, U., S.-L. Chan, B. B. McCain, J. T. Landahl, M. H. Schiewe, R. C. Clark Jr., D. W. Brown, M. S. Myers, M. M. Krahn, W. D. Gronlund and W. D. MacLeod Jr.** 1989. National Benthic Surveillance Project: Pacific Coast, Part II, Technical Presentation of the Results for Cycles I to III (1984–1986). NOAA Tech. Memo. NMFS/F/NWC-170. US Department of Commerce, Seattle, WA.
7. **Safe, S.** 1984. Polychlorinated biphenyls (PCBs) and polybrominated biphenyls (PBBs): biochemistry, toxicology, and mechanism of action. *CRC Crit. Rev. Toxicol.* **13**: 319–395.
8. **Waid, J. S. (ed.)** 1986. *PCBs and the Environment*, Vol. II. CRC Press, Boca Raton, FL.
9. **Stein, J. E., T. Hom and U. Varanasi.** 1984. Simultaneous exposure of English sole (*Parophrys vetulus*) to sediment-associated xenobiotics: I. Uptake and disposition of [^{14}C]-polychlorinated biphenyls and [^{3}H]-benzo[*a*]pyrene. *Mar. Environ. Res.* **13**: 97–119.
10. **Varanasi, U., W. L. Reichert, J. E. Stein, D. W. Brown and H. R. Sanborn.** 1985. Bioavailability and biotransformation of aromatic hydrocarbons in benthic organisms exposed to sediment from an urban estuary. *Environ. Sci. Technol.* **19**: 836–841.
11. **Stein, J. E., T. Hom, E. Casillas, A. Friedman and U. Varanasi.** 1987. Simultaneous exposure of English sole (*Parophrys vetulus*) to sediment-associated xenobiotics: II. Chronic exposure to an urban estuarine sediment with added ^{3}H-benzo[*a*]pyrene and ^{14}C-polychlorinated biphenyls. *Mar. Environ. Res.* **22**: 123–149.
12. **McEwen, F. L. and G. R. Stephenson.** 1979. *The Use and Significance of Pesticides in the Environment.* John Wiley and Sons, New York, NY.
13. **Riseborough, R. W.** 1986. Pesticides and bird populations. *Current Ornithol.* **3**: 397–427.
14. **Holden, A. V.** 1972. The effects of pesticides on life in fresh waters. *Proc. Royal Soc. Found. B* **180**: 383–394.
15. **Lee, R. F., R. Sanerheber and G. H. Dobbs.** 1972. Uptake, metabolism and discharge of polycyclic aromatic hydrocarbons by marine fish. *Mar. Biol.* **17**: 201–208.
16. **Varanasi, U., D. J. Gmur and P. A. Treseler.** 1979. Influence of time and mode of exposure on biotransformation of naphthalene by juvenile starry flounder (*Platichthys stellatus*) and rock sole (*Lepidopsetta bilineata*). *Environ. Contam. Toxicol.* **8**: 673–692.
17. **Boroujerdi, M., H. King, A. G. E. Wilson and M. W. Anderson.** 1981. Metabolism and DNA binding of benzo[a]pyrene *in vivo* in the rat. *Cancer Res.* **41**: 951–957.
18. **Stegeman, J. J.** 1981. Polynuclear aromatic hydrocarbons and their metabolism in the marine environment. In H. V. Gelboin and P. O. P. Ts'O (eds), *Polycyclic Aromatic Hydrocarbons.* Academic Press, Inc., New York, NY, pp. 1–60.
19. **Melancon, M. J. and J. J. Lech.** 1983. Dose-effect relationship for induction of hepatic monooxygenase activity in rainbow trout and carp by Aroclor 1254. *Aquat. Toxicol.* **4**: 51–61.
20. **Varanasi, U., J. E. Stein and M. Nishimoto.** 1989. Biotransformation and disposition of polycyclic aromatic hydrocarbons in fish. In U. Varanasi (ed.), *Metabolism of Polycyclic Aromatic Hydrocarbons in the Aquatic Environment.* CRC Press, Boca Raton, FL, pp. 93–149.
21. **James, M. O.** 1989. Biotransformation and disposition of PAH in aquatic invertebrates. In U. Varanasi (ed.), *Metabolism of Polycyclic Aromatic Hydrocarbons in the Aquatic Environment.* CRC Press, Boca Raton, FL, pp. 69–91.
22. **Livingstone, D. R.** 1991. Organic xenobiotic metabolism in marine invertebrates. In R. Gilles (ed.). *Advances in Comparative and Environmental Physiology*, Vol. 7. Springer-Verlag, Berlin, Heidelberg, pp. 46–185.
23. **National Research Council.** 1985. *Oil in the Sea, Inputs, Fates, and Effects.* National Academy Press, Washington, DC.

24. **Dipple, A., R. C. Moschel and C. A. H. Bigger.** 1984. Polynuclear aromatic hydrocarbons. In C. E. Searle (ed.), *Chemical Carcinogens.* American Chemical Society, Washington, DC, pp. 41–163.
25. **Hendricks, J. D., T. R. Meyers, D. W. Shelton, J. L. Castee and G. S. Bailey.** 1985. Hepatocarcinogenicity of benzo[a]pyrene to rainbow trout by dietary exposure and intraperitoneal injection. *J. Natl. Cancer Inst.* **74**: 839–851.
26. **Lucier, G. W. and G. E. R. Hook.** (eds.) 1987. *Environ. Health Perspect.* **74**.
27. **Stein, Z. and M. Hatch.** 1987. Biological markers in reproductive epidemiology: prospects and precautions. *Environ. Health Perspect.* **74**: 67–75.
28. **Thomas, P.** 1989. Effects of Aroclor 1254 and cadmium on reproductive endocrine function and ovarian growth in Atlantic croaker. *Mar. Environ. Res.* **28**: 499–503.
29. **Spies, R. B., D. W. Rice Jr., P. A. Montagna and R. R. Ireland.** 1985. Reproductive success, xenobiotic contaminants and hepatic mixed-function oxidase (MFO) activity in *Platichthys stellatus* populations from San Francisco Bay. *Mar. Environ. Res.* **17**: 117–121.
30. **Johnson, L. L., E. Casillas, T. K. Collier, B. B. McCain and U. Varanasi.** 1988. Contaminant effects on ovarian development in English sole (*Parophrys vetulus*) from Puget Sound, Washington. *Can. J. Fish. and Aquat. Sci.* **45**: 2133–2146.
31. **Casillas, E., D. A. Misitano, L. L. Johnson, L. D. Rhodes, T. K. Collier, J. E. Stein, B. B. McCain and U. Varanasi.** 1991. Inducibility of spawning and reproductive success of female English sole (*Parophrys vetulus*) from urban and nonurban areas of Puget Sound, Washington. *Mar. Environ. Res.* **31**: 99–122.
32. **McCain, B. B., S.-L. Chan, M. M. Krahn, D. W. Brown, M. S. Myers, J. T. Landahl, S. M. Pierce, R. C. Clark Jr. and U. Varanasi.** 1989. Results of the National Benthic Surveillance Project (Pacific Coast): 1987. In *Proceedings*, Oceans '89. IEEE, Washington, DC, pp. 590–596.
33. **Reichert, W. L., B.-T. L. Eberhart and U. Varanasi.** 1985. Exposure of two species of deposit-feeding amphipods to sediment-associated ^{3}H-benzo[*a*]pyrene: uptake, metabolism and covalent binding to tissue macromolecules. *Aquat. Toxicol.* **6**: 45–56.
34. **Widdows, J., S. L. Moore, K. R. Clark and P. Donkin.** 1983. Uptake, tissue distribution and elimination of [1-^{14}C]naphthalene in the mussel *Mytilus edulis. Mar. Biol.* **76**: 109–114.
35. **Varanasi, U. and D. J. Gmur.** 1981. Hydrocarbons and metabolites in English sole (*Parophrys vetulus*) exposed simultaneously to [^{3}H]benzo[*a*]pyrene and [^{14}C]naphthalene in oil-contaminated sediment. *Aquat. Toxicol.* **1**: 49–67.
36. **Thomas, R. E. and S. D. Rice.** 1982. Metabolism and clearance of phenolic and mono-, di-, and polynuclear aromatic hydrocarbons by Dolly Varden char. In J. Vernberg, A. Calabrese, F. P. Thurberg and W. B. Vernberg (eds), *Physiological Mechanisms of Marine Pollutant Toxicity.* Academic Press, New York, NY, pp. 161–176.
37. **Varanasi, U., M. Uhler and S. Stranahan.** 1978. Uptake and release of naphthalene and its metabolites in skin and epidermal mucus of salmonids. *Toxicol. Appl. Pharmacol.* **44**: 277–289.
38. **Thomas, R. E. and S. D. Rice.** 1981. Excretion of aromatic hydrocarbons and their metabolites by freshwater and seawater Dolly Varden char. In F. J. Vernberg, A. Calabrese, F. P. Thurberg and W. B. Vernberg (eds), *Biological Monitoring of Marine Pollutants.* Academic Press, New York, NY, pp. 425–428.
39. **Goldstein, J. A.** 1979. The structure–activity relationships of halogenated biphenyls as enzyme inducers. *Ann. N.Y. Acad. Sci.* **320**: 164–178.
40. **Safe, S., A. Parkinson, L. Robertson, R. Cockerline, L. Safe, S. Bandiera and A. Okey.** 1982. PCBs and AHH inducers. In O. Hutzinger, R. W. Frei, E. Merian and F. Pocchiari (eds), *Chlorinated Dioxins and Related Compounds.* Pergamon Press, New York, NY, pp. 383–392.
41. **Clarke, J. U.** 1986. Structure-activity relationships in PCBs: Use of principal components analysis to predict inducers of mixed-function oxidase activity. *Chemosphere* **15**(3): 275–287.
42. **Connor, M. S.** 1984. Comparison of the carcinogenic risks from fish vs. groundwater contamination by organic compounds. *Environ. Sci. Technol.* **18**: 628–630.
43. **DeWitt, T. H., R. C. Swartz and J. O. Lamberson.** 1989. Measuring the acute toxicity of estuarine sediments. *Environ. Toxicol. Chem.* **8**: 1035–1048.
44. **Swartz, R. C., W. A. DeBen, K. A. Sercu and J. O. Lamberson.** 1982. Sediment toxicity

and the distribution of amphipods in Commencement Bay, Washington, USA. *Mar. Pollut. Bull.* **13**: 359–364.

45. **Swartz, R. C., P. F. Kemp, D. W. Schults, G. R. Ditsworth and R. J. Ozretich.** 1989. Acute toxicity of sediment from Eagle Harbor, Washington to the infaunal amphipod *Rhepoxynius abronius*. *Environ. Toxicol. Chem.* **8**: 215–222.
46. **Mearns, A. J., R. C. Swartz, J. M. Cummins, P. A. Dinnel, P. Plesha and P. M. Chapman.** 1986. Inter-laboratory comparison of a sediment toxicity test using the marine amphipod, *Rhepoxynius abronius*. *Mar. Environ. Res.* **19**: 13–37.
47. **Krahn, M. M., L. D. Rhodes, M. S. Myers, L. K. Moore, W. D. MacLeod Jr. and D. C. Malins.** 1986. Associations between metabolites of aromatic compounds in bile and the occurrence of hepatic lesions in English sole (*Parophrys vetulus*) from Puget Sound, Washington. *Arch. Environ. Contam. Toxicol.* **15**: 61–67.
48. **Gronlund, W. D., S.-L. Chan, B. B. McCain, R. C. Clark Jr., M. S. Myers, P. D. Plesha, J. E. Stein, D. W. Brown, J. T. Landahl, M. M. Krahn, W. L. Reichert and U. Varanasi.** 1991. Multidisciplinary assessment of pollution in Long Island Sound. *Estuaries* **14**: 299–305.
49. **Birnbaum, L. S.** 1985. The role of structure in the disposition of halogenated aromatic xenobiotics. *Environ. Health Perspect.* **61**: 11–20.
50. **Albaiges, J., A. Farran, M. Soler and A. Gallifa.** 1987. Accumulation and distribution of biogenic and pollutant hydrocarbons, PCBs and DDT in tissues of western Mediterranean fishes. *Mar. Environ. Res.* **22**: 1–18.
51. **Pizza, J. E. and J. M. O'Connor.** 1983. PCB dynamics in Hudson River striped bass. II. Accumulation from dietary sources. *Aquatic Toxicol.* **3**: 313–327.
52. **Van Veld, A. P., D. J. Westbrook, B. R. Wookin, R. C. Hale, C. L. Smith, R. J. Huggett and J. J. Stegeman.** 1990. Induced cytochrome P-450 in intestine and liver of spot *Leiostomus xanthurus* from a polycyclic aromatic hydrocarbon contaminated environment. *Aquatic Toxicol.* **17**: 119–132.
53. **McElroy, A. E.** 1985. *In vivo* metabolism of benz(a)anthracene by the polychaete *Nereis virens*. *Mar. Environ. Res.* **17**: 133–136.
54. **Farrington, J. W., E. D. Goldberg, R. W. Risebrough, J. H. Martin and V. T. Bowen.** 1983. US "Mussel Watch" 1976–1978: An overview of the trace-metal, DDE, PCB, hydrocarbon, and artificial radionuclide data. *Environ. Sci. Technol.* **17**: 490–496.
55. **McElroy, A. E., J. W. Farrington and J. M. Teal.** 1989. Bioavailability of polycyclic aromatic hydrocarbons in the aquatic environment. In U. Varanasi (ed.), *Metabolism of Polycyclic Aromatic Hydrocarbons in the Aquatic Environment*. CRC Press, Boca Raton, FL, pp. 1–39.
56. **Varanasi, U., J. E. Stein, M. Nishimoto, W. L. Reichert and T. K. Collier.** 1987. Chemical carcinogenesis in feral fish: uptake, activation, and detoxication of organic xenobiotics. *Environ. Health Perspect.* **71**: 155–170.
57. **Landrum, P. F.** 1989. Bioavailability and toxicokinetics of polycyclic aromatic hydrocarbons sorbed to sediments for the amphipod *Pontoporeia hoyi*. *Environ. Sci. Technol.* **22**: 588–595.
58. **Plummer, J. L., B. R. Smith, L. M. Ball and J. R. Bend.** 1980. Metabolism and biliary excretion of benzo[*a*]pyrene 4,5-oxide in the rat. *Drug Metabl. Dispos.* **8**: 68–72.
59. **Preston, B. D., J. A. Miller and E. C. Miller.** 1983. Non-arene oxide aromatic ring hydroxylation of 2,2′,5,5′-tetrachlorobiphenyl as the major metabolic pathway catalyzed by phenobarbital-induced rat liver microsomes. *J. Biol. Chem.* **258**: 8304–8311.
60. **Smith, L. M., T. R. Schwartz, K. Feltz and T. J. Kubiak.** 1990. Determination and occurrence of AHH-active polychlorinated biphenyls, 2,3,7,8-tetrachloro-*p*-dioxin and 2,3,7,8-tetrachlorodibenzofuran in Lake Michigan sediment and biota. The question of their relative toxicological significance. *Chemosphere* **21**: 1063–1085.
61. **Tanabe, S., N. Kannan, M. Ono and R. Tatsukawa.** 1989. Toxic threat to marine mammals: Increasing toxic potential of non-*ortho* and mono-*ortho* coplanar PCBs from land to ocean. *Chemosphere* **18**: 485–490.
62. **Haglund, P., L. Asplund, U. Jarnberg and B. Jansson.** 1990. Isolation of toxic polychlorinated biphenyls by electron donor-acceptor high-performance liquid chromatography on a 2-(1-pyrenyl)ethyldimethylsilylated silica column. *J. Chromatogr.* **507**: 389–398.
63. **Krahn, M. M.** 1991. (in preparation).

64. **Bailey, G. E., D. E. Goeger and J. Hendricks.** 1989. Factors influencing experimental carcinogenesis in laboratory fish models. In U. Varanasi (ed.), *Metabolism of Polycyclic Aromatic Hydrocarbons in the Aquatic Environment*. CRC Press, Boca Raton, FL, pp. 253–268.
65. **Krahn, M. M., D. G. Burrows, J. W. D. MacLeod and D. C. Malins.** 1987. Determination of individual metabolites of aromatic compounds in hydrolyzed bile of English sole (*Parophrys vetulus*) from polluted sites in Puget Sound, Washington. *Arch. Environ. Contam.* **16**: 511–522.
66. **Varanasi, U., J. E. Stein, L. L. Johnson, T. K. Collier, E. Casillas and M. S. Myers.** 1991. Evaluation of bioindicators of contaminant exposure and effects in coastal ecosystems. In *Proceedings* International Symposium on Ecological Indicators. Elsevier Applied Science, Essex, England (in press).
67. **Collier, T. K. and U. Varanasi.** 1991. Hepatic activities of xenobiotic metabolizing enzymes and biliary levels of xenobiotics in English sole (*Parophrys vetulus*) exposed to environmental contaminants. *Arch. Environ. Contam. Toxicol.* (in press).
68. **Varanasi, U., W. L. Reichert, B.-T. Eberhart and J. E. Stein.** 1989. Formation and persistence of benzo[a]pyrene-diolepoxide-DNA adducts in liver of English sole (*Parophrys vetulus*). *Chem.-Biol. Interact.* **69**: 203–216.
69. **Maccubbin, A. E., J. J. Black and B. P. Dunn.** 1990. ^{32}P-postlabeling detection of DNA adducts in fish from chemically contaminated waterways. *Sci. Tot. Environ.* **94**: 89–104.
70. **Randerath, K., M. V. Reddy and R. C. Gupta.** 1981. ^{32}P-labeling test for DNA damage. *Proc. Natl. Acad. Sci. USA* **78**: 6126–6129.
71. **Gupta, R. C. and K. Randerath.** 1988. Analysis of DNA adducts by ^{32}P-labeling and thin layer chromatography. In E. C. Friedberg and P. H. Hanawalt (eds), *DNA Repair*, Vol. 3. Marcel Dekker, Inc., New York, NY, pp. 399–418.
72. **Stein, J. E., T. K. Collier, W. L. Reichert, E. Casillas, T. Hom and U. Varanasi.** 1991. Bioindicators of contaminant exposure and sublethal effects: Studies with benthic fish in Puget Sound, WA. *Environ. Toxicol. Chem.* **11**: 701–714.
73. **Reichert, W. L., B. L. French, J. E. Stein and U. Varanasi.** 1991. ^{32}P-postlabeling analysis of the persistence of bulky hydrophobic xenobiotic-DNA adducts in liver of English sole (*Parophrys vetulus*), a marine fish. In *Proceedings* Eighty-second Annual Meeting of the American Association for Cancer Research, Houston, TX, pp. 87.
74. **Stein, J. E., W. L. Reichert, M. Nishimoto and U. Varanasi.** 1989. ^{32}P-postlabeling of DNA: A sensitive method for assessing environmentally-induced genotoxicity. In *Proceedings* Oceans '89. IEEE, Washington, DC, pp. 385–390.
75. **Varanasi, U., W. L. Reichert and J. E. Stein.** 1989. ^{32}P-postlabeling analysis of DNA-adducts in liver of wild English sole (*Parophrys vetulus*) and winter flounder (*Pseudopleuronectes americanus*). *Cancer Res.* **49**: 1171–1177.
76. **Landahl, J. T.** *et al.* National Benthic Surveillance Project: Pacific Coast Organic Chemical Contaminants (1984–1988) NOAA Tech. Memo (in preparation).

Part IV

Marine Mammals

CHAPTER 6

The Toxicokinetics of PCBs in Marine Mammals with Special Reference to Possible Interactions of Individual Congeners with the Cytochrome P450-dependent Monooxygenase System: an Overview*

JAN P. BOON, EVELINE VAN ARNHEM, STEPHAN JANSEN

Netherlands Institute for Sea Research (NIOZ), PO Box 59, 1790 AB Den Burg, Texel, The Netherlands

NARAYANAN KANNAN, GERT PETRICK, DETLEF SCHULZ, JAN C. DUINKER

Institut für Meereskunde an der Universität Kiel, Duesternbrookerweg 20, 2300 Kiel 1, Germany

PETER J. H. REIJNDERS

Institute for Forestry and Nature Research (IBN-DLO), PO Box 167, 1740 AD Den Burg, Texel, The Netherlands

and

ANDERS GOKSØYR

Laboratory of Marine Molecular Biology, University of Bergen, Bergen High Technology Centre, N-5020 Bergen, Norway

INTRODUCTION

Several organochlorines, such as polychlorinated biphenyls[1] and organochlorine insecticide residues, are hydrophobic compounds concentrating strongly in lipid tissues of marine organisms. There is ample

[1] Throughout this chapter the abbreviation PCB is used when the contribution of individual congeners to the mixture is not considered. The term Σ-PCB is used when total CB levels were calculated as the sum of all individual congeners measured. In all other cases the abbreviation CB is used.

* This is publication No. 20 of the Applied Science Project of the Netherlands Institute for Sea Research (Be WON – N10Z)

evidence that certain CBs and DDT and its environmental metabolites are at least partly responsible for reproductive and immunological abnormalities observed in some marine mammal populations (Fuller and Hobson, 1986; Anon., 1987). This has specifically been reported for California sealions[2] in coastal waters (De Long *et al.*, 1973; Gilmartin *et al.*, 1976), for harbour seals and harbour porpoises of the Dutch Wadden Sea (Reijnders, 1978, 1979, 1980, 1986), for grey seals, ringed seals and harbour seals in the Baltic (Hook and Johnels, 1972; Olsson *et al.*, 1975; Helle *et al.*, 1976a,b; Helle, 1980, 1981, 1986; Bergman *et al.*, 1981; Almkvist, 1982; Olson, 1986) and for belugas in the St. Lawrence river estuary (Beland *et al.*, 1987; Martineau *et al.*, 1987).

Equilibrium partitioning between body lipids, blood and ambient water is the dominant mechanism for elimination of apolar compounds in invertebrates and fish (Hamelink *et al.*, 1971). This mechanism, involving exchange between blood and water via the gills, is not available for marine mammals. This accounts, at least partially, for the considerably higher concentrations of some apolar compounds in marine mammals than in lower organisms.

Differences in the CB patterns between tissues of marine mammals and potential food sources such as fish and shellfish were already reported in the days when organochlorine analyses were still performed with packed column gas chromatography (GC); especially peaks representing clusters of compounds with a lower degree of chlorination showed relatively low contributions to the CB pattern in marine mammals (Jansson *et al.*, 1975).

Laboratory studies on rodents have shown that metabolism is of vital importance for the elimination of lipophilic substances in mammals (Abdel-Hamid *et al.*, 1981; Gage and Holm, 1976; Sipes *et al.*, 1980).

In a recent literature review on organochlorines in marine mammals it was concluded that (i) information on organochlorine levels in tissues other than blubber is scarce, (ii) much less information is available on cetaceans than on seals and (iii) many data have been obtained with the use of packed column GC with technical PCB mixtures (e.g. Aroclor and Clophen mixtures) as reference materials for quantification (Reuthergårdh and Knap, 1987). The resulting data on concentrations of PCB cannot be used to discriminate between the behaviour of individual CB congeners. As different congeners have different physicochemical properties, a more accurate insight into the mechanisms of uptake, distribution between tissues and elimination can be gained from studies of individual congeners in various tissues. The major depot tissue (blubber) as well as possible target organs should be considered.

[2] Latin names of species: sealion = *Zalophus californianus*; harbour seal = *Phoca vitulina*; harbour porpoise = *Phocoena phocoena*; grey seal = *Halichoerus grypus*; ringed seal = *Phoca hispida*; beluga = *Delphinapterus leucas*.

The present paper intends to present an overview of important processes involved in the toxicokinetics of individual CB congeners in marine mammals with some discussion of structurally related hydrocarbons. Results from studies on rodents will also be considered wherever appropriate. Also, the influence of some processes is illustrated by data on PCB concentrations, since no data on individual congeners were available. Because of the major role of the cytochrome P450 monooxygenase enzyme system, an important part of this paper is devoted to its general characterization in marine mammals.

BIOCHEMICAL CHARACTERISTICS OF THE CYTOCHROME P450 MONOOXYGENASE ENZYME SYSTEM OF MARINE MAMMALS

The Cytochrome P450 System

General Aspects and Molecular Properties

The cytochrome P450 monooxygenase system is a central catalyst in the oxidative "phase I" metabolism and disposition of organic xenobiotics. It is located in the smooth endoplasmic reticulum. This multicomponent membrane-bound electron transport chain receives reducing equivalents from NADPH (or in some cases NADH), which are then passed to the flavoprotein NADPH–cytochrome P450 reductase (or NADH–cyt b_5 reductase) and subsequently directly to clusters of cytochrome P450 (in the case of NADH as electron donor, this occurs via cytochrome b_5). In combination with molecular oxygen and a suitable organic substrate, P450 forms a trimolecular complex, the substrate is oxygenated and a molecule of water is formed.

The system comprises a superfamily of more than 150 individual forms of cytochrome P450, each belonging to one of the 27 gene families identified so far (Nebert *et al.*, 1991). The P450s are hemoproteins with a single polypeptide chain and molecular weights in the range 45,000–60,000 daltons. Each form catalyses the oxidative metabolism of a relatively specific group of lipophilic substrates. A gene family (or subfamily) may respond to a characteristic group of foreign chemicals or endogenous regulators. A common nomenclature for these families has recently been established (Nebert *et al.*, 1989a, 1991). In essence, the gene is indicated by an italicized root symbol *CYP*, representing *CY*tochrome *P*450, followed by an Arabic number denoting the family, a letter denoting the subfamily (when more than one exists), and an Arabic numeral representing the individual gene within the subfamily (e.g. *CYP1A1*). For mRNA, (sub-)families of proteins and individual proteins, CYP is not italicized, while the other rules follow those for

TABLE 6.1 *Nomenclature of Selected Cytochrome P450 Gene Subfamilies, Their Most Prominent Inducers and Characteristic Model Substrates Used to Assess Their Activity*

P450 gene subfamily	Prominent inducers	Model reactions
CYP 1A	Aromatic hydrocarbons, β-naphthoflavone, chlorinated dibenzodioxins and dibenzofurans, certain PBBs & PCBs (see Table 6.2)	EROD, AHH, ECOD
CYP 2B	Phenobarbital, certain PBBs & PCBs (see Table 6.2)	PROD, EMND, APND, AE
CYP2E	Ethanol	EO
CYP3A	Pregnenolone-16α-carbonitrile glucocorticoids	EMND, T6H
CYP4A	Clofibrate, phtalate esters, 2,4,5-T	FAH

Data from Nebert *et al.* (1989b); Gibson (1990) and Estabrook (1990).
Abbreviations: EROD: 7-ethoxyresorufin *O*-deethylase; AHH: Aryl hydrocarbon hydroxylase; ECOD: Ethoxycoumarin *O*-deethylase; PROD: Pentoxyresorufin *O*-deethylase; EMND: Ethylmorphine N-demethylase; APND: Aminopyrine N-demethylase; AE: Aldrin epoxidase; EO: Ethanol oxidase; T6H: Testosterone 6β-hydroxylase; FAH: Fatty acid Ω-hydroxylase.
(Note: These activities are not necessarily catalysed solely by the indicated P450 subfamily.)

the genes (e.g. CYP1A1). Individual proteins can also be written as P450 XYZ, where X, Y and Z indicate family, subfamily and the number of the protein within the subfamily (e.g. P450 1A1). The most important families and subfamilies of cytochrome P450 are presented in Table 6.1 with their characteristic inducers.

Interactions between CBs and Cytochrome P450

There are several ways in which CB congeners interact with the cytochrome P450 system: they can act as inducers, substrates, and inhibitors. In Table 6.2, CB congeners for which data are available have been divided into different groups with respect to induction type for the cytochrome P450 system and environmental occurrence (based on data of Safe, 1984; Safe *et al.*, 1985; McFarland and Clarke, 1989; Schulz *et al.*, 1989). Congeners with the same induction pattern, e.g. 2,3,7,8-tetra-chloro-dibenzo-*p*-dioxin (2,3,7,8-TCDD) and 3-methylcholanthrene (3-MC), induce the CYP1A subfamily only. Congeners with the same induction pattern as the model compound phenobarbital (PB) induce only the CYP2B subfamily. Congeners with a mixed 1A/2B induction type show the same reaction as a mixture of 3-MC and PB. Other congeners do not show an inducing potency towards any form of P450. For a pure CYP1A inducer, chlorine substitution should occur at least at both *para* positions and two *meta* positions at the different rings of

TABLE 6.2 *Groups of CB Congeners According to P450 Induction Potential and Environmental Occurrence*

CYP1A

Nr.	Cl-subst.	nr. ortho-Cl
A:	none	
B:	none	
C:		
77	(3,3′,4,4′)	0
126	(3,3′,4,4′,5)	0
169	(3,3′,4,4′,5,5′)	0

Mixed CYP1A/2B

Nr.	Cl-subst.	nr. ortho-Cl
A:		
105	(2,3,3′,4,4′)	1
118	(2,3′,4,4′,5)	1
128	(2,2′,3,3′,4,4′)	2
138	(2,2′,3,4,4′,5′)	2
170	(2,2′,3,3′,4,4′,5)	2
B:		
123	(2′,3,4,4′,5)	1
156	(2,3,3′.4,4′5)	1
158	(2,3,3′,4,4′,6)	2
C:		
37	(3,4,4′)	0
81*	(3,4,4′,5)	0
114	(2,3,4,4′,5)	1
157	(2,3,3′,4,4′,5′)	1
167	(2,3′,4,4′,5,5′)	1
189	(2,3,3′,4,4′,5,5′)	1
119	(2,3′,4,4′,6)	2
166*	(2,3,4,4′,5,6)	2
168*	(2,3′,4,4′,5′,6)	2

CYP2B

Nr.	Cl-subst.	nr. ortho-Cl
A:		
87	(2,2′,3,4,5′)	2
99	(2,2′,4,4′,5)	2
101	(2,2′,4.5,5′)	2
153	(2,2′,4,4′,5,5′)	2
180	(2,2′,3,4,4′,5,5′)	2
B:		
15	(4,4′)	0
194	(2,2′,3,3′,4,4′,5,5′)	2
183	(2,2′,3,4,4′,5′,6)	3
C:		
100	(2,2′,4,4′,6)	3
140*	(2,2′,3,4,4′,6′)	3
147*	(2,2′,3,4′,5,6)	3
154*	(2,2′,4,4′,5,6′)	3
182*	(2,2′,3,4,4′,5,6′)	3

Non-/Weak Inducers

Nr.	Cl-subst.	nr. ortho-Cl
A:		
70	(2,3′,4′,5)	1
74	(2,4,4′,5)	1
18	(2,2′,5)	2
44	(2,2′,3,5′)	2
49	(2,2′,4,5′)	2
52	(2,2′,5,5′)	2
151	(2,2′,3,5,5′,6)	3
187	(2,2′,3,4′,5,5′,6)	3
B:		
177	(2,2′,3,3′,4′,5,6)	3
201	(2,2′,3,3′,4′,5,5′,6)	3

Sources: Safe, 1984; Safe *et al.*, 1985; McFarland and Clarke, 1989.
Categories A-C represent an estimation of environmental occurrence, calculated as cumulative contribution of the individual congeners (CBs) to the series of Aroclor (1016–1260) and Clophen (A30–A60) technical mixtures (ΣCBs = 819.9%; Schulz *et al.*, 1989). A: CBs contributing >7.5%; B: 2.5< CBs >7.5%; C: CBs <2.5%.
* CB not detected in any of the technical mixtures. Within each category, CBs are given in order of increasing number of *ortho*-Cl atoms. Systematic numbering of CB congeners after Ballschmiter and Zell (1980).

the biphenyl skeleton. Since planarity of the molecule is an important factor, this toxicity type of CBs decreases with an increasing degree of ortho substitution. Yet, the mono-*ortho* derivatives of the CYP1A inducers still show a mixed type of induction pattern. The di-*ortho* derivatives show a mixed induction pattern or only CYP2B induction (Safe, 1984; Safe *et al.*, 1985). In general, only compounds which can easily assume a planar configuration (resembling the 2,3,7,8-TCDD structure) show a pure CYP1A induction (Lewis *et al.*, 1986).

The same structure–effect relationships as described above for the chlorinated biphenyls were found for the closely related group of brominated biphenyls (PBBs, used as fire-retardants). Bromine substitution seems to cause a stronger effect than chlorine substitution at the same positions (Safe, 1984).

As inducers, CBs will also affect the metabolism and toxicokinetics of other contaminants. Depending on which P450 subfamily is induced, CB induction may lead to increased formation of reactive intermediates, giving increased toxicity or carcinogenicity (additive or synergistic effects), or increased detoxification (antagonistic effects). These phenomena have been demonstrated in several experimental models, using both mammals and fish (e.g. Bailey *et al.*, 1987).

CBs are also substrates for the P450 monooxygenase reaction. The structure/activity relationship for biotransformation of CBs in marine mammals is treated extensively in the section on CB Patterns of this chapter (p. 133). In *in vitro* studies, working with the non-chlorinated biphenyl molecule, the P450 system inserts a hydroxyl group predominantly at the 4-position, but in some species (mouse and hamster) also at the 2-position (Kulkarni and Hodgson, 1980). In minke whale liver, only the 4-hydroxylase activity with biphenyl was observed (Goksøyr *et al.*, 1988).

Recently, several studies have indicated the ability of some CBs to act as inhibitors of P450 monooxygenase activities studied *in vitro*. This effect has, for example, been observed with CB-77 on the P450 1A-mediated 7-ethoxyresorufin *O*-deethylase (EROD) activity in fish (Gooch *et al.*, 1989), in studies with PCB mixtures (Clophen A40; Boon *et al.*, 1992) and in field studies (Elskus *et al.*, 1989), resulting in situations where the catalytic P450 1A activity measured may not reflect the true induction state of environmentally exposed organisms (Goksøyr *et al.*, 1991).

Pinnipeds (Seals)

Only a limited number of studies on the cytochrome P450 system of pinnipeds exist. Addison *et al.* (1986) have investigated the cytochrome P450 system of the harbour seal (*Phoca vitulina*) and characterized the

P450-mediated EROD activity in the grey seal (*Halichoerus grypus*; Addison and Brodie, 1984). In both of these cases, post-mitochondrial supernatants were prepared from the samples (liver and kidney) shortly after death to avoid deactivation due to freezing. Engelhardt (1982) studied aryl hydrocarbon hydroxylase (AHH) activity in ringed seals (*Phoca hispida*), in frozen tissue samples. The method of freezing, however, is unspecified. In a more recent study, Goksøyr and co-workers (1992) have studied the cytochrome P450 system of harp seal (*Phoca groenlandica*) and hooded seal (*Cystophora cristata*) caught during research cruises in the West Ice under the auspices of the Marine Mammal Programme of the Norwegian Fisheries Research Council. In 1989, samples were immediately frozen on liquid nitrogen on board the ship, but still suffered from a pronounced degradation of cytochrome P450 to inactive P420. In 1990, postmitochondrial supernatants were prepared on board ship before freezing in liquid nitrogen. Still, a conversion to P420 seemed to have occurred, unless the high P420 peak is a general feature of marine mammals, or represents some other pigment present in these animals (see also the discussion on cetaceans below). However, the higher cytochrome P450 levels reported by Addison *et al.* (1986) for harbour seal compared to harp and hooded seals (Table 6.3), together with the higher cytochrome b_5 levels (which is resistant to degradation), suggest that the differences in P450 levels are more probably species-specific than due to the method of sampling.

The cytochrome P450-mediated monooxygenase activities measured in liver samples from the four seal species most extensively studied are presented in Table 6.4. The different substrates used reflect to a

TABLE 6.3 *Levels of Cytochrome P450 and Microsomal Electron Transport Components Reported in Liver Samples from Adults of Different Pinniped Species*

Parameter	Harbour seal (9)[a]	Harp seal (12)[b]	Hooded seal (7)[b]
P450[c]	0.27 ± 0.11	0.11 ± 0.04	0.15 ± 0.10
cytb5[d]	0.33 ± 0.13	0.11 ± 0.05	0.09 ± 0.02
NADPH-red.[e]	n.m.[g]	47 ± 56	20 ± 12
NADH-red.[f]	n.m.	64 ± 38	54 ± 31

Values indicate mean ± standard deviation. The numbers of animals used is given between brackets

[a] Data reported by Addison *et al.* (1986).

[b] Data from Goksøyr *et al.* (1992).

[c] Cytochrome P450 in nmol mg^{-1} microsomal protein.

[d] Cytochrome b5 in nmol mg^{-1} microsomal protein.

[e] NADPH cytochrome P450 reductase in nmol cytochrome-c reduced min^{-1} mg^{-1}protein.

[f] NADH cytochrome b5 reductase in nmol cytochrome-c reduced min^{-1} mg^{-1} protein.

[g] n.m. = not measured.

certain degree the activities of different subfamilies of cytochrome P450 (CYP). EROD activity, representing the CYP1A subfamily, was the only activity studied in all these species. Female harp seals seem to have lower activities than the other species. 7-Ethoxycoumarin *O*-deethylase (ECOD), 7-methoxycoumarin *O*-demethylase (MCOD) and 7-pentoxyresorufin *O*-deethylase (PROD) can only be compared between harp and hooded seal, where the only difference seems to be the low, but detectable, levels of PROD in harp seal, which is not seen in hooded seal.

TABLE 6.4 *Comparison of Monooxygenase Activities Reported in Liver Samples from Adults of Different Pinniped Species*

Reaction	Grey Seal[a]		Harbour seal[b]		Harp seal[c]	Hooded seal[c]	
	F(5)	M(3)	F(9)	M(1)	F(10)	F(11)	M(7)
EROD	24 ± 20	21 ± 29	28 ± 17	8.2	2.1 ± 1.4	7.1 ± 4.4	30 ± 7
ECOD	n.m.[d]	n.m.	n.m.	n.m.	15 ± 13	24 ± 12	22 ± 13
MCOD	n.m.	n.m.	n.m.	n.m.	11 ± 11	8.8 ± 5.2	2.5 ± 5.5
PROD	n.m.	n.m.	n.m.	n.m.	0.13 ± 0.12	0.6 ± 0.7	1.3 ± 0.4

All values represent means ± standard deviation in pmol min^{-1} mg^{-1} post-mitochondrial supernatant (PMS) protein. F = female; M = male. The number of individuals is given between brackets. Explanation of abbreviations: Table 6.1.
[a] Data from Addison and Brodie (1984).
[b] Data from Addison *et al.* (1986), transformed into pmol min^{-1} mg^{-1} PMS protein.
[c] Data from Goksøyr *et al.* (1992).
[d] n.m. = activity not measured.

Sex differences in monooxygenase activities were not reported by Addison and Brodie (1984) or Addision *et al.* (1986), but were observed in the EROD activity of adult hooded seals (female lower than male) Goksøyr *et al.* (1992). The results are summarized in Table 6.4.

Monooxygenase activities were also measured in pups of four pinniped species; these data are shown in Table 6.5. Most activities in pups are lower than in adults, and hooded seal pups have even undetectable levels of EROD, ECOD and PROD. However, both harp seal EROD and MCOD, and hooded seal MCOD, were higher in pups than in adults. In addition, both NADPH and NADH cytochrome c reductase activities were higher in the pups of harp and hooded seals (Goksøyr *et al.*, 1992). Sex differences were observed in the pups as well: MCOD activity was higher in male harp seal pups and EROD activity was higher in male grey seal pups (Table 6.5).

Without any specific information on purified isozymes of P450 from pinniped species, it is difficult to comment much on the molecular properties of these forms. However, to a certain extent the application of specific antibody probes against P450 isozymes of other species (mammals and fish) can give information on structural relationships

TABLE 6.5 *Comparison of Monooxygenase Activities Reported in Liver Samples from Pups of Different Pinniped Species*

Reaction	Grey seal[a] F(4)	Grey seal[a] M(3)	Harbour seal[b] F+M(6)	Harp seal[c] F(5)	Harp seal[c] M(5)	Hooded seal[c] F(3)	Hooded seal[c] M(4)
EROD	3.6 ± 1.5	8.9 ± 3.7	6.3 ± 4.6	3.5 ± 3.2	7.1 ± 4.1	n.d.	n.d.
ECOD	n.m.	n.m.	n.m.	4.1 ± 9.1	11.0 ± 11.0	n.d.	n.d.
MCOD	n.m.	n.m.	n.m.	8.5 ± 16.6	66.7 ± 26.7	17.5 ± 30.3	20.6 ± 15.9
PROD	n.m.	n.m.	n.m.	n.d.	n.d.	n.d.	n.d.

All values represent means ± standard deviation in pmol min^{-1} mg^{-1} post-mitochondrial supernatant (PMS) protein. F = female; M = male. The number of individuals is given between brackets. Explanation of abbreviations: Table 6.1.

[a] Data from Addision and Brodie (1984).

[b] Data from Addison *et al.* (1986). Values from pups with detectable levels (>24 hr old) have been transformed into pmol min^{-1} mg^{-1} PMS protein.

[c] Data from Goksøyr *et al.* (1992).

with well-characterized forms in known subfamilies of the P450 system. In Western blotting, antibodies against cod P450 1A1 cross-reacted with two bands of proteins with molecular weights of 49 and 50 kD from liver samples of adult individuals of both harp and hooded seals (Goksøyr *et al.*, 1992). These results reflect the gene duplication event of the *CYP1A* subfamily before the divergence of mammals 100–130 million years ago, as discussed by Nebert and Gonzalez (1987). Antibodies against a dog P450 2B form (anti-dog PBD-2 IgG) gave a single band around 52 kD in both these species, strongest in male pups (Goksøyr *et al.*, 1992). This form is reported to be an effective catalyst of CB–153 metabolism (Graves *et al.*, 1990), which is only apparent in dogs (Schnellmann *et al.*, 1985). Dogs and seals both belong to the order Carnivora, and thus evolutionary relationships may give some indication to molecular and catalytic properties of P450 forms in the pinnipeds, and to their ability to metabolize certain CBs.

There are, of course, numerous reasons why marine mammals have not been employed to any extent in experimental exposure studies, including ethical issues and purely logistical problems of working with such large animals. Engelhardt (1982) fed ringed seals (*Phoca hispida*) with crude oil-contaminated fish and studied the CYP1A-mediated AHH activity. He found elevated activities compared to control seals, especially in kidney samples, reflecting induction of CYP1A isozymes. In a recent study with harp and hooded seal, single female pups from each species were treated intravenously with phenobarbital (Goksøyr *et al.*, 1992). This experiment was a part of the Norwegian Marine Mammal Programme, and was approved by the Norwegian National Board for Experiments with Animals. The treatment resulted in increased levels of EROD activity in both species, whereas MCOD, ECOD and PROD activities were induced only in harp seal. The results demonstrate that seals do respond to PB treatment, but typical CYP2B activities (e.g. PROD) were much lower than observed in other species even after PB treatment, which is in accordance with the postulations of Tanabe *et al.* (1988; see also section on CB Patterns of this chapter, p. 133).

Cetaceans (Whales)

The cytochrome P450 system of cetaceans is even more sparsely studied than that of pinnipeds. The cetaceans can be divided into the baleen whales (mysticeti) and the toothed whales (odontoceti). Most of the studies concern the family Delphinidae of the odontoceti. A report by Geraci and St. Aubin (1982) described the presence of spectrophotometrically measurable cytochrome P450 in liver samples of the bottlenose dolphin (*Tursiops truncatus*), the white-sided dolphin (*Lagenorhychnus acutus*) and the harbour porpoise (*Phocoena phocoena*); the levels were comparable to those found in other mammals (0.19–1.77 nmol P450/mg microsomal protein). Since then, the most comprehensive characterization of xenobiotic metabolizing enzymes in Delphinidae has been performed by Watanabe *et al.* (1989) on the cytochrome P450 system of the short-finned pilot whale (*Globicephala macrorhynchus*) and a more limited set of samples of the striped dolphin (*Stenella coeruleoalba*) and the killer whale (*Orcinus orca*).

Goksøyr and co-workers have conducted the only studies on the P450 system of a baleen whale species; i.e. the minke whale (*Balaenoptera acutorostrata*) which is the smallest species of the baleen whales (Goksøyr *et al.* 1985, 1986a, 1988, 1989).

In these studies, samples were taken from tissues within 30 minutes after death on board a whaling vessel—at the time of commercial whaling—and stored on liquid nitrogen until processing in the laboratory. Even so, a partial degradation (inactivation) of cytochrome P450 to P420 could be observed (Goksøyr *et al.*, 1986a; Watanabe *et al.*, 1989), unless this phenomenon is due to other factors as discussed above. The levels of microsomal electron transport components (NADPH-cytochrome P450 reductase, NADH cytochrome b5 reductase, cytochrome b5 and cytochrome P450) found in liver of foetuses and adults are presented in Table 6.6.

It should be noted that inactive P420 is not included in the P450 numbers. In the adult minke whale samples these levels varied from one-third to about the same as the P450 level, whereas in the foetal samples it was three to four times higher than the P450 level (Goksøyr *et al.*, 1986a). In the toothed whale samples the P420 levels ranged from 0.2 to 0.4 nmol mg^{-1} protein (Watanabe *et al.*, 1989). Otherwise, the levels of microsomal electron transport components compare well with

TABLE 6.6 *Levels of Cytochrome P450 and Microsomal Electron Transport Components Reported in Liver Samples from Different Cetacean Species*

Parameter	Striped dolphin[a] adult (5)	Killer whale[a] adult (3)	Short-finned pilot whale[a] foetus (1)	Short-finned pilot whale[a] adult (33)	Minke whale[b] foetus (4)	Minke whale[b] adult (10)
P450	0.19 ± 0.02	0.21 ± 0.07	0.045	0.17 ± 0.05	0.06 ± 0.02	0.35 ± 0.09
cytb5	n.m.	n.m.	n.m.	n.m.	0.07 ± 0.03	0.25 ± 0.03
NADPH red.	93 ± 18	35 ± 15	29	64 ± 13	21 ± 5	76 ± 24
NADH red.	n.m.	n.m.	n.m.	n.m.	145 ± 22	282 ± 83

For explanations and units, see Table 6.3.
[a] Data reported by Watanabe *et al.* (1989).
[b] Data reported by Goksøyr *et al.* (1988); see also Goksøyr *et al.* (1986a).

TABLE 6.7 *Comparison of Liver Microsomal Monooxygenase Activities Reported in Liver Samples from Different Cetacean Species*

Model reaction	Striped dolphin[a] adult (5)	Killer whale[a] adult (3)	Short-finned pilot whale[a] foetus (1)	Short-finned pilot whale[a] adult (33)	Minke whale[b] foetus (4)	Minke whale[b] adult (10)
EROD[c]	0.19 ± 0.015	0.61 ± 0.37	n.d.	0.04 ± 0.04	0.16 ± 0.10	0.74 ± 0.17
AHH[d]	7.2 ± 1.6	27.4 ± 6.1	0.15	7.8 ± 4.2	14 ± 5	220 ± 60
ECOD[e]	n.m.	n.m.	n.m.	n.m.	0.08 ± 0.02	0.53 ± 0.11
AE[f]	0.17 ± 0.05	0.04 ± 0.01	n.d.	0.06 ± 0.05	n.m.	n.m.
AH[g]	0.29 ± 0.10	0.22 ± 0.05	0.03	0.15 ± 0.06	n.m.	n.m.
APDM[h]	n.m.	n.m.	n.m.	n.m.	n.d.	5.6 ± 0.9
E-2OH[i]	n.m.	n.m.	n.m.	n.m.	16 ± 5	122 ± 12

All values represent means ± standard deviation in nmol min^{-1} mg^{-1} microsomal protein, unless indicated otherwise below. F = female; M = male. The number of individuals is given between brackets.
[a] Data reported by Watanabe *et al.* (1989).
[b] Data reported by Goksøyr *et al.* (1986a; 1988).
[c] EROD: 7-ethoxyresorufin *O*-deethylase.
[d] AHH: Aryl hydrocarbon hydroxylase (pmol min^{-1} mg^{-1} microsomal protein).
[e] ECOD: Ethoxycoumarin *O*-deethylase.
[f] AE: Aldrin epoxidase.
[g] AH: Aniline hydroxylase.
[h] APDM: Aminopyrine N-demethylase.
[i] E-2OH: Estradiol 2-hydroxylase (pmol min^{-1} mg^{-1} microsomal protein).

each other, as well as with those found in other mammals (e.g. Kulkarni *et al.*, 1976) and with those observed in seals (see above).

The cytochrome P450-mediated monooxygenase activities measured in three toothed whale species and the minke whale are presented in Table 6.7. EROD and AHH activities, representing the CYP1A subfamily, were the only activities measured in both minke whale and the toothed whales. However, AHH activity was measured by different methods, and the large difference in the results may be a reflection of this (Table 6.7). The highest EROD activity seems to occur in the adult killer whale and minke whale, whereas short-finned pilot whales display the lowest activities. The activities of EROD, AHH, ECOD and E-2OH were lower in foetal than in adult minke whale. However, foetal preparations from minke whale possess high capabilities for certain monooxygenase reactions, when expressed as turnover per nmol total P450 (Goksøyr *et al.*, 1988). Expressed in this way, foetal EROD turnover was higher than in adult minke whale livers, ECOD was similar, whereas AHH and biphenyl 4-hydroxylase was about half of the adult rate. Minke whale kidney microsomes, although containing a substantial level of cytochrome P450 (0.13 ± 0.05 nmol mg^{-1}), showed much lower activities and turnover numbers with these substrates (Goksøyr *et al.*, 1988).

Of the steroid transformation activities that have been investigated, estradiol 2-hydroxylase, when expressed as turnover numbers, was almost as high in foetal minke whales as in adults. Androstendione, when incubated with foetal liver, adult liver and adult kidney microsomes, showed characteristic metabolite patterns with the different tissues (Goksøyr, Andersson, Förlin, Snowberger, Woodin and Stegeman, manuscript in preparation).

Although no purified isoenzymes of cytochrome P450 have been reported from cetacean sources, Goksøyr and co-workers (1988, 1989, in prep.) have studied the molecular properties of the cytochrome P450 system in the minke whale with antibody probes against several mammalian and fish cytochromes P450, belonging to different sub-families. In these studies, polyclonal antibodies against cod P450 1A1 cross-reacted with two protein bands in adult livers, one strong band around 51-52,000 daltons, and a weaker band around 54 kD. The same two bands were barely detectable in foetal livers, whereas kidneys showed no cross-reactions. Monoclonal antibodies against another fish P450 1A1 (scup P450E), stained only one band in minke whale livers (weakly in foetals), probably the same 54 kD band. This pattern is similar to that found in rats induced with β-naphthoflavone (Goksøyr *et al.*, 1991c), and in seals (see above) with these antibodies. This reflects the duplication of the *CYP1A* gene into two gene products (the proteins P450 1A1 and P450 1A2) before the radiation of the mammals, but after the divergence of fish (Nebert and Gonzalez, 1987).

Polyclonal antibodies against a constitutive P450 form from rainbow trout, P450 con (Celander *et al.*, 1989), also cross-reacted with two protein bands in minke whale liver (approximately 51 kD and 53 kD) in the study by Goksøyr *et al.* (1989). The strongest of these bands (51 kD) was weakly present also in kidneys. It was observed that this band co-migrated with the 51–52 kD band detected by the anti-P450IA antibodies, and that it also cross-reacted with a monoclonal anti-rabbit P450 3c (3C2) antibody, generously supplied by Dr. Eric F. Johnson (ibid.). These results indicate the presence of products of the *CYP3* gene family in the minke whale. In the same studies, no cross-reactions were observed with anti-rat P450 2B1 antibodies, indicating either low levels of this phenobarbital-inducible form, or a strong divergence of the molecular properties of the minke whale P450 2B1 homologue. Based on CB congener patterns, Tanabe *et al.* (1988) postulated that cetaceans lack the CYP2B activities, which is in accordance with the former of these possibilities (see also section on CB Patterns of this chapter, p. 133).

Antibodies against three rat cytochrome P450 forms, i.e. P450 1A1, P450 1A2 and P450 2B1 (formerly P450c, d and b respectively), were used by Watanabe *et al.* (1989) to investigate the effect of these on the monooxygenase activities in liver microsomes from the short-finned pilot whale. They observed a pronounced inhibitory effect of the anti-rat P450 1A1 IgC on both EROD, AHH and metabolic activation on the promutagen MeIQ. A weaker effect was observed by the anti-rat P450 1A2 IgG, whereas anti-rat P450 2B1 did not affect these activities at all. Similar studies on the effect of anti-cod P450 1A1 IgG on EROD activity in minke whale liver microsomes gave no inhibition, but—for reasons unknown—rather a 20% elevation at 40 mg IgG/nmol P450 (Goksøyer *et al.*, 1988). Such results reflect the changes in antigenic epitopes involved in catalytic activity that have taken place during the course of evolution.

In Vitro Biotransformation of Polyaromatic Compounds

The aryl hydrocarbon hydroxylase (AHH) activity observed in different whale species (See Table 6.7) reflect the ability of the cytochrome P450 system to convert (poly-)aromatic compounds into polar metabolites. Regrettably no data have been reported *in vitro* biotransformation of CB congeners. Therefore, we summarize the results obtained with representatives of the related class of unhalogenated polyaromatic hydrocarbons (PAHs).

Goksøyr *et al.* (1986a) incubated liver microsomes from minke whale with the three-ring aromatic hydrocarbon phenanthrene and the metabolites formed were analysed by gas chromatography–mass spectrometry (GC-MS). Liver microsomes from adult whales preferentially attacked at the 9,10-position (K-region) of phenanthrene, giving 9,10-dihydro-9,10-dihydroxy-phenanthrene as 50–65% of the metabolites, while 10–35% was attacked at the 1,2-position. In contrast, foetal liver microsomes metabolized only 25-35% at the 9,10-position and 38-50% at the 1,2-position. The turnover of phenanthrene was around 1 nmol/min/nmol P450 in adult liver microsomes, four to five times the levels in foetal samples (Goksøyr *et al.*, 1986a). In the study of Jacob *et al.* (1982), liver microsomes from untreated male rats converted phenanthrene solely to the 9,10-diol. However, in the comparative study of Chaturapit and Holder (1978), untreated male guinea pigs, rats and mice showed microsomal hydroxylation of phenanthrene to the 1,2-diol at percentages of 22, 15 and 9 respectively. In both these studies, and in the study with fish liver microsomes by Goksøyr *et al.* (1986b), a larger proportion of 1,2-diol could only be obtained by *in vivo* treatment of the animal with inducers of the CYP1A subfamily. The results with minke whale liver microsomes again suggest that CYP1A isozymes are present in relatively large proportions in foetal samples.

The metabolic profile of benzo(a)pyrene conversion with minke whale liver microsomes has also been studied (Goksøyr, Andersson, Förlin, Snowberger, Woodin and Stegeman, manuscript in preparation). With adult liver samples, 3-OH B(a)P formation dominated (20–50%), but a substantial amount of activity was observed also at the 7,8- (10–30%) and 9,10-positions (7–10%), which are the bay region positions of B(a)P (Jerina and Daly, 1974). In foetal samples, 4,5-dihydrodiols of B(a)P were dominating (30% of total extractable metabolites).

CB PATTERNS

Methods for Calculation

The behaviour of individual compounds can be studied from data on absolute or relative concentrations. The latter can be expressed as a ratio of the concentrations of compound X and a single reference compound Y. Alternatively, the concentration of each individual compound can be expressed as the fraction of the sum of concentrations of all compounds measured. Both methods yield a reduction of the variability of the data compared to the use of absolute concentrations. For example, when the concentrations for all compounds measured differ by a factor 10 between samples A and B, the relative concentrations are still identical.

In the method where all compounds are expressed as the ratio to a chosen reference compound Y, the choice of the latter is critical and should meet the following criteria: (i) it must be present in all samples in concentrations well above the detection limit, (ii) it must be well separated from all other compounds in GC analysis and (iii) it must be a metabolically inert compound. The following CB congeners have been used for this purpose: CB-153 (2,2′,4,4′,5,5′-hexaCB; Boon *et al.*, 1987; Duinker *et al.*, 1989), CB-138 (2,2′,3,4,4′,5-hexaCB; Boon *et al.*, 1989) and CB-180 (2,2′,3,4,4′,5,5′-heptaCB; Tanabe *et al.*, 1988). The advantage of CB-153 is its presence at often the highest concentrations in environmental compartments with hydrophobic moieties, such as biota and particles. In this case, the concentrations of all other congeners can be expressed as a fraction of unity. However, CB-132 and CB-105 may be partly co-eluting after single column capillary GC. In the case of CB-138, CB-158 and CB-160 are co-eluting and may contribute about 10% to this peak (Schulz *et al.*, 1989). CB-180 is usually well separated, but occurs in lower concentrations than both other potential reference compounds. Throughout this overview, CB-153 is used as the reference compound, because it has commonly been used as such.

The alternative method for the calculation of relative concentrations expresses the concentration of each compound as a percentage of the total concentration of all compounds belonging to a structurally related group, e.g. the concentration of each CB congener as a percentage of Σ-PCB (being the sum of concentrations of all congeners measured; Duinker *et al.*, 1989). This is less sensitive to problems that arise in cases of inaccurate date on the reference compound. On the other hand, a relative decrease in some compounds due to, for example, biotransformation, automatically results in an increase in values of more persistent compounds, the total always remaining 100%.

Thus, for each goal the method of choice may differ, but its choice should always be well considered and justified.

Comparison of CB Patterns in Fish and Marine Mammals

In contrast to the situation in gill-breathing animals, food is virtually the only source of uptake for organochlorines in marine mammals. The elevated PCB concentrations in the Dutch part of the Wadden Sea have been correlated with a decreased reproduction of the resident harbour seal population (Reijnders, 1978, 1979, 1980, 1986). In order to evaluate the role of diet contamination on the reproduction rate, two groups of females were held in captivity at the Research Institute for Nature Management at Texel, The Netherlands. One group was fed with a diet of contaminated fish from the Dutch Wadden Sea (WSD). It received a 5–7.5 times higher daily dose of CBs compared to another group fed with a relatively clean diet from the Atlantic (AD). Concentrations of individual CBs were measured in the cellular fraction of seal blood obtained from one of the rear flippers. The first comparison of the reproductive success between both groups of seals was made in 1983, when CB concentrations in the WSD group were about five times higher than in the AD group (Boon *et al.*, 1987). The organochlorines pentachloro- and hexachlorobenzene, *p,p′*-DDE, γ-hexachlorocyclohexane and dieldrin were significantly higher in the WSD group of 1983, but the concentrations of the *p,p′* isomers of DDD and DDT did not differ (Boon, Reijnders, Dols, Wensvoort and Hillebrand, unpublished data).

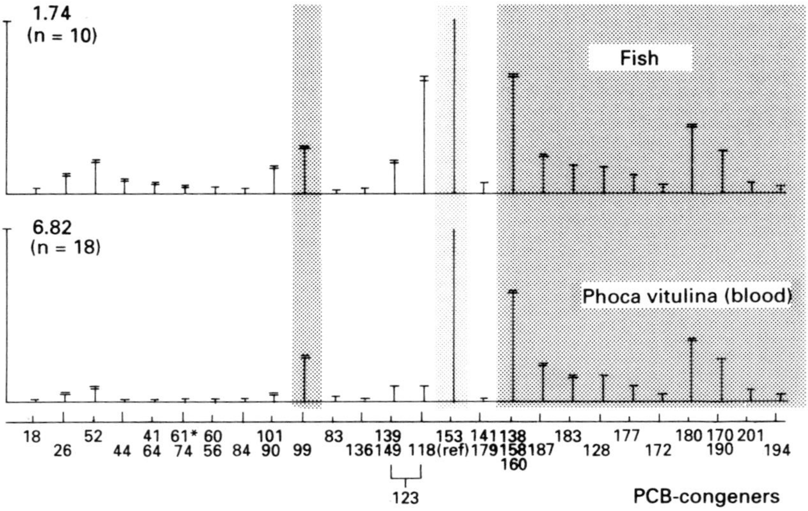

FIG. 6.1 Mean concentrations of CB-153 in μg g^{-1} pentane extractable lipid (PEL) in whole fish from the Dutch Wadden Sea and the cellular fraction of the blood of harbour seals. Numbers of CBs are given in order of elution from the GC column by their systematic numbers according to IUPAC rules as proposed by Ballschmiter and Zell (1980). Molecular structures of most CB congeners are also given in Table 6.2. All concentrations are proportional to height of bar. The numbers a/(b) in the upper left-hand corner of each histogram represent, a: The mean concentration of CB-153, b: the numbers of chromatograms used for these calculations. (Source: Boon *et al.*, 1987.)

The levels of diet contamination with CBs and organochlorine pesticides correlated well with a decrease in reproductive success from 10:12 (83%) in the AD to 4:12 (33%) in the WSD. In the case of a healthy population in the field, about 80% of the mature females deliver pups, but in the population of the western Wadden Sea this is only about 50% (Reijnders, unpublished results).

The CB patterns in fish and seal blood are given in Fig. 6.1 as concentration ratios CB-X/CB-153.

In this particular case, CB-153 was well separated from CB-105. CB-132 was below detection limit in seal blubber samples analysed by two-dimensional gas chromatography with electron capture detection (GC-ECD; not shown), but its presence in the fish samples is uncertain. The CB patterns in the diet and blood are characterized by two different parts, one involving very similar ratios of different congeners (dashed part of Fig. 6.1), but in the other part the congeners show lower relative concentrations in the seal blood than in the diet. This contrasts with the often mentioned food chain accumulation model from fish to seals.

All CBs with lower relative contribution in seals possess vicinal H atoms (Fig. 6.2).

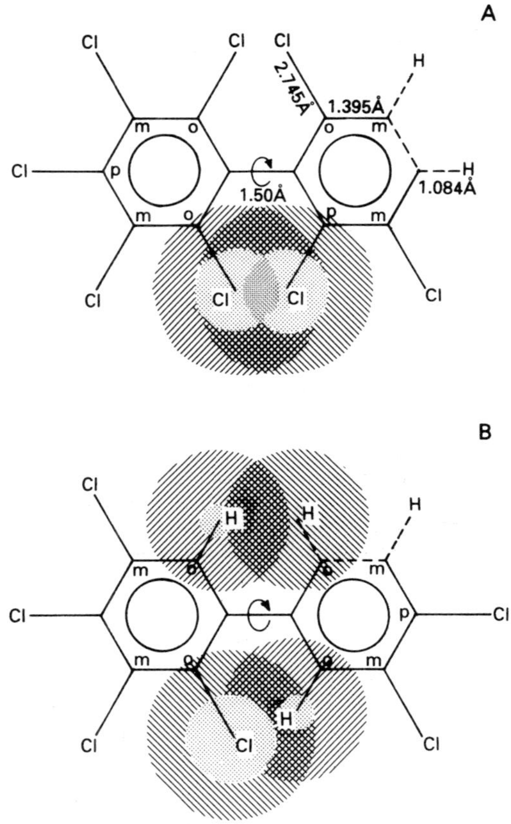

FIG. 6.2 Structural features of CB congeners influencing enzymatic metabolism. Areas where the principal enzymatic reaction occurs are given by broken lines. Bond lengths are taken from McKinney *et al.* (1983). For atoms in the *ortho* position, the outer circle represents the area within the van der Waals radius of an atom; the dotted inner circle represents the part of this area which is also within the single bond covalent radius. The van der Waals' radius indicates the maximum distance for any possible influence of an atom (H = 1.45 Å; Cl = 1.90 Å; Huheey, 1975). The covalent radius represents the minimum distance to which atoms can approach each other (H = 0.37 Å; Cl = 0.99 Å; Huheey, 1975).
A: Vicinal atoms in the *meta* and *para* positions. Overlapping covalent radii for two *ortho*-Cl show that a planar configuration is highly improbable when three or four *ortho*-Cl are present.
B: Vicinal H atoms in the *ortho* and *meta* positions. Non-overlapping covalent radii for *ortho*-Cl and *ortho*-H show that a planar configuration causes a much lower energy barrier when chlorine atoms do not oppose each other. (Source: Everaarts *et al.*, 1990.)

Laboratory experiments with rats, mice, dogs and monkeys have shown that this feature increases the rates of enzyme mediated biotransformation of CBs (Gage and Holm, 1976; Matthews and Tuey, 1980; Sipes *et al.*, 1980; Abdel-Hamid *et al.*, 1981; Matthews and Dedrick, 1984; Safe, 1984; Safe *et al.*, 1985).

All CB molecules show an energy minimum with rings at an angular configuration. With an increasing degree of *ortho*-Cl substitution, both the inter-ring angle of minimum energy and the energy barrier for a planar configuration increase (Table 6.8; Cullen and Kaiser, 1984).

TABLE 6.8 *Rotational Barriers of PCB Molecules in Relation to Inter-Ring Torsion (INDO Method)*

Ortho-Cl substitution pattern	Energy minimum (·)	Barrier height planar configuration (kcal mol^{-1})
Non-	28.5	1.94
Mono-	56	6.9
2,2′ di-	70	11.3
2,6 di-	68	12.3
tri-, tetra-	90	Repulsive (>30)

Source: Cullen and Kaiser (1984).

With three or four *ortho*-Cl atoms present, a planar configuration will be highly improbable at physiological temperatures of 35–45°C, since the covalent radii of two *ortho*-chlorine atoms at different rings will always overlap in that case (Table 6.8; Fig. 6.2).

With vicinal H atoms in *meta* and *para* (*m,p*) positions, the congeners were always lower in seal blood than in fish, irrespective of the degree of *ortho*-chlorination. However, with vicinal H atoms in the *ortho* and *meta* (*o,m*) positions, congeners were only lower when (not more than) one *ortho*-chlorine was present. From the general structure–activity data on biotransformation, induction of the cyt P450 monooxygenase enzyme system (see p. 122), and the fact that metabolites have been isolated from harbour seals (Jansson *et al.*, 1975; Jensen and Jansson, 1976), we conclude that the ability to assume a planar configuration is also an important prerequisite for biotransformation when vicinal H atoms are present in the *o,m* positions of CBs. With vicinal H atoms in the *m,p* position, a planar configuration is apparently unnecessary for biotransformation.

In general, the CYP1A subfamily shows a high affinity for planar molecules, but the CYP2B family shows a high affinity for globular molecules (Lewis *et al.*, 1986). The structure–effect relationship for *in vitro* metabolism of brominated biphenyl congeners (BBs) containing 0–2 *ortho*-Cl atoms, was investigated very extensively in microsomal systems of the rat by Mills *et al.* (1985). The 3-MC induced CYP1A subfamily attacked BBs via vicinal H atoms at the *o,m* positions, while the PB-induced CYP2B subfamily attacked the molecules via vicinal H atoms at the *m,p* positions. BBs without any vicinal H atoms were persistent. In this study, congeners with vicinal H atoms in the *o,m* position with two *ortho*-Br substituents were still metabolized, which appears to be in contrast to the situation in seals and cetaceans. BBs with more than two *ortho*-Br atoms were not investigated. Although studied less extensively, the *in vitro* data for metabolism of CBs in rats appear consistent with those of BBs (Kato *et al.*, 1980; Preston and Allen, 1980; Shimada and Sato, 1980; Shimada and Sawabe, 1983; Millis *et al.*, 1985).

Voorman and Aust (1988) reported the very important observation, that although BB-169 induced both P450 1A1 and 1A2 in rats, binding occurred only to the 1A2 form. The BB-169/protein complex appeared to be more stable than the P450 1A2 by itself. The binding between BB-169 and P450 1A2 was non-covalent and could be broken by extraction with dichloromethane. BB-169 also inhibited the estradiol-2-hydroxylase activity of purified P450 1A2 in a non-competitive way (Voorman and Aust, 1987).

The question whether the CB pattern in seal blood represents the entire animal cannot be answered directly, since no other tissue types were analysed in the experiment discussed above. However, a comparison is possible for six congeners between the blood samples of the experimental animals and blubber samples of harbour seals and grey

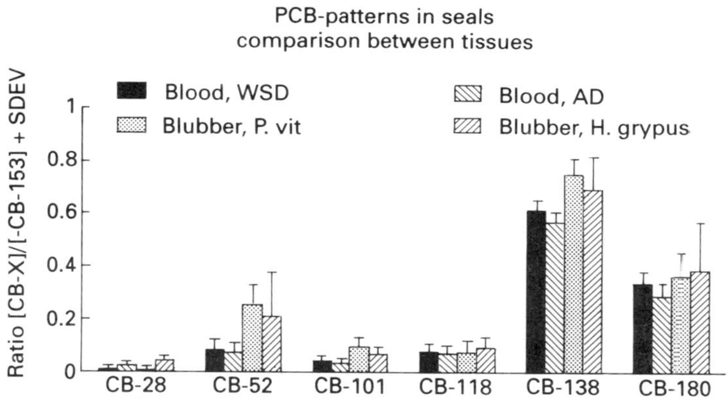

FIG. 6.3 A comparison between CB patterns in blood of captive harbour seals fed with fish from the Wadden Sea (WSD) and the Atlantic Ocean (AG) and blubber of harbour seal (*P. vitulina*) and grey seals (*H. grypus*) found dead at the English east coast (data of Law *et al.*, 1989). CBs are represented by their systematic numbers according to IUPAC rules as proposed by Ballschmiter and Zell (1980). Molecular structures of most CBs are also given in Table 6.2.

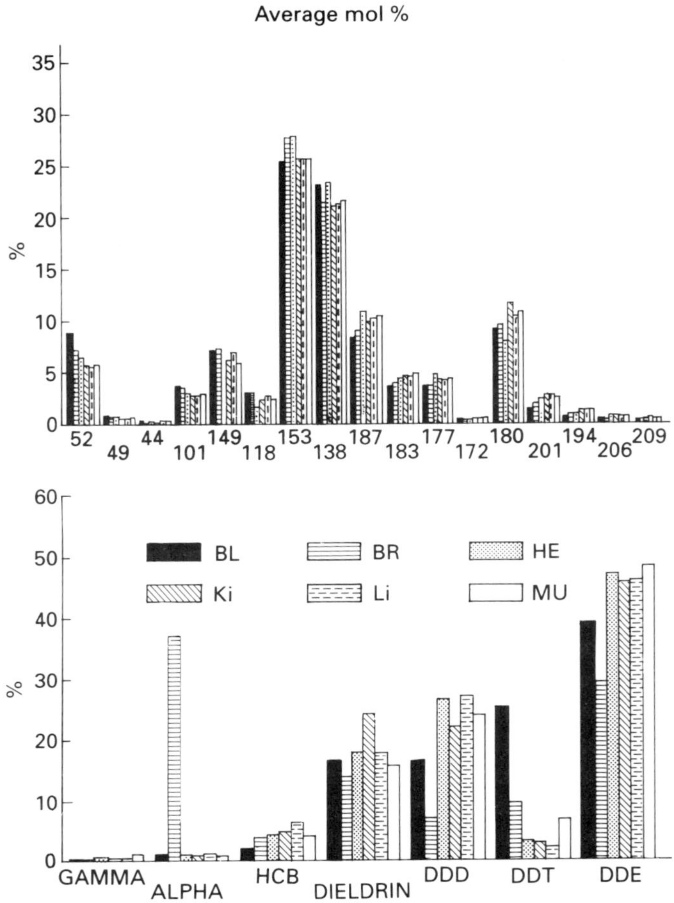

FIG. 6.4 Average mol% contributions (based on concentrations in nmol g^{-1} hexane extractable lipid) of CB congeners (identified by systematic numbers) to their mixtures in blubber (BL), brain (BR), heart (HE), kidney (Ki), liver (Li) and muscle (MU) in harbour porpoises (n = 11). (Source: Duinker *et al.*, 1989.) Systematic numbering of CB congeners according to Ballschmiter and Zell (1980). Molecular structures of most CBs are also given in Table 6.2.

seals (*Halichoerus grypus*) from the English North Sea coast (Fig 6.3; Law *et al.*, 1989).

The following conclusions can be drawn:

(i) The pattern in blood of the seals fed with a diet from the Wadden Sea (WSD) was virtually identical to that of the group fed with a diet from the Atlantic Ocean (AD). Thus, concentrations of individual CB congeners co-varied in spite of a factor of 5 difference in absolute concentrations (Boon *et al.*, 1987).

(ii) The pattern in blubber of grey seals is virtually identical to that of harbour seals.

(iii) The CB patterns of blubber and blood are very similar. This is striking when considering that the blubber samples originated from different animals from another area of the North Sea.

A similar agreement in CB patterns was also found between different organs of harbour porpoises (Duinker *et al.*, 1989). In Fig. 6.4, these CB patterns are represented as the percentage of the contribution of each congener to Σ-PCB (as the sum of all congeners determined). Each cluster of bars represents a single CB congener in different organs. A surprisingly similar conclusion was already reached by Jansson *et al.* (1975), even with data obtained from packed column GC!

However, a different distribution between organs was found for some organochlorine pesticides: e.g. α-HCH accumulated preferentially in brain tissue (Duinker *et al.*, 1989).

The results all show that CBs are free to migrate between different organs and tissues within an animal. Such behaviour can be described by the mammillary model applied in pharmacology and toxicology (Robinson and Roberts 1968; Robinson *et al.*, 1969; Moriarty, 1975, 1978). Here, blood plays the role of the central transport compartment, acting as a bridge between the outside world and a series of peripheral compartments (organs) which are more or less independent from each other while maintaining a dynamic equilibrium with the blood compartment (Matthews and Dedrick, 1984; Aguilar, 1985).

When we assume that cetaceans eat fish with the same CB pattern as present in Wadden Sea fish, we can make the same comparison between CB patterns of fish and an organ of the cetaceans as previously performed with the pattern in seal blood. Figure 6.5 compares the pattern for the CBs that were persistent in our seal study. Although the similarity between fish and cetacean samples is less good than between fish and blood of the WSD seals, we found the similarity good enough to compare groups of CBs which were metabolized by harbour seals. Figure 6.6a shows the situation for CBs with vicinal H atoms in the *o,m* positions in combination with one *ortho*-Cl. Ratios CB-X/CB-153 are much lower in all types of mammals than in fish.

A different situation exists for CBs with *m,p* vicinal H atoms: in seals, these congeners are much lower than in fish, but this is not the case in both groups of cetaceans, where the ratios are similar to fish (Fig. 6.6b). This indicates a low biotransformation capacity for this type of CBs in cetaceans. Metabolism of this class of CBs is catalysed by proteins of the CYP2B subfamily, which were shown to be present in seals but not in cetaceans (see pp. 127, 131). The ratios CB-44/CB-153 and CB-49/CB-153 are given in Fig. 6.6c. According to our model, both CB congeners are expected to be persistent in cetaceans. However,

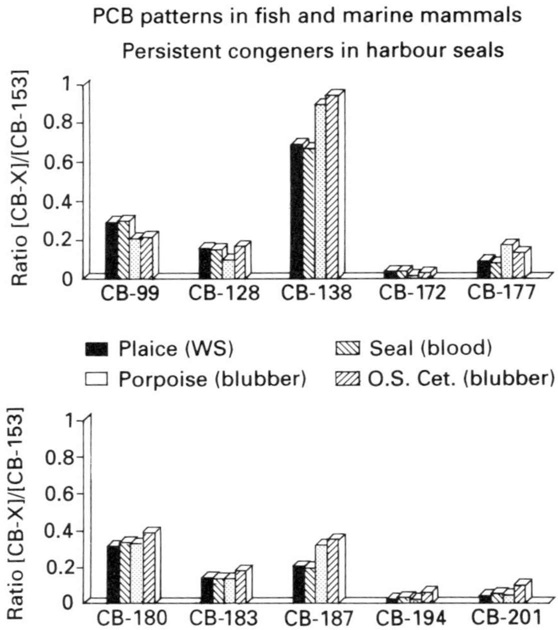

FIG. 6.5 CB patterns expressed as ratio of [CB-X]/[CB-153] in fish, blood of harbour seals fed with this fish and blubber of harbour porpoises and mixed cetacean species from the open North Sea and the Atlantic Ocean (O.S. Cet.; molar basis). All CB congeners given were persistent in harbour seals (data of Duinker *et al.*, 1989). Systematic numbering of CB congeners according to Ballschmiter and Zell (1980). Molecular structures of most CBs are also given in Table 6.2.

instead they show similar concentration ratios as in harbour seals. This may indicate that the model presented above needs some refinement; e.g. Kannan *et al.* (1989b) reported that of the carbon atoms in the *meta* and *para* positions, the atom with the highest frontier (π) electron density is most readily hydroxylated when unsubstituted with chlorine. If chlorine is bound to the carbon atom with the highest frontier (π) electron density, either a chlorine replacement occurs (NIH shift) or the carbon with the next highest electron density is involved in enzymatic metabolism. For carbon atoms in the *ortho* positions, steric hindrance prevents enzymatic attack, even in the case of a favourable frontier (π) electron density.

The study of Muir *et al.* (1988) allows a comparison of CB pattern in an Arctic food chain: Arctic cod (*Boreogadus saida*), ringed seal (*Phoca hispida*) and the polar bear (*Ursus maritimus*). The pattern in cod is similar to that in fish from the Wadden Sea (Fig. 6.7a). The differences between the ringed seal and its diet is somewhat different from those of the seal study described above: some congeners that were

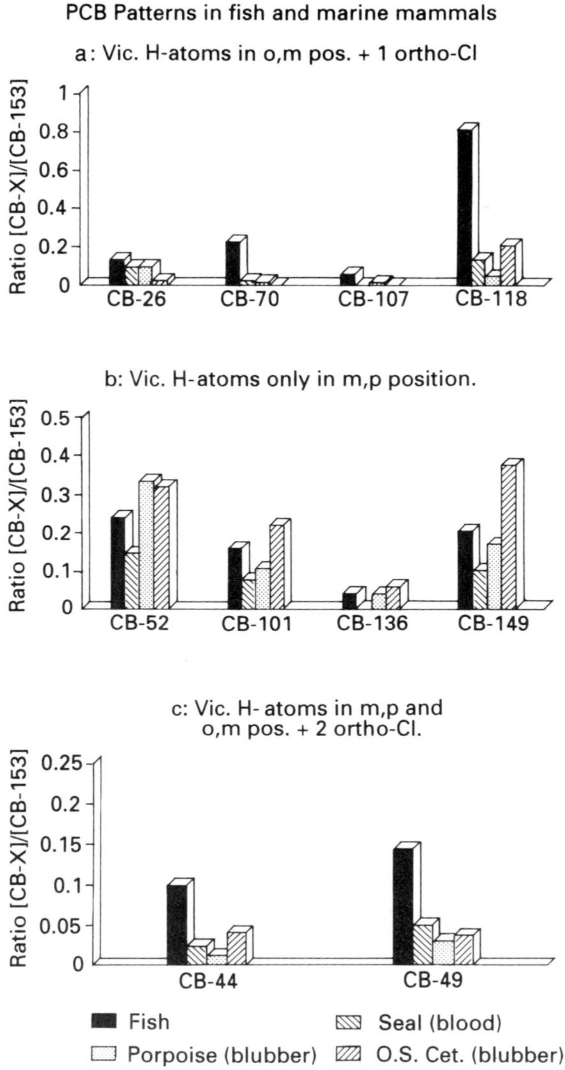

FIG. 6.6 CB patterns expressed as ratio of [CB-X]/[CB-153] in fish, blood of harbour seals fed with this fish and blubber of harbour porpoises and mixed cetacean species from the open North Sea and the Atlantic Ocean (O.S. Cet.; molar basis). All CB congeners given were metabolizable by harbour seals (data of Duinker *et al.*, 1989). Systematic numbering of CB congeners according to Ballschmiter and Zell (1980). Molecular structures of most CBs are also given in Table 6.2.

persistent in the study of Boon *et al.* (1987) were much lower in the ringed seal compared to cod; this applies to CB-177, CB-183, CB-187, CB-194 and CB-201. Other congeners were persistent in both studies: CB-99, CB-138 and CB-180. The ratios CB-105/CB-153 and CB-118/

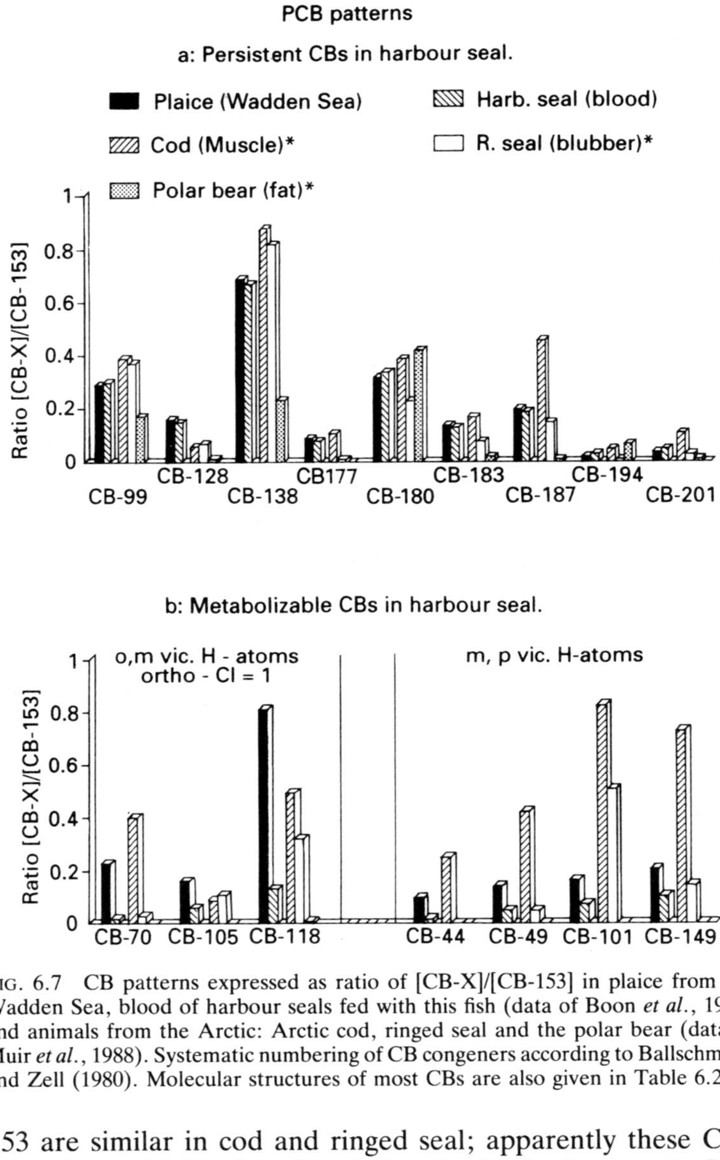

FIG. 6.7 CB patterns expressed as ratio of [CB-X]/[CB-153] in plaice from the Wadden Sea, blood of harbour seals fed with this fish (data of Boon *et al.*, 1987) and animals from the Arctic: Arctic cod, ringed seal and the polar bear (data of Muir *et al.*, 1988). Systematic numbering of CB congeners according to Ballschmiter and Zell (1980). Molecular structures of most CBs are also given in Table 6.2.

CB-153 are similar in cod and ringed seal; apparently these CBs are more persistent in the ringed seal than in the harbour seal. CB congeners with *m,p* vicinal H atoms behaved similarly in all seal species investigated, i.e. they were easily metabolized (Fig. 6.7b).

The largest scope for biotransformation seems to be present in the polar bear. (i) All congeners that were metabolized by the harbour seal were also metabolized by the polar bear. (ii) Since the presence of *o,m* vicinal H atoms in combination with two (CB-99, CB-128, CB-138) or even three *ortho*-Cl atoms (CB-177) still allows for biotransformation, a planar configuration does not appear to be necessary for biotransformation of this type of CBs in this species. However, this structure–metabolism relationship is not always valid, since CB-170 appears persistent. In some cases, the presence of one unsubstituted *para* position also allows for biotransformation, even in the absence of any vicinal H atoms (CB-187 and CB-201). CB-153, CB-180 and CB-194 are persistent congeners even in the polar bear; they lack vicinal H atoms and are substituted at both *para* positions. However, CB-183, which also possesses these structural features, appears to be metabolizable.

The apparent ability of marine mammals to metabolize CBs in general, but especially the category metabolized by CYP1A (with vicinal H atoms in the *o,m* positions and a maximum of one *ortho*-Cl), may be very important where toxicological effects are concerned (see p. 145). In contrast to marine mammals, seabirds may not be able to metabolize this category of CBs (Borlakoglu *et al.*, 1988; Boon *et al.*, 1989; Walker, 1990).

In summary, it can be stated that the harbour seal, cetaceans and the polar bear appear to be able to metabolize CB congeners with *o,m* vicinal H atoms in combination with at maximum one *ortho*-chlorine atom. This capacity may be less developed in the ringed seal. All seal species investigated and the polar bear appear to have a larger capacity than cetaceans to metabolize CB congeners with vicinal H atoms in the *m,p* positions; metabolism occurred irrespective of the *ortho*-Cl substitution pattern. Biotransformation of the first category of CBs is associated with the CYP1A subfamily and a planar configuration is necessary. Biotransformation of the second category of CBs is catalysed by CYP2B, whose presence has been shown in seals but not in cetaceans.

The polar bear shows an ability to metabolize CB congeners without any vicinal H atoms which has not been reported for any other marine mammal species, but the structural requirements for metabolism are still partly unclear.

The pattern of CBs appears to be virtually identical in different organs and tissues of the same animal. The data fit best a model where blood serves as the central transport compartment which will strive to be at a partitioning equilibrium with all organs and tissues. This is not true for all organic contaminants, viz. the selective accumulation of α-HCH in brain.

Patterns of CBs with Special Reference to Congeners with a "TCDD-type" of Toxicity: the Toxic Equivalency Concept

The toxicity of CB congeners is determined by their chlorine substitution pattern. Some CBs show the same toxicological mechanism as 2,3,7,8-tetrachlorodibenzodioxin (TCDD), i.e. initial binding to the cytosolic Ah receptor and transport of the CB receptor complex to the cellular nucleus, where it leads to transcription of DNA for CYP1A. Thus the toxicity of this type of CBs can be determined by their induction capacity for the CYP1A subfamily (Safe, 1984; Safe *et al.*, 1985, 1989).

The normal procedure to quantify individual CBs, using single capillary columns, does not allow the determination of several of the most toxic "TCDD-type" CBs, since these are still not separated from other congeners. The necessary extra separation capacity can be supplied by the use of two different GC columns in series in a special two-dimensional gas chromatograph (2DGC; Duinker *et al.*, 1988a, b). Alternatively, CBs can be separated according to molecular planarity on active carbon columns prior to GC-ECD or GC-MS. The method was reported by Tanabe *et al.* (1987) for the separation of the three pure 3-MC type inducers (with numbers CB-77, CB-126 and CB-169) from other congeners.

The performances of carbon separation and 2 DGC were compared recently (Kannan *et al.*, 1990; 1991): commercial charcoals behaved differently in their separation efficiency towards CBs without *ortho*-Cl substitution (non-*ortho* CBs). Despite high separation efficiencies, the amounts of interfering compounds remaining on the column were still high enough to interfere in the determination of the toxic CBs by single column GC-ECD or even GC-MS. Also, it appeared to be difficult to obtain sufficiently clean charcoals. However, when blank chromatograms are thoroughly checked, charcoals could be used to remove more than 95% of the CBs with two or more *ortho*-Cl (*ortho* CBs). This increased the sensitivity of the method with more than an order of magnitude for the determination of the non-*ortho* CBs using

TABLE 6.9 *Relative Concentrations (expressed as their ratios to CB-153 ± SEM;* n=5 *for all congeners) and Relative Range of Toxic Equivalents (TE) of Some Non-*ortho*-Cl Substituted PCBs and Their Mono-*ortho *Substituted Derivatives in Blubber of Harbour Seals (*Phoca vitulina*).*

CB no.	Cl-substitution pattern	Concentration ratio CB-X / CB-153 ± SEM	TEF	TE range
77	3,3′,4,4′	$1.4 \pm 0.4 \times 10^{-5}$	1×10^{-2}	$1\text{–}2 \times 10^{-7}$
126	3,3′,4,4′,5	$2.1 \pm 0.6 \times 10^{-4}$	1×10^{-1}	$1\text{–}3 \times 10^{-5}$
169	3,3′,4,4′,5,5′	$7.3 \pm 2.3 \times 10^{-5}$	5×10^{-3}	$2\text{–}5 \times 10^{-7}$
105	2,3,3′,4,4′	$4.6 \pm 0.5 \times 10^{-2}$	1×10^{-4}	$4\text{–}5 \times 10^{-6}$
118	2,3′,4,4′,5	$7.4 \pm 2.0 \times 10^{-2}$	5×10^{-5}	$3\text{–}5 \times 10^{-6}$
123	2′,3,4,4′,5	$1.3 \pm 0.6 \times 10^{-2}$	5×10^{-5}	$0.4\text{–}1 \times 10^{-6}$
156	2,3,3′,4,4′,5	$8.3 \pm 2.2 \times 10^{-2}$	5×10^{-4}	$3\text{–}5 \times 10^{-5}$

The toxic equivalency factors (TEFs) are established by the Dutch group of experts on TEFs. The mean concentration of CB-153 was 3.4 ± 1.6 μg g^{-1} wet weight ± SEM.

2DGC-ECD. These combined techniques were used to determine the concentrations of the non-*ortho* CBs in seal blubber (Table 6.9). The *ortho* CBs were determined with 2DGC-ECD only.

The relative concentrations of classes of CBs with regard to their degree of *ortho*-Cl substitution were inversely proportional to their relative "TCDD-type" toxicity, i.e. di-*ortho* > mono-*ortho* > non-*ortho* CBs. A ranking of the potential hazard of several toxic congeners in toxic equivalents (TEs) was made by multiplying their concentration ratio to CB-153 by their toxicities relative to 2,3,7,8-TCDD [toxic equivalency factors (TEFs) as determined in rats]. CB-126 and CB-156 showed TEs in order of 10^{-5}, CB-105, CB-118 and CB-123 in the order of 10^{-6} and CB-77 and CB-169 in the order of 10^{-7}. Values of di-*ortho* CBs could not be given, since no toxic equivalency factors (TEFs) have been determined so far.

Surprisingly, CB-118 is the only congener mentioned in Table 6.9 being involved in international monitoring programmes in the marine environment, such as the Joint Monitoring Programme.

A comparison based on TE values of the relative importance of CBs, chlorinated dioxins and the closely related dibenzofurans is only available for Japanese waters: in contrast to the terrestrial situation, CBs contribute much more to the sum of toxic equivalents in marine mammals than the dioxins and dibenzofurans (Kannan *et al.*, 1989a; Tanabe *et al.*, 1989).

NATURE AND TOXIC EFFECTS OF THE METABOLITES FORMED

Already in the seventies, Jansson *et al.* (1975) and Jensen and Jansson (1976) had reported the occurrence of hydroxylated and methyl sulfone metabolites of PCBs and DDE in grey seals (*Halichoerus grypus*). Since no further reports have appeared on this matter in marine mammals, we have to rely on results obtained with "classical" laboratory mammals (rats, mice and guinea pigs) with regards to routes of metabolism. Mio and Sumino (1985) have reviewed the metabolic pathways leading to sulfonated metabolites: the first step in all cases is the formation of an active intermediate by the cytochrome P450 monooxygenase system. The formation of an epoxide may occur, but is not necessarily the first step (Bush and Träger, 1985; Träger, 1989), leading to either a hydroxylated compound (phenolic metabolite) or binding of a reactive intermediate (epoxide) to glutathione (GSH). Via the mercapturic acid pathway, this conjugate is subsequently degraded to an –SH containing metabolite, which is then methylated to a methylsulfide ($-SCH_3$) group and ultimately oxidized to a methylsulfinyl ($-SOCH_3$) or sulfonyl ($-SO_2CH_3$) side chain of the aromatic moiety.

A hydroxylated metabolite of CB-77 caused lower plasma levels of vitamin A and thyroxine in rats and mice. This effect was observed for the metabolite, but not for the parent compound (Brouwer *et al.*, 1986). In the very same group of captive seals as used for the study on the effects of contaminated food on reproductive success, significantly lowered plasma levels of retinol (vitamin A) and thyroid hormones (total and free thyroxin and triiodothyronin) were established (Brouwer *et al.*, 1989). Thus, this may well be an effect of CB metabolites that already occurs under field circumstances.

At least some methylsulfonyl metabolites can specifically accumulate in the lungs of rodents and humans (Andersson *et al.*, 1987), where they associate with a specific binding protein (Lund *et al.*, 1987). This complex can be secreted into the airway lumen and spread over the entire surface lining of the lungs. The complex can also be transported with the mucociliary escalator to the pharynx and swallowed. In the gut, the metabolite–protein complex may even be released, reabsorbed and recirculated to the lung (Brandt and Bergman, 1987), resulting in an almost continuous recycling within the body. Accumulation of hydroxylated and sulfonated metabolites in the uterine fluid of mice has also been reported (Brandt and Bergman, 1987), which may indicate that another binding site of sulfonated CBs may be present in the uterus. This may be an important finding with respect to the reproductive failure of the seal population in the Dutch Wadden Sea.

DEVELOPMENT OF ABSOLUTE CONCENTRATIONS

The concentrations of CBs built up in marine mammals is mainly determined during the phases of uptake, kinetics and deposition by the chemical properties of CBs and by ecological and physiological factors (Reijnders, 1988a). The influence of biotransformation on the kinetics is elaborately treated on p. 133 of this chapter and will not be discussed here. Regrettably, most of the literature cited in this section is based on PCB concentrations and differences between congeners are not recognized. It can be assumed, however, that PCB concentrations involve mostly the behaviour of the group of metabolically inert congeners, since the metabolizable CBs do not contribute more than about 20% to PCB.

Chemical Properties

The low solubilities of CBs lead to the build up of relatively high concentrations in lipid-rich tissues. Since blubber is the body compartment with the highest proportion of lipids, the highest levels of PCBs are found there (e.g. Reijnders, 1980; Gaskin, 1982; Duinker *et al.*,

1989). Generally, CBs are distributed over the different organs relative to their fat content. However, the retention capacity of the respective organ is influenced by its lipid composition. This explains why in marine mammals the CB residue levels in brain tissue—basically constituted of phospholipids—did not follow the general pattern of a ratio tissue: blubber close to 1, when expressed on lipid based concentrations (Reijnders, 1980; Kawai and Fukushima, 1981; Kawai *et al.*, 1988; Duinker *et al.*, 1988). However, the CB pattern in brain does not differ from other organs, since the concentrations of all congeners are lower by the same factor. Thus, the blood–brain barrier does not operate selectively between different CB congeners. In contrast, α-hexachlorocyclohexane is accumulated selectively in brain (Kawai *et al.*, 1988; Duinker *et al.*, 1989).

A further complicating factor is the stratification of lipid composition, especially in large whales with a thick blubber layer (Aguilar, 1985). Comparison of CB levels in blubber between animals, even from one species, should account for this potential artifact.

Ecological Factors

CBs are mainly taken up by marine mammals via ingested prey. Amplification through food chains is obviously resulting in bioaccumulation of those compounds. However, it is clear that the expected tenfold magnification in every subsequent trophic level is not a general phenomenon (Drescher, 1979; Tanabe *et al.*, 1984; Hidaka *et al.*, 1984). Equilibrium partitioning between body lipids, blood and ambient water is the dominant mechanism for elimination of apolar compounds in invertebrates and fish (Hamelink *et al.*, 1971). This mechanism, involving exchange between blood and water via the gills, is not available for marine mammals. This accounts at least partially for the considerably higher concentrations of some apolar compounds in marine mammals than in lower organisms. Thus, bioaccumulation factors (ratio of concentration of PCB in the organism to that in its food) range from about 1 to over 10^7 (wet weight basis). Irrespective of the mechanism behind the biocaccumulation processes, it can be concluded that at least persistent CBs exhibit bioaccumulative properties.

One of the factors that can influence residue levels in marine mammals is seasonal differences in feeding behaviour. These are either caused by their migratory behaviour or by their opportunistic feeding strategy. Depending on the differences in residue levels in their prey, the seasonal differences in uptake of CBs by marine mammals will be reflected in their tissue burdens. It is difficult to provide a general quantitative concept for this process. Changes in diet can also influence nutritional state of the animals concerned. This will be discussed further on.

However, some indication is given by the experiment described by Reijnders (1988b). Within a period of about 8 months, the average concentration of CBs in blood from harbour seals dropped slightly in the group receiving "clean" fish and increased by about a factor of 2 in the group receiving more contaminated fish. This was caused by a difference in diet concentration of about 7-fold. This occurred without changes in the CB pattern, i.e. the metabolizable and the persistent congeners decreased by the same factor (Boon *et al.*, 1987).

Physiological Factors

Changes in the physiological state of marine mammals can have a considerable effect on their PCB burden.

Growth

In most marine mammal species so far studied, PCB levels increase with age in both sexes until sexual maturity is reached (Helle *et al.*, 1976b; Aguilar, 1987; Clausen and Andersen, 1988). In some cases, Arctic ringed seals (Addison and Smith, 1974) and harbour seals from the north-eastern part of the Wadden Sea (Reijnders, 1980), concentrations in subadults were equal or lower than in newborn animals. This can probably be explained by the fact that during that period relatively more fat was deposited than CBs and consequently a dilution effect occurred.

Nutritional State

Changes in nutritional state will be directly reflected in the thickness of the blubber layer. This is very well documented for baleen whales that carry out seasonal migration from highly productive to oligotrophic waters. The difference in average daily food intake between summer and winter respectively is about tenfold (Tomilin, 1957; Best, 1967), resulting in a total decrease in body weight of 25–50%, mostly caused by fat mobilization in blubber (Lockyer and Brown, 1981). This is expected to be responsible for the observed differences of lipophilic compounds, including CBs, in blubber between baleen and toothed whales (Aguilar, 1985).

Changes in diet during migration between more and less contaminated areas resulted in changes in PCB and Σ-DDT (i.e. *p,p'*–DDT + *p.p'*–DDE + *p,p'*–DDD) body burdens of southern minke whales (Tanabe *et al.*, 1984).

Next to changes in diet, circumannual cycles in blubber layer thickness do occur as a consequence of changes in feeding intensity. Drescher and Huschenbeth (1977) demonstrated that the PCB concentration in blubber of harbour seals was inversely related to blubber layer thickness. It would be interesting to know whether such differences in blubber concentrations correlate with similar changes in other organs, but this has not been studied.

Reproductive State

The reproductive state of marine mammals can clearly influence the CB concentrations in their tissues. In order to address this aspect adequately, the available literature data should not be generalized (as e.g. in Addison, 1989) but treated according to their area of origin. To that end, several categories of areas relative to the levels of pollution have to be distinguished. Irrespective of the category, levels of PCB are usually higher or equal in subadults compared to newborn animals, irrespective of sex.

Difference between the sexes do generally occur after the animals have become mature. These differences appear to depend on the degree of contamination of the area from where the animals originate. In seals originating from areas with relative high levels of PCB (Wadden Sea), no differences in PCB levels between adult males and females have been observed (Drescher *et al.*, 1977; Reijnders, 1980). However, in porpoises as well as seals from areas with medium levels of PCB (North Sea and Baltic), blubber PCB levels increase with age in mature males but not in mature females (Helle *et al.*, 1976b; Born *et al.*, 1981; Donkin *et al.*, 1981; Clausen and Andersen, 1988). In marine mammals from areas with low levels of PCB (Arctic, N.W. Atlantic, Pacific), blubber PCB levels increase with age in mature males but remain constant or decrease slightly in females (Frank *et al.*, 1973; Addison and Smith, 1974; Calambokidis *et al.*, 1978; Subramanian *et al.*, 1988). In order to indicate these conceptions, Figs 6.8 and 6.9 are provided. These are schematic curves, based on the former cited literature, and only intended to demonstrate the different trends corresponding to the categories identified.

As the diet for both sexes in each of the different studies is supposed to be the same, the difference in tissue burden must be a consequence of different physiological processes in males and females. The two routes to eliminate metabolically inert CBs are the production of offspring and the subsequent lactation. Transfer of PCB from female lipid tissue to foetus via placental-transport and to offspring via lactation is demonstrated in many marine mammals (e.g. Addison and Brodie, 1977; Duinker and Hillebrand, 1979; Reijnders, 1980; Wagemann and Muir, 1984; Subramanian *et al.*, 1988). In harbour seals, the concentration of PCB in blubber of stillborn pups ranged from 1 to 4% of the

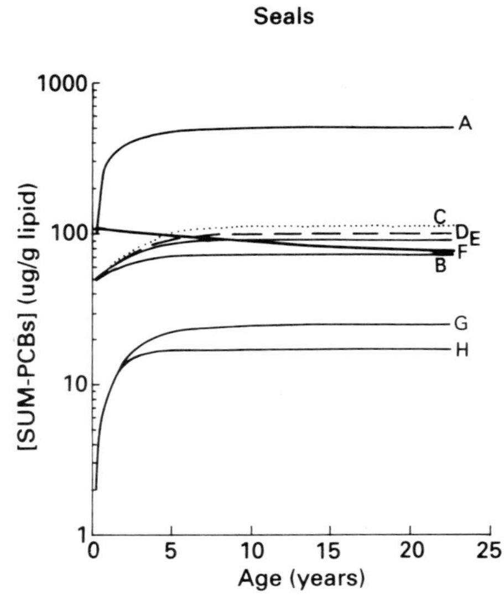

FIG. 6.8 PCB concentrations (μg g^{-1} lipid) in seal blubber. A,B: males and females from respectively the Dutch part and the north-eastern part of the Wadden Sea (Reijnders, 1980); C–F: respectively non-pregnant abnormal females, males, non-pregnant normal females and pregnant females from the Baltic (Helle *et al.*, 1976b); G,H: males resp. females from Washington State waters (Calambokidis *et al.*, 1978).

mean value in adult females (Reijnders, 1980). In striped dolphins, the transfer rate of some organochlorines through parturition ranged from 4 to 9% (Tanabe *et al.*, 1982). Through lactation, this rate ranged even from 80 to 90% (Tanabe *et al.*, 1980). It is evident from these data that the main excretion route for persistent CBs in marine mammals is the mobilization of large amounts of lipids and the associated CBs during lactation.

Subramanian *et al.* (1988) concluded from decreasing ratios of [CB–X]/[CB–153], that the tetra- and penta-chlorinated congeners were eliminated more rapidly during parturition and lactation than higher chlorinated congeners from adult female Dall's porpoises (*Phocoenoides dalli*). This was also observed for CB–52 (2,2′,5,5′-tetraCB), which is metabolically inert in cetaceans, since it is lacking *o,m* vicinal H atoms. After the cessation of parturition, the ratios of [CB–52]/[CB–153] slowly increased again to pre-maturity levels.

The large influence of the reproductive state of female marine mammals on their PCB blubber burden is underlined by the observations on ringed seals (Helle *et al.*, 1976b): increasing concentrations have been found in the series pregnant females < non pregnant (normal) females < males < non-pregnant females with pathological reproductive disorders.

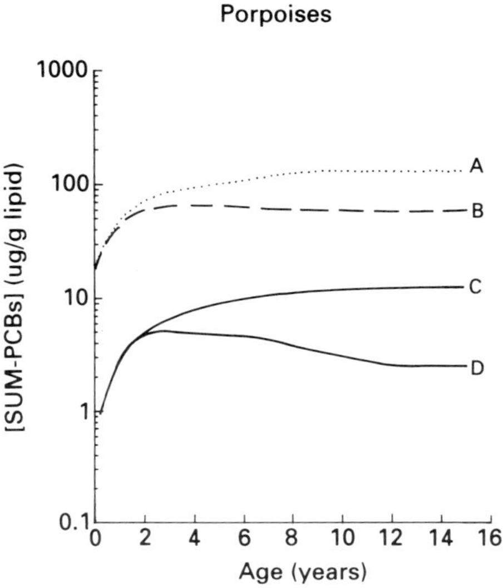

FIG. 6.9 PCB concentrations ($\mu g\ g^{-1}$ lipid) in porpoise blubber. A,B: males resp. females from Danish waters (Clausen and Andersen, 1988); C,D: males resp. females from the north-western Pacific (Subramanian *et al.*, 1988).

It is emphasized that the former discussions have been carried out on data obtained from literature and our own research. Due to differences in sampling procedures and analytical techniques those data can only be used to indicate relative trends within the respective areas. They should not be interpreted as absolute values for comparison between different areas, and only serve to indicate the order of magnitude.

In summary; for seals in relatively highly contaminated areas, lifetime PCB levels develop equally in both sexes. Here, CB deposition in females exceeds excretion within an annual sexual cycle, despite the elimination of CBs with parturition and lactation. In areas with medium to low contamination levels of CBs, PCB levels are generally higher in mature males than in females. This is true for both seals and porpoises. Thus, parturition and lactation are mainly responsible for this sexual difference. The transfer rate of individual congeners from mother to foetus can differ between congeners.

CONCLUSIONS

In order to understand the ability of marine mammals to dispose of (halogenated) polyaromatic contaminants such as CBs and related compounds, a thorough knowledge at the molecular level of the interaction between xenobiotic compounds and biotransformation enzyme systems of these animals is necessary. Although limited in number, the studies reported indicate differences in marine mammals from the model mammals studied mostly (i.e. rodents), as well as differences between seals, cetaceans and the polar bear. Indeed, even differences between seal species may occur, pointing to the dangers of extrapolation between species.

The cytochrome P450 dependent monooxygenase system of seals and cetaceans possesses two proteins of the CYP1A subfamily, which is the usual case for mammalian systems. The CYP2B subfamily appears underdeveloped compared to rodents, especially in cetaceans.

The harbour seal, cetaceans and the polar bear appear to be able to metabolize CB congeners with *o,m* vicinal H atoms in combination with at maximum one *ortho*-chlorine atom. This capacity may be less developed in the ringed seal. All seal species investigated and the polar bear, appear to have a larger capacity than cetaceans to metabolize CB congeners with vicinal H atoms in the *m,p* positions; metabolism occurred irrespective of the *ortho*-Cl substitution pattern. The first category of CBs is probably metabolized by the CYP1A subfamily and a planar configuration is necessary. The second category of CBs is metabolized by the CYP2B subfamily; here planarity is unnecessary for metabolism. The polar bear shows a thus far unique capacity among marine mammals to metabolize CB congeners without any vicinal H atoms, but the structural requirements for metabolism are still partly unclear.

In marine mammals, CBs can potentially exert their toxicity in at least three different ways: as parent compounds and as hydroxylated or sulphonated metabolites.

The pattern of CBs appears to be virtually identical in different organs and tissues of the same animal. Absolute concentrations largely differ as a function of lipid contents of the organs and tissues, but lipid based concentrations in brain are lower than in other organs. This is due to the composition of the lipids; in brain mostly phospholipids and sterols are present, which have a lower affinity for CBs than free fatty acids and triacylglycerols. The blood/brain barrier does not cause a difference in the CB pattern between brain and the other organs; i.e. the concentrations are lowered by the same factor for all congeners in brain. The data fit best a model where CBs are taken up from food with blood serving as the central transport compartment which will strive to be at a partitioning equilibrium with all organs and tissues. This is not necessarily true for all organic contaminants, viz. the selective accumulation of α-HCH in brain.

For seals in relatively highly contaminated areas, lifetime PCB levels develop equally in both sexes. Here, PCB deposition in females exceeds excretion within an annual sexual cycle, despite the elimination of PCB with parturition and lactation. In areas with medium to low contamination levels of PCB, the levels are generally higher in mature males than in females. This is true for both seals and porpoises. Thus, parturition and lactation are mainly responsible for this sexual difference.

Migration and nutritional status are other factors influencing PCB concentrations in marine mammals.

DIRECTIONS FOR FUTURE RESEARCH

In order to understand the ability of marine mammals to dispose of polyaromatic contaminants such as CBs and related compounds, a thorough knowledge of their kinetics at the level of individual congeners is necessary. This should specifically include the interactions between these xenobiotic compounds and the biotransformation enzyme systems of these animals. Despite the interest in marine mammals in the context of marine sciences, the few reports reviewed in this paper represent our knowledge about these systems today. Although limited in volume, these studies indicate differences from the model mammals (i.e. rodents) studied mostly, as well as differences between seals, cetaceans and the polar bear. Indeed, even differences between seal species may occur, pointing to the dangers of extrapolation from one species to another.

Acknowledgements The work of A.G. was supported by the Marine Mammal Programme of the Norwegian Fisheries Research Council (NFFR). Dr. J. M. Everaarts and both editors of this book are kindly acknowledged for their critical reading of the original manuscript.

REFERENCES

Abdel-Hamid, F. M., J. A. Moore and H. B. Matthews. 1981. A comparative study of 3,4,3′,4′-tetrachlorobiphenyl in male and female rats and in female monkeys. *J. Toxicol. Environ. Health* 7: 181–191.

Addison, R. F. 1989. Organochlorines and marine mammal reproduction. *Can. J. Fish. Aquat. Sci.* **46**: 360–368.

Addison, R. F. and P. F. Brodie. 1977. Organochlorine residues in maternal blubber, milk and pup blubber from grey seals (*Halichoerus grypus*) from Sable Island, Nova Scotia. *J. Fish. Res. Bd. Can.* **34**: 937–941.

Addision, R. F. and P. F. Brodie. 1984. Characterization of ethoxyresorufin *O*-deethylase in grey seal *Halichoerus grypus*. *Comp. Biochem. Physiol.* **79C**: 261–263.

Addison, R. F., P. F. Brodie, A. Edwards and M. C. Sadler. 1986. Mixed function oxidase activity from the harbour seal (*Phoca vitulina*) from Sable Is., N.S. *Comp. Biochem. Physiol.* **85C**: 121–124.

Addison, R. F. and T. G. Smith. 1974. Organochlorine residue levels in Arctic ring seals: Variation with age and sex. *Oikos* **25**: 335–337.

Aguilar, A. 1985. Compartmentation and reliability of sampling procedures in organochlorine pollution surveys of cetaceans. *Res. Rev.* **95**: 91–114.

Aguilar, A. 1987. Using organochlorine pollutants to discriminate marine mammal populations: a review and critique of methods. *Mar. Mamm. Sci.* **3**(3): 242–262.

Almkvist, L. 1982. Baltic marine mammals—A status report. ICES CM. 1982/N:16: 19 pp.

Andersson, O., J. Lund and E. Ripe. 1987. A binding protein for the PCB metabolite 4,4′–bis(methylsulfonyl)–2,2′,5,5′–tetrachlorobiphenyl: Characterization in broncho-alveolar lavage from healthy smokers and non-smokers. *Chemosphere* **16**: 1667–1670.

Anonymous. 1987. ICES/IOC/UNEP Review of contaminants in marine mammals. ICES CM. 1987/N:8.

Baily, G. S., D. Selivonchick and J. Hendricks. 1987. Initiation, promotion, and inhibition of carcinogenesis in rainbow trout. *Environ. Health Perspect.* **71**: 147–153.

Ballschmiter, K. and M. Zell. 1980. Analysis of polychlorinated biphenyls (PCB) by glass capillary gas chromatography. Composition of technical Aroclor and Clophen-PCB mixtures. *Z. Analyt. Chem. 302*: 20–31.

Beland, J., R. Michaud and D. Martineau. 1987. Recensements de la population de belugas du Saint-Laurent en 1985 par embarcations. *Rapp. tech. can. sci. halieut. aquat.* No. 1545.

Bergman, A., M. Olsson and L. Reuthergårdh. 1981. Lowered reproduction rate in seal population and PCB concentration. ICES CM. 1981/N:10: 12 pp.

Best, P. B. 1967. Distribution and feeding habits of baleen whales of the province. Rep. S.A. Div. Sea Fish. Invest. Rep. 57.

Boon, J. P., P. J. H. Reijnders, J. Dols, P. Wensvoort and M. T. J. Hillebrand. 1987. The kinetics of individual polychlorinated biphenyl congeners in female harbour seals (*Phoca vitulina*), with evidence for structure-related metabolism. *Aquat. Toxicol.* **10**: 307–324.

Boon, J. P., F. Eijgenraam, J. M. Everaarts and J. C. Duinker. 1989. A structure-activity relationship (SAR) approach towards metabolism of PCBs in marine animals from different trophic levels. *Mar. Environ. Res.* **27**: 159–176.

Boon, J. P., J. M. Everaarts, M. T. J. Hillebrand, M. L. Eggens, J. Peijnenburg and A. Goksøyr. 1992. Changes in levels of hepatic biotransformation of enzymes and haemoglobin in female plaice (*Pleuronectes platessa*) after oral administration of a technical PCB mixture (Clophen A40). *The Science of the Total Environment*, **114**: 113–133.

Borlakoglu, J. T., J. P. G. Wilkins and C. H. Walker. 1988. Polychlorinated biphenyls in fish-eating seabirds: Molecular features and metabolic interpretations. *Mar. Environ. Res. 24*: 15–19.

Born, E. W., I. Kraul and T. Kristensen. 1981. Mercury, DDT and PCB in the Atlantic walrus (*Odobenus rosmarus rosmarus*) from the Thole district, Northern Greenland. *Arctic* **34**: 255–260.

Brandt, I. and A. Bergman. 1987. PCB methyl sulphones and related compounds: Identification of target cells and tissues in different species. *Chemosphere* **16**: 1671–1676.

Brouwer, A., A. J. van den Berg, W. S. Blaner and D. S. Goodman. 1986. Transthyretin (prealbumin) binding in PCBs, a model for the mechanism of interference with vitamin A and thyroid hormone metabolism. *Chemosphere* **15**: 1699–1706.

Brouwer, A., P. J. H. Reijnders and J. H. Koeman. 1989. Polychlorinated biphenyl (PCB)-contaminated fish induces vitamin A and thyroid hormone deficiency in the common seal (*Phoca vitulina*). *Aquat. Toxical.* **15**: 99–106.

Bush, E. D. and W. F. Träger. 1985. Substrate probes for the mechanism of aromatic hydroxylation catalysed by cytochrome P–450: selectively deuterated analogues of warfarin. *J. Medicin. Chem.* **28**: 992–996.

Calambokidis, J., K. Bowman, S. Carter, J. Cubbabe, P. Dawson, T. Fleischner, J. Schuett-Hames, J. Skidmore, B. Taylor and S. G. Herman. 1978. Chlorinated hydrocarbon concentrations and the ecology and behavior of harbor seals in Washington State waters. Final report to National Science Fundation, The Evergreen State College, Olympia, WA. 121 pp.

Celander, M., M. Ronis and L. Förlin. 1989. Initial characterization of a constitutive cytochrome P–450 isoenzyme in rainbow trout liver. *Marine Environ. Res.* **28**: 9–13.

Chaturapit, S. and G. M. Holder. 1978. Studies on the hepatic microsomal metabolism of (14)C-phenanthrene. *Biochem. Pharmacol.* **27**: 1865–1871.

Clausen, B. and S. Andersen. 1988. Evaluation of by catch and health status of the harbour porpoise (*Phocoena phocoena*) in Danish waters. *Danish Rev. Game. Biol.* **13**(5): 1–20.

Cullen, J. M. and K. L. E. Kaiser. 1984. An examination of the role of rotational barriers in the toxicology of PCBs. In K. L. E. Kaiser (ed.), *QSAR in Environmental Toxicology*. Reidel Publishing Company, pp. 39–66.

De Long, R. L., W. G. Gilmartin and J. G. Simpson. 1973. Premature births in California sea lions: Association with high organochlorine pollutant residue levels. *Science* **181**: 1168–1170.

Donkin, P., S. V. Mann and E. I. Hamilton. 1981. Polychlorinated biphenyl, DDT and dieldrin residues in grey seal (*Halichoerus grypus*) males, females and mother-foetus pairs sampled at the Farne Islands, during breeding season. *Sci. Tot. Environ.* **19**: 121–142.

Drescher, H. E. 1979. *Biologie, Ökologie und Schutz der Seehunde im Schleswig-Holsteinischen Wattenmeer*. Wiss. Sch. I. Landesjagdverband Schleswig-Holstein, Meldorf, FRG.

Drescher, H. E., V. Harms and E. Huschenbeth. 1977. Organochlorines and heavy metals in the harbour seal (*Phoca vitulina*) from the German North Sea coast. *Mar. Biol.* **41**: 99–106.

Duinker, J. C. and M. T. J. Hillebrand. 1979. Mobilization of organochlorines from female lipid tissue and transplacental transfer to fetus in a harbour porpoise (*Phocoena phocoena*) in a contaminated area. *Bull. Environ. Contam. Toxicol.* **23**: 728–732.

Duinker, J. C., D. E. Schulz and G. Petrick. 1988a. Multidimensional gas chromatography with electron capture detection for the determination of toxic congeners in polychlorinated biphenyl mixtures. *Anal. Chem.* **60**: 478–482.

Duinker, J. C., D. E. Schulz and G. Petrick. 1988b. Selection of chlorinated biphenyl congeners for analysis in environmental samples. *Mar. Poll. Bull.* **19**: 19–25.

Duinker, J. C., T. Zeinstra, M. Th. J. Hillebrand and J. P. Boon. 1989. Individual chlorinated biphenyls and pesticides in tissues of some cetacean species from the North Sea and the Atlantic Ocean: Tissue distribution and biotransformation. *Aquat. Mammals* **15**: 95–124.

Elskus, A. A., J. J. Stegeman, L. C. Susani, D. Black, R. J. Pruell and S. J. Fluck. 1989. Polychlorinated biphenyl concentration and cytochrome P–450E expression in winter flounder from contaminated environments. *Marine Environ. Res.* **28**: 25–30.

Engelhardt, F. R. 1982. Hydrocarbon metabolism and cortisol balance in oil-exposed ringed seals, *Phoca hispida*. *Comp. Biochem. Physiol.* **72C**: 133–136.

Estabrook, R. W. 1990. Future developments with the cytochrome P–450: promises for the applied chemical and medical sciences. *Biochem. Soc. Trans.* **18**: 34–36.

Everaarts, J. M., A. de Buck, M. T. J. Hillebrand and J. P. Boon. 1991. Residues of chlorinated biphenyl congeners and pesticides in brain and liver of the oystercatcher (*Haematopus ostralegis*) in relation to age, sex and biotransformation capacity. *Science Tot. Environ.* **100**: 483–499.

Frank, R., K. Ronald and H. E. Braun. 1973. Organochlorine residues in harp seals (*Pagophilus groenlandicus*) caught in eastern Canadian waters. *J. Fish. Res. Bd. Can.* **30**: 1053–1063.

Fuller, G. B. and W. C. Hobson. 1986. Effects of PCBs on reproduction in mammals. In J. S. Waid (ed.), *PCBs and the Environment*. CRC Press, Vol. II, pp. 101–125.

Gage, J. C. and S. Holm. 1976. The influence of molecular structure on retention and excretion of polychlorinated biphenyls by the mouse. *Toxicol. Appl. Pharmacol.* **36**: 555–560.

Gaskin, D. E. 1982. *The Ecology of Whales and Dolphins*. Heinemann, London.

Geraci, J. R. and D. J. St. Aubin. 1982. Study on the effects of oil on cetaceans. Report for US Department of the Interior, Bureau of Land Management, Contract No. AA–551–CT9–29, 274 pp.

Gibson, G. G. 1990. The cytochrome P450IV family: constitutive and inducible haemoproteins. *Biochem. Soc. Trans.* **18**: 14–15.

Gilmartin, W. G., R. L. DeLong, W. Smith, J. C. Sweeney, B. W. DeLappe, R. W. Risebrough, L. A. Griner, M. D. Dailey and D. B. Peakall. 1976. Premature parturition in the California sea lion. *J. Wildl. Diseases* **12**: 104–115.

Goksøyr, A., J. Tarlebø, J. E. Solbakken and J. Klungsøyr. 1985. Characteristics of the hepatic microsomal cytochrome P–450 system of the minke whale (*Balaenoptera acutorostrata*). *Marine Environ. Res.* **17**: 113–116.

Goksøyr, A., J. E. Solbakken, J. Tarlebø and J. Klungsøyr. 1986a. Initial characterization of the hepatic microsomal cytochrome P–450 system of the piked whale (minke) *Balaenoptera acutorostrata*. *Marine Environ. Res.* **19**: 185–203.

Goksøyr, A., J. E. Solbakken and J. Klungsøyr. 1986b. Regioselective metabolism of phenanthrene in Atlantic cod (*Gadus morhua*): Studies on the effects of monooxygenase inducers and role of cytochrome P–450. *Chem.-Biol. Interact.* **60**: 247–263.

Goksøyr, A., T. Andersson, L. Förlin, J. Stenersen, E. A. Snowberger, B. R. Woodin and J. J. Stegeman. 1988. Xenobiotic and steroid metabolism in adult and foetal piked (minke) whales, *Balaenoptera acutorostrata. Marine Environ. Res.* **24**: 9–13.

Goksøyr, A., T. Andersson, L. Förlin, E. A. Snowberger, B. R. Woodin and J. J. Stegeman. 1989. Cytochrome P–450 monooxygenase activity and immunochemical properties of adult and foetal piked whales, *Balaenoptera acutorostrata.* In I. Schuster (ed.), *Cytochrome P–450, Biochemistry and Biophysics* Taylor & Francis, pp. 698–701.

Goksøyr, A., H. E. Larsen and A.–M. Husøy. 1991a. Application of a cytochrome P–450IA1 ELISA in environmental monitoring and ecotoxicological testing of fish. *Comp. Physiol. Biochem.*, **100**: 157–160.

Goksøyr, A., T. Andersson, D. R. Buhler, J. J. Stegeman, D. E. Williams and L. Förlin. 1991b. Immunochemical cross-reactivity of beta-naphthoflavone-inducible cytochrome P–450 (P–450IA) from different fish species and rat. *Fish Physiology and Biochemistry*, **9**: 1–13.

Goksøyr, A., J. Beyer, H. E. Larsen, T. Andersson and L. Förlin. 1992. Cytochrome P450 in seals: monooxygenase activities, immunochemical cross-reactions and response to phenobarbital treatment. *Marine Environ. Res.*, in press.

Gooch, J. W., A. A. Elskus, P. J. Klopper-Sams, M. E. Hahn and J. J. Stegeman. 1989. Effects of *ortho*- and non-*ortho*-substituted polychlorinated biphenyl congeners on the hepatic monooxygenase system in scup (*Stenotomus chrysops*). *Toxicol. Appl. Pharmacol.* **98**: 422–433.

Graves, P. E., G. A. Elhag, P. J. Ciaccio, D. P. Bourque and J. R. Halpert. 1990. cDNA and deduced amino acid sequences of a dog hepatic cytochrome P450IIB responsible for the metabolism of 2.2′,4,4′,5,5′-hexachlorobiphenyl. *Arch. Biochem. Biophys.* **281**: 106–115.

Hamelink, J. L., R. C. Waybrant and R. C. Ball. 1971. A proposal: exchange equilibria control the degree chlorinated hydrocarbons are biologically magnified in lentic environments. *Trans. Am. Fish. Soc.* **100**: 207–214.

Helle, E. 1980. Lowered reproductive capacity in female ringed seals (*Pusa hispida*) in the Bothnian Bay, northern Baltic Sea, with special reference to uterine occlusions. *Ann. Zool. Fenn.* **17**: 147–158.

Helle, E. 1981. Reproductive trends and occurrence of organochlorines and heavy metals in the Baltic seal populations. ICES CM. 1981/E:37: 13 pp.

Helle, E. 1986. The decrease in the ringed seal population of the Gulf of Bothnia in 1975–1984. *Finnish Game Res.* **44**: 28–32.

Helle, E., M. Olsson and S. Jensen. 1976a. DDT and PCB levels and reproduction in ringed seals from the Bothnian Bay. *Ambio* **5**: 188–189.

Helle, E., M. Olsson and S. Jensen. 1976b. PCB levels correlated with pathological changes in seal uteri. *Ambio* **5**: 261–263.

Hidaka, H., S. Tanabe, M. Kawano and R. Tatsukawa. 1984. Fate of DDTs, PCBs and chlordane compounds in Arctic marine ecosystem. *Mem. Nat. Inst. Polar Res. (Spec. Iss.)* **32**: 153–161.

Hook, O. and A. G. Johnels. 1972. The breeding and distribution of the grey seal (*Halichoerus grypus* Fab.) in the Baltic Sea, with observations on other seals of the area. *Proc. R. Soc. Lond. Ser. B* **182***: 37.*

Huheey, J. E. 1975. *Inorganic Chemistry: Principles of Structure and Reactivity*. Harper & Row, London: 737 pp.

Jacob, J., A. Schmoldt and G. Grimmer. 1982. Influence of monooxygenase inducers on the metabolic profile of phenanthrene in rat liver microsomes. *Toxicology* **25**: 333–343.

Jansson, B., S. Jensen, M. Olsson, L. Renberg, G. Sundström and R. Vaz. 1975. Identification by GC–MS of phenolic metabolites of PCB and *p,p′*–DDE isolated from Baltic guillemot and seal. *Ambio* **4**: 93–97.

Jensen, S. and B. Jansson. 1976. Anthropogenic substances in seals from the Baltic: Methylsulfone metabolites of polychlorinated biphenyls and DDE. *Ambio* **5**: 257–260.

Jerina, D. M. and J. W. Daly. 1974. Arene oxides: A new aspect of drug metabolism. *Science* **185**: 573–582.

Kannan, N., T. Wakimoto and R. Tatsukawa. 1989a. Possible involvement of frontier (π) electrons in the metabolism of polychlorinated biphenyls (PCBs). *Chemosphere* **18**: 1955–1963.

Kannan, N., S. Tanabe, M. Ono and R. Tatsukawa. 1989a. Critical evaluation of polychlorinated biphenyl toxicity in terrestrial and marine mammals: Increasing impact of non-*ortho* and mono-*ortho* coplanar polychlorinated biphenyls from land to ocean. *Arch. Environ. Contam. Toxicol.* **18**: 850–857.

Kannan, N., G. Petrick, D. E. Schulz, J. C. Duinker, J. P. Boon, E. van Arnhem and S. Jansen. 1990. Multidimensional GC and charcoal separation techniques in the determination of non-ortho Cl chlorobiphenyls—A comparison. *Proc. 10th Int. Meeting Dioxin 90, Bayreuth, FRG.* O. Hutzinger, H. Fiedler and G., Addis, (eds). Vol. 2; pp. 165–168.

Kannan, N., G. Petrick, D. E. Schulz, J. C. Duinker, J. P. Boon, E. van Arnhem and S. Jansen. 1991. A comparison between activated charcoals and multidimensional GC in the separation and determination of (non-*ortho* Cl substituted) toxic chlorobiphenyls. *Chemosphere* **23**: 1055–1076.

Kato, S., J. D. McKinney and H. B. Matthews. 1980. Metabolism of symmetrical hexachlorobiphenyl isomer in the rat. *Toxicol. Appl. Pharmacol.* **53**: 389–398.

Kawai, S. and M. Fukushima. 1981. Relation between lipid composition and the concentrations of organochlorine compounds in the various organs of striped dolphins *Stenella coeruleoalba*. In T. Fujiymam (ed.), *Studies on the Levels of Organochlorine Compounds and Heavy Metals in Marine Organisms*. Univ. of Ryukyus.

Kawai, S. M., N. Fukushima, N. Miyazaki and R. Tatsukawa. 1988. Relationships between lipid composition and organochlorine levels in the tissues of striped dolphin. *Mar. Pollut. Bull.* **19**: 129–133.

Kulkarni, A. P., E. Smith and E. Hodgson. 1976. Occurrence and characterization of microsomal cytochrome P–450 in several vertebrate and insect species. *Comp. Biochem. Physiol.* **54B**: 509–513.

Kulkarni, A. P. and E. Hodgson. 1980. Comparative toxiocology. In E. Hodgson and F. E. Guthrie (eds), *Introduction to Biochemical Toxicology*. Elsevier, New York, pp. 106–132.

Law, R. J., C. R. Allchin and John Harwood. 1989. Concentrations of organochlorine compounds in the blubber of seals from eastern and north-eastern England. 1988. *Mar. Pollut. Bull.* **20**: 110–115.

Lewis, D. F. V., C. Ioannides and D. V. Parke. 1986. Molecular dimensions of the substrate binding site of cytochrome P–448. *Biochem. Pharmacol.* **35**: 2179–2185.

Lockyer, C. H. and S. G. Brown. 1981. The migration whales. In: H. Aidley, (ed.), *Animal Migration*. Soc. Exp. Biol. Semin. Ser. 13. Cambridge Univ. Press, 105 pp.

Lund, J., L. Nordlund, T. Devereux, H. Glaumann and J.–A. Gustafsson. 1987. Physiocochemical and immunochemical characterization of a binding-protein for PCB methyl sulfones. *Chemosphere* **16**: 1677–1680.

Martineau, D., P. Beland, C. Desjardins and A. Lagace. 1987. Levels of organochlorine chemicals in tissues of beluga whales (*Delphinapterus leucas*) from the St Lawrence estuary, Quebec, Canada. *Arch. Environ. Contam. Toxicol.* **16**: 137–147.

Matthews, H. B. and R. L. Dedrick. 1984. Pharmacokinetics of PCBs. *Ann. Rev. Pharmacol. Toxicol.* **24**: 85–103.

Matthews, H. B. and D. B. Tuey. 1980. The effect of chlorine position on the distribution and excretion of four hexachlorobiphenyl isomers. *Toxicol. Appl. Pharmacol.* **53**: 377–388.

McFarland, V. A. and J. U. Clarke. 1989. Environmental occurrence, abundance, and potential toxicity of polychlorinated biphenyl congeners: considerations for a congener-specific analysis. *Environ. Health Perspect.* **81**: 225–239.

McKinney, J. D., K. E. Gottschalk and L. Pedersen. 1983. The polarizability of aromatic systems. An application to polychlorinated biphenyls (PCBs), dioxins and polyaromatic hydrocarbons. *J. Molec. Struct.* **105**: 427–438.

Millis, C. D., R. A. Mills, S. D. Sleight and S. D. Aust. 1985. Toxicity of 3,4,5,3′,4′,5′-hexabrominated biphenyl and 3,4,3′,4′-tetrabrominated biphenyl. *Toxicol. Appl. Pharmacol.* **78**: 88–95.

Mills, R. A., C. D. Millis, G. A. Dannan, F. P. Guengerich and S. D. Aust. 1985. Studies on the structure-activity relationships for the metabolism of polybrominated biphenyls by rat liver microsomes. *Toxicol. Appl. Pharmacol.* **78**: 96–104.

Mio, T. and K. Sumino. 1985. Mechanism of biosynthesis of methylsulfones from PCBs and related compounds. *Environ. Health Perspect.* **59**: 129–135.

Moriarty, F. 1975. Exposure and residues. In F. Moriarty (ed.), *Organochlorine Insecticides: Persistent Organic Pollutants*. Academic Press, London, p. 29.

Moriarty, F. 1978. Terrestrial animals. In: G. C. Butler, (ed.), *Principles of Ecotoxicology*. Wiley, New York, p. 169.

Muir, D. C. G., R. J. Norstrom and M. Simon. 1988. Organochlorine contaminants in Arctic marine food chains: accumulation of specific polychlorinated biphenyls and chlordane-related compounds. *Environ. Sci. Technol.* **22**: 1071–1079.

Nebert, D. W. and F. J. Gonzalez. 1987. P450 genes: Structure, evolution, and regulation. *Ann. Rev. Biochem.* **56**: 945–993.

Nebert, D. W., D. R. Nelson, M. Adesnik, M. J. Coon, R. W. Estabrook, F. J. Gonzalez, F. P. Guengerich, I. C. Gunsalus, E. F. Johnson, B. Kemper, W. Levin, I. R. Phillips, R. Sato and M. R. Waterman. 1989a. The P450 superfamily: updated listing of all genes and recommended nomenclature for the chromosomal loci. *DNA* **8**: 1–13.

Nebert, D. W., D. R. Nelson and R. Feyereisen. 1989b. Evolution of the cytochrome P450 genes. *Xenobiotica* **19**: 1149–1160.

Nebert, D. W., D. R. Nelson, M. J. Coon, R. W. Estabrook, R. Feyereisen, Y. Fujii-Kuriyama, F. J. Gonzales, R. P. Guenngerich, I. C. Gunsalus, E. F. Johnson, J. C. Loper, R. Sato, M. R. Waterman and D. J. Waxman. 1991. The P450 superfamily: update on new sequences, gene mapping and recommended nomenclature. *DNA and Cell Biol.* **10**: 1–14.

Olsson, M. 1986. PCBs in the Baltic environment. In J. S. Waid (ed.), *PCBs and the Environment*. CRC Press, Vol. III, pp. 181–208.

Olsson, M., A. G. Johnels and R. Vaz. 1975. DDT and PCB levels in seals from Swedish waters. The occurrence of aborted seal pups. *Nat. Swed. Environ. Prot. Board. PM* **591**: 43–65.

Preston, B. D. and J. R. Allen. 1980. 2,2′,5,5′-tetrachlorobiphenyl: Isolation and identification of metabolites generated by rat liver microsomes. *Drug Metab. Disp.* **8**: 197–204.

Reijnders, P. J. H. 1978. Recruitment in the harbour seal (*Phoca vitulina*) population in the Dutch Wadden Sea. *Neth. J. Sea Res.* **12**: 164–179.

Reijnders, P. J. H. 1979. Organochlorine and heavy metal residues in harbour seals of Schleswig-Holstein plus Denmark and the Netherlands: Their possible effects in relation to reproduction in both populations. ICES CM 1979/N: 18.

Reijnders, P. J. H. 1980. Organochlorine and heavy metal residues in harbour seals from the Wadden Sea and their possible effects on reproduction. *Neth. Sea Res.* **14**: 30–65.

Reijnders, P. J. H. 1986. Reproductive failure in common seals feeding on fish from polluted coastal waters. *Nature* **324**: 456–457.

Reijnders, P. J. H. 1988a. Ecotoxicological perspectives in marine mammalogy: Research principles and goals for a conservation policy. *Mar. Mamm. Sci.* **4**: 91–102.

Reijnders, P. J. H. 1988b. Environmental impact of PCBs in marine environment. In P. J. Newman and A. R. Agg (eds), *Environmental Protection of the North Sea*. Heinemann, Oxford, pp. 85–98.

Reuthergårdh, L. and A. H. Knap. 1987. Overview of organochlorine compound analysis in marine mammals. ICES Mar. Chem. Work. Group, 1987/7.2.2, mimeo: 23pp.

Robinson, J. and M. Roberts. 1968. Accumulation, distribution and elimination of organochlorine insecticides by vertebrates. *S.C.I. Monogr.* **29**: 106.

Robinson, J., M. Baldwin and A. I. T. Walker. 1969. The pharmokinetics of HEOD (Dieldrin) in the rat. *Food Cosmet. Toxicol.* **7**: 317.

Safe, S. 1984. Polychlorinated biphenyls (PCBs) and polybrominated biphenyls (PBBs): Biochemistry, toxicology, and mechanism of action. *CRC Crit. Rev. Toxicol.* **13**: 319–393.

Safe, S., S. Bandiera, T. Sawyer, L. Robertson, L. Safe, A. Parkinson, P. E. Thomas, D. E. Ryan, L. M. Reik, W. Levin, M. A. Denomme and T. Fujita. 1985. PCBs: Structure-function relationships and mechanism of action. *Environ. Health Perspect.* **60**: 47–56.

Safe, S., T. Zacharewski, L. Safe, M. Harris, C. Yao and M. Holcomb. 1989. Validation of the AHH induction bioassay for the determination of 2,3,7,8-TCDD toxic equivalents. *Chemosphere* **18**: 941–946.

Schnellmann, R. G., A. E. M. Vickers and I. G. Sipes. 1985. *Rev. Biochem. Toxicol.* **7**: 247–282.

Schulz, D. E., G. Petrick and J. C. Duinker. 1989. Complete characterization of polychlorinated biphenyl congeners in commercial Aroclor and Clophen mixtures by multidimensional gas chromatography-electron capture detection. *Environ. Sci. Technol.* **23**: 852–859.

Shimada, T. and R. Sato. 1980. Covalent binding of polychlorinated biphenyls to rat liver microsomes *in vitro*: Nature of reactive metabolites and target macromolecules. *Toxicol. Appl. Pharmacol.* **55: 490–500.**

Shimada, T. and Y. Sawabe. 1983. Activation of 3,4,3′,4′-tetrachlorobiphenyl to protein-bound metabolites by rat liver microsomal cytochrome P-448-containing monooxygenase system. *Toxicol. Appl. Pharmacol.* **70**: 486–493.

Sipes, I. G., M. L. Slocumb, D. F. Perry and D. E. Carter. 1980. 4,4′-dichlorobiphenyl: distribution, metabolism and excretion in the dog and the monkey. *Toxicol. Appl. Pharmacol.* **55**: 554–563.

Subramanian, A., S. Tanabe and R. Tatsukawa. 1988. Use of organochlorines as chemical tracers in determining some reproductive parameters in Dalli-type Dall's porpoises *Phocoenoides dalli. Mar. Env. Res.* **25**: 161–174.

Tanabe, S., N. Kannan, M. Ono and R. Tatsukawa. 1989. Toxic threat to marine mammals: Increasing toxic potential of non-ortho coplanar PCBs from land to ocean. *Chemosphere* **18**: 485–490.

Tanabe, S., N. Kannan, T. Wakimoto and R. Tatsukawa. 1987. Method for the determination of three toxic non-orthochlorine substituted coplanar PCBs in environmental samples at the part per trillion levels. *Intern. J. Environ. Anal. Chem.* **29**: 199–213.

Tanabe, S., T. Mori and R. Tatsukawa. 1984. Bioaccumulation of DDTs and PCBs in the southern minke whale (*Balaenoptera acutorostrata*). *Mem. Natl. Inst. Pol. Res. Spec. Iss. 32*: 140–150.

Tanabe, S., H. Tanaka and K. Maruyama. 1980. Bioaccumulation of *Stenella coeruleoalba*: Elimination of chlorinated hydrocarbons from mother striped dolphins through parturition and lactation. In T. Fujiyama (ed.), *Studies on the Level of Organochlorine Compounds and Heavy Metals in the Marine Organisms*. Univ. Ryuks, Okinawa, pp. 115–121.

Tanabe, S., R. Tatsukawa, K. Maruyama and N. Miyazaki. 1982. Transplacental transfer of PCNs and chlorinated hydrocarbon pesticides from pregnant striped dolphin (*Stenella coeruleoalba*) to her fetus. *Agric. Biol. Chem.* **46**: 1249–1254.

Tanabe, S., S. Watanabe, H. Kan and R. Tatsukawa. 1988. Capacity and mode of PCB metabolism in small cetaceans. *Mar. Mammal. Sci.* **4**: 103–124.

Tomilin, A. G. (ed.). 1957. *Mammals of the USSR and Adjacent Countries*, Vol. 9, *Cetacea*. Trs. Prog. Sci. Transl. Ser. 1967, Hepner, V. G. (ed.), Jerusalem, p. 717.

Träger, W. F. 1989. Isotope effects as mechanistic probes of cytochrome P450 catalysed reactions. In Baillie and Jones (eds), *Synthesis and Application of Isotopically Labelled Compounds*. Elsevier.

Voorman, R. and S. D. Aust. 1987. Specific binding of polyhalogenated aromatic hydrocarbon inducers of cytochrome P–450 d to the cytochrome and inhibition of its estradiol 2-hydroxylase activity. *Toxicol. Appl. Pharmacol.* **90**: 69–78.

Voorman, R. and S. D. Aust. 1988. Inducers of cytochrome P450 d: influence on microsomal catalytic activities and differential regulation of enzyme stabilization. *Arch. Biochem. Biophys.* **262**, 76–84.

Wagemann, R. and D. G. C. Muir. 1984. Concentrations of heavy metals and organochlorines in marine mammals of northern waters: overview and evaluation. *Can. Tech. Rep. Fish. Aquat. Sci.* **1279**: 1–97.

Walker, C. H. 1990. Persistent pollutants in fish-eating sea birds—bioaccumulation, metabolism and effects. *Aquat. Toxicol.* **17**: 293–324.

Watanabe, S., T. Shimada, S. Nakamura, N. Nishiyama, N. Yamashita, S. Tanabe and R. Tatsukawa. 1989. Specific profile of liver microsomal cytochrome P–450 in dolphin and whales. *Marine Environ. Res.* **27**: 51–65.

CHAPTER 7

Chemical Modernization and Vulnerability of Cetaceans: Increasing Toxic Threat of Organochlorine Contaminants

SHINSUKE TANABE and RYO TATSUKAWA

Department of Environment Conservation, Ehime University, Tarumi 3-5-7, Matsuyama 790, Japan

INTRODUCTION

The demand for many new materials in modern civilization and the concomitant development of the chemical industry has resulted in the production of man-made chemicals in huge numbers and quantities, and contributed greatly to mankind's achieving a convenient and pleasant life. On the other hand, it should be also noted that many tragic disasters involving environmental quality and human health have emerged as an unexpected outcome of these developments. The undesirable effects of some of these chemicals are suspected in natural ecosystems where the occurrence of immunologic, reproductive and teratogenic dysfunction in various higher animals such as the double crested cormorant (*Phalacrocorax auritus*) and caspian tern (*Hydroprogne caspia*) in the Great Lakes (Kubiak *et al.*, 1989), common and grey seals (*Phoca vitulina* and *Halichoerus grypus*, respectively) in the North and Baltic Seas (Anonymous, 1988), the beluga whale (*Delphinapterus leucas*) in the St. Lawrence estuary (Martineau *et al.*, 1987) give cause for great concern. These abnormalities are believed to be caused by the toxic effect of persistent organochlorines such as PCBs (polychlorinated biphenyls), PCDDs (polychlorinated dibenzo-*p*-dioxins) and related compounds.

The present article gives an overview of the biological contamination by persistent organochlorines and its ecotoxicological implication in higher organisms of the marine environment, particularly dolphins, porpoises and whales. In contrast to large number of papers reporting organochlorine residues in cetaceans (e.g. O'Shea *et al.*, 1980; Wagemann and Muir, 1984), the clear evidence of their toxic manifestations is relatively small in spite of the fact that this group of animals is highly contaminated by these persistent chemicals. However, this is not a conclusive observation arising from comprehensive investigations. It has yet to be established how cetaceans accumulate and excrete persistent contaminants and during this process how they respond to these toxic chemicals.

Cetaceans have a wide variety of organs and tissues with different biological and physiological functions—including the usual mammalian processes such as gestation and lactation as well as specific ones like migration. Such functions influence the uptake and release kinetics of pollutants and their resultant toxic impacts. For assessing the biological effects of environmental contaminants, it is necessary to understand their detailed and specific behaviour according to those biological and physiological processes. The present paper also reviews our current understanding of the specific pattern of bioaccumulation, and possible toxic threats of persistent organochlorines to these animals.

WIDESPREAD CONTAMINATION

Among the large number of man-made chemicals, persistent organochlorines such as PCBs and DDTs are the most widely studied environmental contaminants, with much social concern because of their highly bioaccumulative nature and harmful biological effects. The first evidence of contamination of cetaceans was reported by Holden and Marsden in 1967 who noted DDT and dieldrin residues in the harbour porpoise (*Phocoena phocoena*) from the east coast of Scotland (Holden and Marsden, 1967). Since then, many researchers have documented the widespread contamination of persistent organochlorines in cetaceans and pinnipeds. A previous report by Wagemann and Muir (1984) also provided useful information for understanding the level of contamination in marine mammals of the northern hemisphere.

Dolphins and whales extend the boundaries of their habitat all over the seas and oceans including cold, temperate and tropical zones. However, individual species have their own feeding and migrating grounds within defined geographical areas and hence the residue levels of contaminants in particular species reflect the pattern of contamination in particular habitat ranges. According to the analytical data so far published, the highest concentrations of persistent organochlorines such as PCBs and DDTs have been detected in the coastal species of small cetaceans like harbour porpoise and common dolphin (*Delphinus delphis*) from the Baltic Sea, Gulf of Bothnia, Canadian east coast, US northeast coast and Californian coast where industrial and human activities are very high and point-source contaminations are also conspicuous (Gaskin *et al.*, 1971; Taruski *et al.*, 1975; O'Shea *et al.*, 1980; Wagemann and Muir, 1984). These areas are still suffering from heavy pollution and are of concern due to the long-term exposure of marine organisms to toxic chemicals and their adverse effects.

Besides monitoring of persistent organochlorines in inland and coastal areas, residues of these contaminants were also detected in open ocean cetaceans including the minke whale (*Balaenoptera acutrostrata*), sperm whale (*Physeter catodon*), sei whale (*Balaenoptera borealis*), etc., from the polar regions (Addison *et al.*, 1972; Tanabe *et al.*, 1984a; 1986). Interestingly, in the western North and South Pacific, higher levels of PCBs, DDTs and HCHs (hexachlorocyclohexanes) were recorded in animals from the northern hemisphere than the southern one and the maximum concentrations were found in the Pacific white-sided dolphin (*Lagenorhyncus obliquidence*) that principally inhabits the temperate waters of the North Pacific (Fig. 7.1). This geographical distribution can plausibly be explained by the extensive production and use of these chemicals in developed countries which are mostly located in the mid-latitudes of the northern hemisphere. Moreover, as shown in Fig. 7.1,

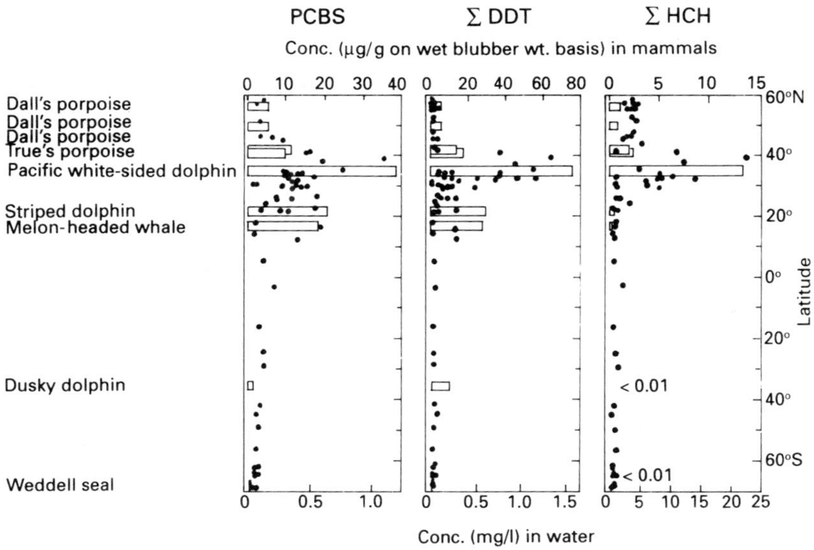

FIG. 7.1 Latitudinal variation of PCB, ΣDDT and ΣHCH concentrations in marine mammals (□) and surface seawater (●) from Bering Sea, western North and South Pacific, and Antarctic Oceans (Tanabe *et al.*, 1983).

a similar pattern of distribution was recognized in water samples collected from the areas inhabited by cetaceans, indicating that water pollution initiates contamination of aquatic ecosystems which leads to substantial levels of pollutants reaching higher organisms at the top of food chains.

SPECIFIC ACCUMULATION

The reason why the persistent organochlorines are ubiquitous in natural ecosystems is primarily their strong tendency to bioaccumulate with passage through food chains as a consequence of their lipophilicity and resistance to biodegradation. In particular, their marked bioaccumulative nature and possible toxic impact have been apparent in diverse aquatic ecosystems since the 1960s (Robinson *et al.*, 1967; Woodwell *et al.*, 1967; Jensen *et al.*, 1969). However, investigations to illustrate the exact status and mechanism of bioaccumulation of persistent organochlorines in various trophic stages of marine food chain, including cetaceans, are still rare. To our knowledge, the study conducted in the western North Pacific is the only work describing the whole picture of bioaccumulation in the marine ecosystem from the above point of view (Tanabe *et al.*, 1984b).

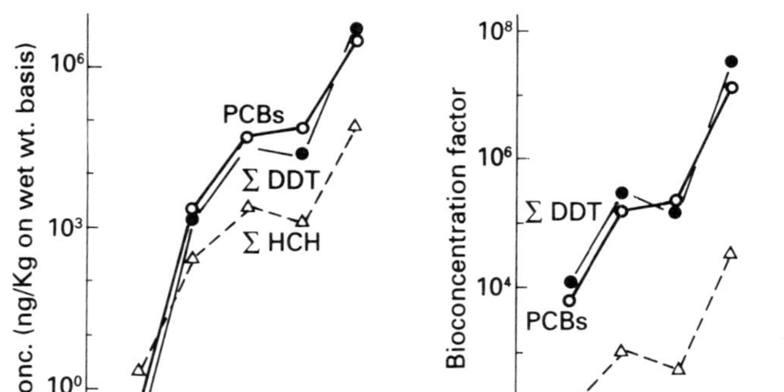

FIG. 7.2 Concentrations and bioconcentration factors of persistent organochlorines in the western North Pacific ecosystem (Tanabe *et al.*, 1984b). Wa: water, Zo: zooplankton (copepods), My: myctophid (*Diaphus suborbitalis*), Sq: squid (*Todarodes pacificus*), St: striped dolphin (*Stenella coeruleoalba*).

As shown in Fig. 7.2, this study clearly indicates the preferential biomagnification of persistent organochlorines from lower to higher organisms. It is well exemplified by the striped dolphin (*Stenella coeruleoalba*), the highest predator of this food chain which revealed high bioconcentration factors (concentration ratio of contaminants in

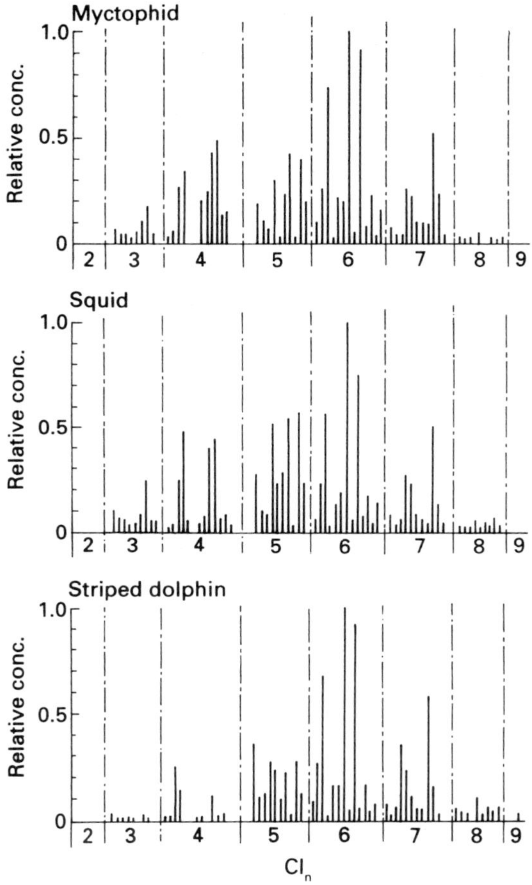

FIG. 7.3 PCB isomer and congener compositions in striped dolphin and its food organisms from the western North Pacific (Tanabe *et al.*, 1984b). Each vertical bar represents the relative concentration of an individual isomer. Relative concentration shows the ratio of individual PCB concentrations to the concentration of the maximum peak (IUPAC No. 153) which was treated as 1.

organisms to water) of PCBs and DDTs exceeding 1×10^7. When the bioconcentration factors were compared, PCBs and DDTs revealed higher values than HCHs; the former compounds are more lipophilic and biochemically stable than the latter and showed a greater tendency to bioconcentrate in aquatic ecosystems. In this food chain, interesting phenomena were observed in the case of myctophid and squid samples which showed similar bioconcentration factors, although these two animals occupy different niches of feeding (Fig. 7.2). At the same time, a clear correlation between bioconcentration factors and water solubilities of isomers and congeners of PCBs, DDTs and HCHs was noted in these animals (Tanabe *et al.*, 1984b). Such a result gives credence to the fact that the law of equilibrium partitioning can determine the degree of bioaccumulation of lipophilic organochlorines by gill-breathing aquatic animals like finfish and shellfish and its degree depends on their physicochemical properties such as lipophilicity and water solubility. On the other hand, higher organisms have a different pattern of bioaccumulation from the lower trophic organisms. In striped dolphins from the western North Pacific, the residual pattern of PCB isomers and congeners was quite different from those in their food organisms such as myctophid and squid where the percentage compositions of lower chlorinated members are smaller in striped dolphins (Fig. 7.3). Furthermore, it was noted in this animal that the percentage composition of biodegradable members of DDT compounds and HCH isomers was apparently lower than those in myctophid and squid (Tanabe *et al.*, 1984b). These observations suggest that metabolism is the primary factor controlling the bioaccumulation process of contaminants in striped dolphins. Additionally, this animal was found to transfer large quantities of persistent organochlorines from mother to offspring through lactation, as evidenced by the fact that the residual levels of PCBs were much lower in adult females than adult males and high values were observed in newborn calves (Fig. 7.4). This remarkable sex difference in accumulation of organochlorine residues has also been documented in other species such as Dall's porpoise (*Phocoenoides*

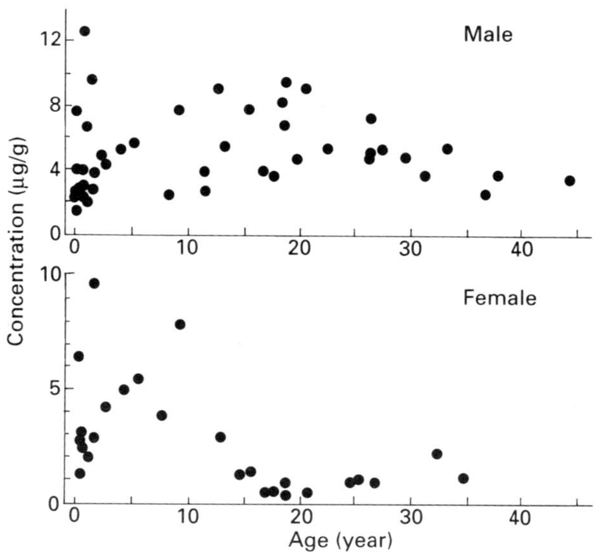

FIG. 7.4 Age trends of PCB concentrations (wet whole body weight basis) in male and female striped dolphins from the western North Pacific (Tanabe, 1985).

dalli) from the northern North Pacific (Subramanian *et al.*, 1986), the short-finned pilot whale (*Globicephala macrorhynchus*) from the Pacific coast of Japan (Tanabe *et al.*, 1987a) and the minke whale from the Antarctic Ocean (Tanabe *et al.*, 1986), indicating that this is one of the important processes determining the levels of contaminants in cetaceans.

Taking account of information so far obtained, the schematic outline of the bioaccumulation of man-made chemicals in the marine ecosystem can be summarized as shown in Fig. 7.5. It has been emphasized by many authors that the transport pathway of persistent organochlorines to open ocean environment is primarily via the atmosphere (e.g. Atlas and Giam, 1981; Bidleman *et al.*, 1981). During this process, the surface water is contaminated through the mode of air-to-water exchange. The contaminants in the water phase are taken into the plankton as a first step of bioaccumulation. Subsequently, these chemicals are accumulated from lower to higher organisms such as plankton – small fish – large fish – birds or mammals following the prey–predator relationship in the ecosystem. In the case of gill-breathing aquatic animals, direct uptake of contaminants also takes place from water. In such a process of bioaccumulation, the uptake and release of persistent organochlorines in lower organisms like plankton and fish can be explained by their adsorption–desorption kinetics on the body surface or exchange kinetics through gills, the intake of contaminants through feeding being of little significance. This suggests that the concentrations of persistent organochlorines in lower aquatic organisms are controlled by a physicochemical process like equilibrium partitioning between organisms (lipid) and ambient medium (water) (Shneider, 1982; Tanabe *et al.*, 1987b). By contrast, higher organisms such as marine birds and mammals principally accumulate the contaminants via food and excrete them through metabolism, since these animals do not have gills and are not able to take up pollutants directly from water to any significant extent. In general, persistent organochlorines in foods are efficiently absorbed into the animal body through digestive tracts (Tanabe *et al.*, 1981, 1982). The contaminants thus incorporated into the body are subjected to metabolism, during the course of which water-soluble and biodegradable chemicals are excreted relatively readily, whereas lipophilic and less biodegradable ones are retained, indicating that the biochemical factors principally determine the levels of contaminants in higher organisms. As mentioned earlier, marine mammals such as cetaceans and pinnipeds have the additional biological process of lactation, which complicates the uptake and release kinetics of persistent contaminants. The fact that large quantities of persistent organochlorines are transferred from one generation to the next through lactation implies their ability to cause contamination

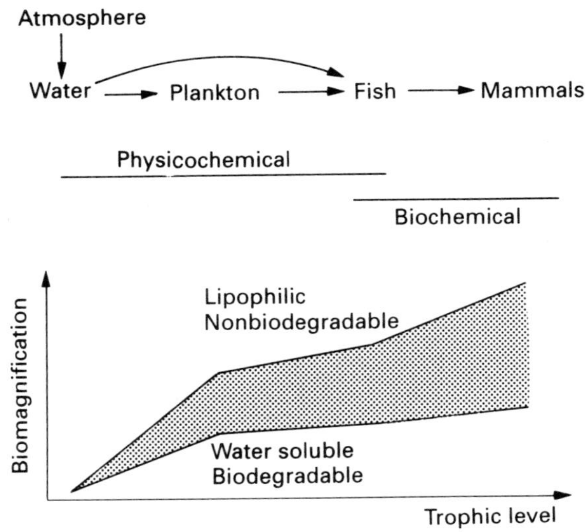

FIG. 7.5 Schematic representation of the bioaccumulation of persistent man-made chemicals in the marine ecosystem.

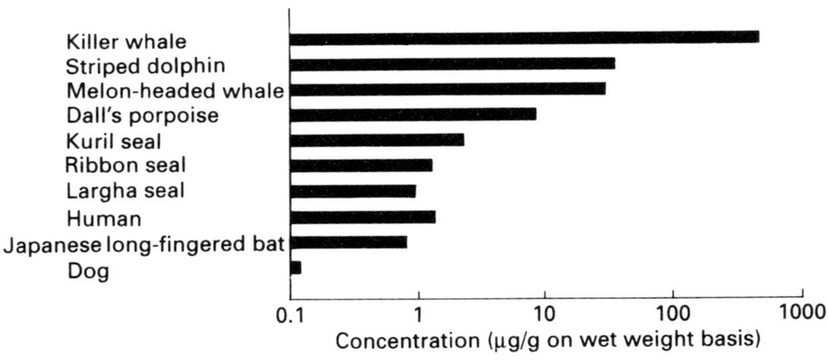

FIG. 7.6 Comparison of PCB concentrations in fat tissue of cetaceans from North Pacific, and coastal and terrestrial mammals from several areas in Japan (data from Tanabe *et al.*, 1988 and Kannan *et al.*, 1989).

of these species in the longer term, with the attendant risk of toxic effects.

CAPACITY OF DEGRADATION

There are some other indications which lead us to suspect the toxic effects of persistent organochlorines on cetaceans. As shown in Fig. 7.6, cetaceans inhabiting pristine oceans far from areas of industrial and human activities were found to contain much higher concentrations

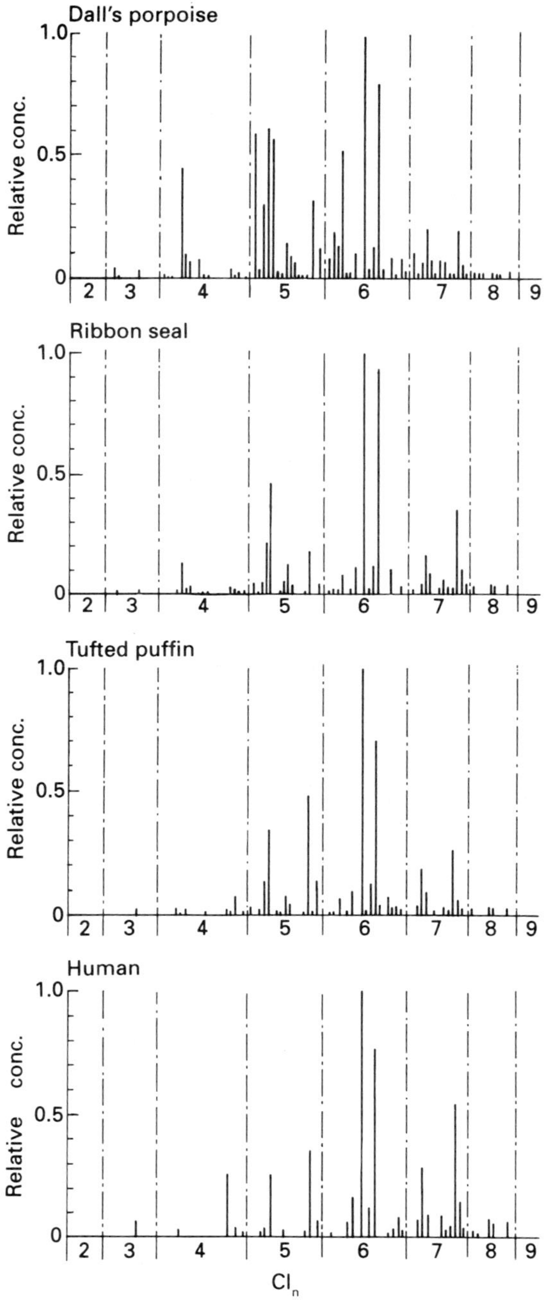

FIG. 7.7 PCB isomer and congener compositions in several species of higher animal (Tanabe *et al.*, 1988). For the explanation of bars see the legend of Fig. 3.

of PCBs than coastal and terrestrial mammals much closer to the pollution sources. This is contrary to the common assumption that the level of contaminants decreases with increasing distance from source. Moreover, it was also noted that PCB compositions in cetaceans consist of larger numbers of isomers and congeners and greater proportions of lower chlorinated members than those of coastal and terrestrial higher organisms (Fig. 7.7). These observations indicate the role of some additional factor in determining the extent of bioaccumulation in cetaceans, which is associated with the degradation of contaminants by metabolism. In fact, subsequent investigations noted that cetaceans have a much smaller capacity to metabolize a group of PCB isomers with adjacent non-chlorinated *meta* and *para* carbons in biphenyl rings than do other mammals such as humans, bat and dog (Tanabe *et al.*, 1988). The comparative approach to examine PCB isomer and congener compositions in a wide range of higher animals further suggests that drug-metabolizing enzyme systems in cetaceans have a smaller function of MC (3-methylcholanthrene)-inducible P–450 enzymes and no function of PB (phenobarbital)-inducible P–450 enzymes (Fig. 7.8). Recent studies using cetacean liver microsomes demonstrated that these animals have much lower activities of aldrin epoxidase (PB-inducible P–450 enzymes) than the rat, whereas activities of 7-ethoxyresorufin *O*-deethylase (MC-inducible P–450 enzymes) are comparable to those of the rat (Table 7.1), supporting the above observations from PCB isomer and congener compositions (Watanabe *et al.*, 1989).

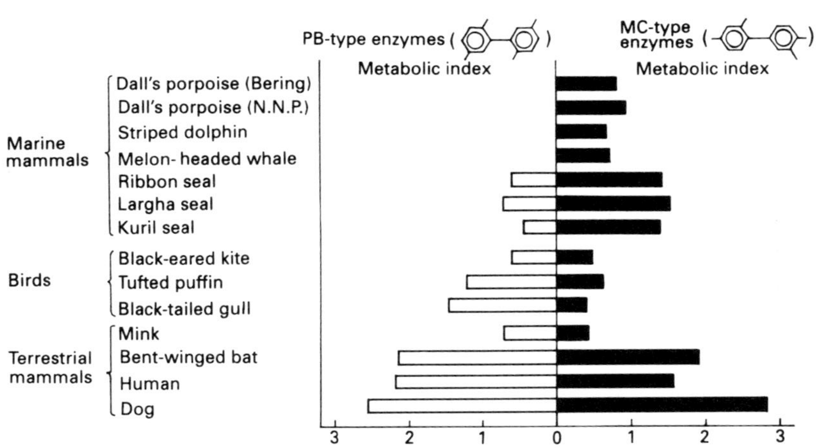

FIG. 7.8 PB (phenobarbital)-inducible and MC (3-methylcholanthrene)-inducible enzyme activities in higher animals estimated by Metabolic Index (MI) of 2,2′,5,5′- and 2,3′,4,4′-tetrachlorobiphenyl isomers which show the capacity of PCB metabolism. For more details of MI see Tanabe *et al.* (1988).

TABLE 7.1 *Comparison of Liver Microsomal Monooxygenase Activities in Short-finned Pilot Whales (*Blobicephala macrorhynchus*), Striped Dolphins (*Stenella coeruleoalba*) and Killer Whales (*Orcinus orca*) (Watanabe* et al., *1989)*

	Short-finned pilot whale	Striped dolphin	Killer whale	Rat (uninduced)
Number of specimens	33	5	3	1
Protein[a]	20.2 ± 2.55	22.4 ± 2.18	21.8 ± 2.47	13.7
NADPH-cytochrome *c* reductase[b]	64.3 ± 13.2	92.9 ± 17.5	35.1 ± 14.8	205
Cytochrome P-450[c]	0.171 ± 0.050	0.186 ± 0.021	0.211 ± 0.068	0.83
Aldrin epoxidase[d]	0.064 ± 0.048 (0.356)	0.172 ± 0.048 (0.928)	0.041 ± 0.011 (0.200)	1.58 (1.90)
Aniline hydroxylase[d]	0.151 ± 0.059 (0.959)	0.294 ± 0.104 (1.62)	0.217 ± 0.050 (1.10)	2.24 (2.70)
Benzo[a]pyrene hydroxylase[e]	7.81 ± 4.23 (44.2)	7.22 ± 1.59 (39.3)	27.4 ± 6.10 (134)	42.0 (50.6)
Ethoxyresorufin *O*-deethylase[d]	0.042 ± 0.042 (0.225)	0.191 ± 0.015 (1.04)	0.612 ± 0.373 (2.71)	0.123 (0.148)

[a] mg/g liver.
[b] nmol cytochrome *c* reduced/min/mg protein.
[c] nmol/mg protein.
[d] nmol product formed/min/mg protein.
[e] pmol product formed/min/mg protein.
Data shows the mean and standard deviation.
Values in parentheses indicate the monooxygenase activities based on cytochrome P–450 contents.

In higher organisms, drug-metabolizing enzyme systems have a significant role in detoxication of hydrophobic xenobiotics from exposed animals, especially by means of catalytic oxidation into polar metabolites. As seen in Fig. 7.8, metabolic indices suggested lower levels of PB-inducible or MC-inducible enzyme activities in piscivorous cetaceans, seals and marine birds than those in other animals which were not piscivores. This could be seen not only in marine organisms but also in terrestrial animals like mink. The low level of hepatic microsomal monooxygenase activities in certain marine animals like fish-eating birds was pointed out earlier (Walker and Knight, 1981). Such an animal specificity is considered to have arisen from the different evolutionary development of drug-metabolizing enzyme systems to facilitate the excretion and detoxication of natural toxicants. Brodie and Maickel (1962) hypothesized that fish need not develop an efficient system of drug-metabolizing enzymes due to an excellent function of the gill which excretes directly the lipophilic xenobiotics into the ambient water from their bodies. Hence, it is believed that in the case of fish-eating birds there is no need to develop detoxication systems because their prey do not accumulate the lipophilic xenobiotics unlimitedly (Walker *et al.*, 1975; Knight *et al.*, 1981). This is also a plausible explanation why cetaceans have only a low activity of certain oxidative enzymes of hepatic microsomes.

The fact that cetaceans have relatively small capacity to degrade persistent organochlorines highlights the risk of long-term bioaccumulation of hazardous chemicals in their bodies and the dangers that they face in an ecotoxicological context.

ECOLOGICAL RISK

Regarding the long-term toxicity of environmental contaminants to humans and wildlife, much more work needs to be done on their carcinogenicity, teratogenicity, immunologic and reproductive effects which are often associated with drug-metabolizing enzyme systems. These enzyme systems are known to modify some xenobiotics into toxic intermediates and to control the critical balance of endogenous molecules such as steroids, fatty acids and prostaglandins (Conney and Kuntzman, 1971; Kupfer, 1982; Kamataki *et al.*, 1983; Dippel *et al.*, 1984). Some of the persistent organochlorines can initiate toxic effects, since they are potent inducers of cytochrome P-450 drug-metabolizing enzymes (Campbell *et al.*, 1983; Safe, 1984).

In natural biological systems, the persistent organochlorines have been reported to cause reproductive abnormalities, especially in pinnipeds (e.g. Delong *et al.*, 1973; Helle *et al.*, 1976a; 1976b; Duinker *et al.*, 1979; Reijinders, 1986; Anonymous, 1988) and birds (e.g. Helander *et al.*, 1982; Newton *et al.*, 1982; Wiemeyer *et al.*, 1984; Hoffman *et al.*, 1986; Peakall and Fox, 1987; Kubiak *et al.*, 1989). It is also known that the mink is extremely sensitive to the toxicity of persistent organochlorines, particularly on its reproductive system (e.g. Jensen *et al.*, 1977; Bleavins *et al.*, 1980; Aulerich *et al.*, 1985). Interestingly, these animals are estimated to have smaller capacity of either PB-inducible or MC-inducible enzymes (Fig. 7.8), suggesting that the animals with a lower ability to metabolize the contaminants are most vulnerable to their reproductive toxic effects. If this is true, cetaceans are expected to be susceptible to the reproductive impairment by persistent organochlorines because of deficiency of PB-inducible enzyme systems and also smaller capacity of MC-inducible enzymes (Fig. 7.8 and Table 7.1).

However, in the case of cetaceans, no direct evidence is so far available on the toxic symptoms caused by organochlorine residues. This might be partly due to the difficulty of collecting a number of samples large enough to conduct an epidemiological survey for determining various abnormalities, because most of the cetaceans are open ocean inhabitants. Notwithstanding the lack of clear evidence, the suggestive indication of organochlorine toxicity to cetaceans is increasing. For example, organochlorines were suggested to be the prime causative factor for a declining population of beluga whales in the St. Lawrence estuary, Canada (Martineau *et al.*, 1987). Furthermore, in Dall's porpoise from the northwestern North Pacific, the residue levels of PCBs and DDE in the blubber were found to have a negative correlation with the testosterone levels in blood (Fig. 7.9). This may well be a causal relationship, since testosterone is metabolized by certain P450-based monooxygenases which may be susceptible to induction by organochlorine compounds. It is also noteworthy that a significant

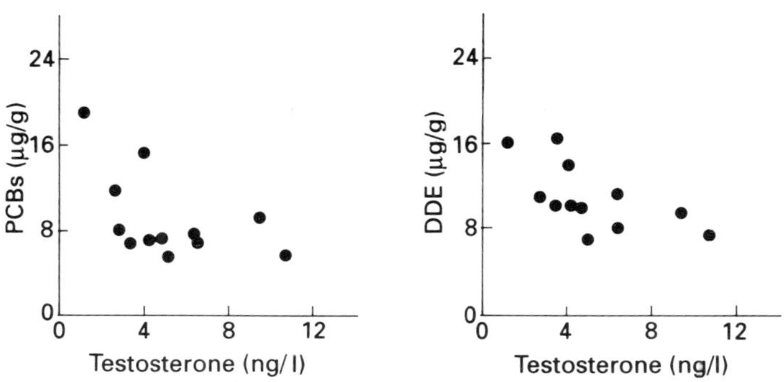

FIG. 7.9 Relationship between the concentrations of PCBs and DDE in blubber and levels of testosterone in blood of adult male Dall's porpoises collected from the northwestern North Pacific (Subramanian *et al.*, 1987).

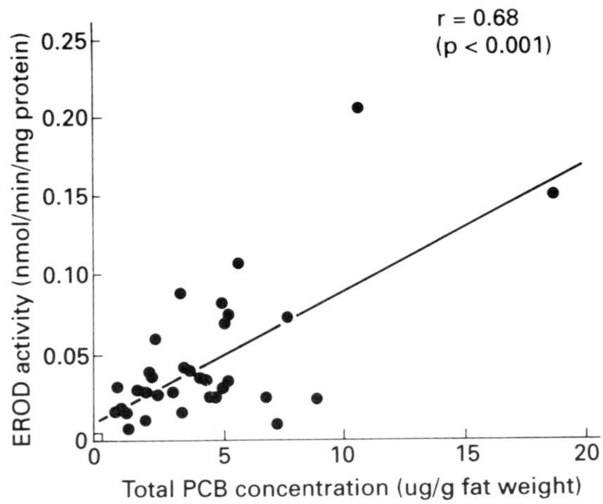

FIG. 7.10 Relationship between PCB concentrations and 7-ethoxyresorufin *O*-deethylase activities in the liver of short-finned pilot whale collected from the western North Pacific (Watanabe *et al.*, unpublished data).

correlation between 7-ethoxyresorufin *O*-deethylase activities and PCB residue levels was recorded in the liver of the short-finned pilot whale from the western North Pacific, indicating the possible P–450 enzyme induction by PCBs (Fig. 7.10). These recent observations suggest that the present residue levels of these persistent contaminants are enough to cause the induction of certain forms of cytochrome P–450 isozymes in cetaceans. Further studies are required to obtain more evidence on this point for protecting marine ecosystems from the toxic threat of hazardous man-made chemicals.

CONCLUSIONS

The present paper reviews the plight of cetaceans in the light of the ecotoxicology of persistent organochlorines in the marine environment. The cetaceans are recognized as one of the animal groups building up some of the highest recorded concentrations of these hazardous man-made chemicals arising from a worldwide contamination. They can strongly bioaccumulate organochlorine contaminants through feeding and also pass them in substantial quantities from one generation to the next through lactation. Unfortunately, these animals have a low capacity for degradation of these contaminants resulting from a specific mode of cytochrome P–450 drug-metabolizing enzyme systems. Moreover, the interaction of organochlorine pollutants with these enzyme systems may lead to effects upon reproduction.

Considering all these facts, it may be concluded that the marine mammals, particularly cetaceans, are one of the most vulnerable target organisms in forthcoming long-term toxicity of hazardous man-made chemicals.

REFERENCES

Addison, R. F., M. E. Zinck, and R. G. Ackman. 1972. Residues of organochlorine pesticides and polychlorinated biphenyls in some commercially produced Canadian marine oils. *J. Fish. Res. Board. Can.* **29**: 349–355.

Anonymous. 1988. The plague summer. *The Pilot* **1**: 1.

Aulerich, R. J., S. J. Bursian, W. J. Breslin, B. A. Olsson and R. K. Ringer. 1985. Toxicological manifestation of 2,4,5,2′,4′,5′–, 2,3,6,2′,3′,6′– and 3,4,5,3′,4′,5′–hexachlorobiphenyls and Arocolor 1254 in mink. *J. Toxicol. Environ. Health* **15**: 63–79.

Atlas, E. and C. S. Giam. 1981. Global transport of organic pollutants: ambient concentrations in the remote marine atmosphere. *Science* **211**: 163–165.

Bidleman, T. F., E. J. Christensen, W. N. Billings and R. Leonard. 1981. Atmospheric transport of organochlorines in the North Atlantic gyre. *J. Mar. Res.* **39**: 443–464.

Bleavins, M. R., R. J. Aulerich and R. K. Ringer. 1980. Polychlorinated biphenyls (Aroclor 1016 and 1242): effects on survival and reproduction in mink and ferrets. *Arch. Environ. Contam. Toxicol.* **9**: 627–635.

Brodie, B. B. and R. P. Maickel. 1962. Comparative biochemistry of drug metabolism. *Proc. 1st Pharmacol. Meeting* pp. 229–324.

Campbell, M. A., J. Gyorkos, B. Leece, K. Homonko and S. Safe. 1983. The effects of twenty-two organochlorine pesticides as inducers of the hepatic drug-metabolizing enzymes. *Gen. Pharmacol.* **14**: 445–454.

Conney, A. H. and R. Kuntzman. 1971. Metabolism of normal body constituents by drug-metabolizing enzymes in liver microsomes. In B. B. Brodie and J. Gillete (eds), *Concept in Biochemical Pharmacology, Handbook of Environmental Pharmacology Series.* Springer-Verlag, Berlin, pp. 401–421.

Delong, R., G. W. Gilmartin and J. G. Simpson. 1973. Premature birth California sea lions: association with high organochlorine pollutant residue levels. *Science* **181**: 1168–1169.

Dippel, A., R. C. Moschel and C. A. H. Bigger. 1984. Polynuclear aromatic carcinogens. In C. E. Searle (ed.), *Chemical Carcinogens* pp. 41–163.

Duinker, J. C., M. Th. J. Hillebrand and R. F. Nolting. 1979. Organochlorines and metals in harbour seals (Dutch Wadden Sea). *Mar. Pollut. Bull.* **10**: 360–364.

Gaskin, D. E., M. Holdrinet and R. Frank. 1971. Organochlorine pesticide residues in harbour porpoises from the Bay of Fundy region. *Nature* **233**: 499–500.

Helander, B., M. Olsson and L. Reutergardh. 1982. Residue levels of organochlorine and mercury compounds in unhatched eggs and the relationships to breeding success in white-tailed sea eagles *Haliaeetus albicilla* in Sweden. *Holarc. Ecol.* **5**: 349–366.

Helle, E., M. Olsson and S. Jensen. 1976a. DDT and PCB levels and reproduction in ringed seal from the Bothnian Bay. *Ambio* **5**: 188–189.

Helle, E., M. Olsson and S. Jensen. 1976b. PCB levels corrected with pathological changes in seal uteri. *Ambio* **5**: 261–263.

Hoffman, D. J., B. A. Rattner, C. M. Bunck, A. Krynitsky, H. M. Ohlendorf and R. W. Lowe. 1986. Association between PCBs and lower embryonic weight in black-crowned night herons in San Francisco Bay. *J. Toxicol. Environ. Health* **19**: 383–391.

Holden, A. V. and K. Marsden. 1967. Organochlorine pesticides in seals and porpoises. *Nature* **216**: 1274–1276.

Jensen, S., G. Johnels, M. Olsson and G. Otterlind. 1969. DDT and PCB in marine animals from Swedish waters. *Nature* **224**: 247–250.

Jensen, S., J. E. Kihlstrom, M. Olsson, C. Lundberg and G. Orberg. 1977. Effects of PCB and DDT on mink (*Meustela vison*) during the reproductive season. *Ambio* **6**: 239.

Kamataki, T., K. Maeda, Y. Yamazoe, N. Matsuda, K. Ishii and R. Kato. 1983. A high-spin form of cytochrome P–450 highly purified from polychlorinated biphenyl-treated rats. *Mol. Pharmacol.* **24**: 146–155.

Kannan, N., S. Tanabe, M. Ono and R. Tatsukawa. 1989. Critical evaluation of polychlorinated biphenyl toxicity in terrestrial and marine mammals: increasing impact of non-*ortho* and mono-*ortho* coplanar polychlorinated biphenyls from land to ocean. *Arch. Environ. Contam. Toxicol.* **19**: 850–857.

Knight, G. C., C. H. Walker, D. C. Cabot and M. P. Harris. 1981. The activity of two hepatic microsomal enzymes in sea birds. *Comp. Biochem. Physiol.* **68C**: 127–132.

Kubiak, T. G., H. J. Harris, L. M. Smith, T. R. Schwartz, D. L. Stalling, J. A. Trick, L. Sileo, D. E. Docherty and T. C. Erdman. 1989. Micro-contaminants and reproductive impairment of the Foster's tern on Green Bay, Lake Michigan—1983. *Arch. Environ. Contam. Toxicol.* **18**: 706–727.

Kupfer, D. 1982. Endogenous substrates of monooxygenases: fatty acids and prostaglandins. In J. B. Schenman and D. Kupfer (eds), *Hepatic Cytochrome P-450 Monooxygenase System*, Pergamon Press, Oxford, pp. 157–187.

Martineau, D., P. Beland, C. Desjardins and A. Lagace. 1987. Levels of organochlorine chemicals in tissues of beluga whales (*Delphinapterus leucas*) from the St. Lawrence estuary, Quebec, Canada. *Arch. Environ. Contam. Toxicol.* **16**: 137–147.

Newton, I., J. Bogan, E. Meak and B. Little. 1982. Organochlorine compounds and shell thinning in British merlins *Falco columbarius*. *Ibis* **124**: 328–335.

O'Shea, T. G., R. L. Brownell Jr., D. R. Clark Jr., W. A. Walker, M. L. Gay and T. G. Lamont. 1980. Organochlorine pollutants in small cetaceans from the Pacific and South Atlantic Oceans, November 1968–June 1976. *Pest. Monit. J.* **14**: 35–46.

Peakall, D. B. and G. A. Fox. 1987. Toxicological investigations of pollutant-related effects in Great Lakes gulls. *Environ. Health Perspect.* **71**: 187–193.

Reijnders, P. J. H. 1986. Reproductive failure in common seals feeding on fish from polluted waters. *Nature* **324**: 456–457.

Robinson, J., A. Richardson, A. N. Crabtree, J. C. Coulson and C. R. Potts. 1967. Organochlorine residues in marine organisms. *Nature* **214**: 1307–1311.

Safe, S. 1984. Polychlorinated biphenyls (PCBs) and polybrominated biphenyls (PBBs): biochemistry, toxicology, and mechanism of action. *CRC Crit. Rev. Toxicol.* **13**: 319–393.

Shneider, R. 1982. Polychlorinated biphenyls (PCBs) in cod tissues from the western Baltic: significance of equilibrium partitioning and lipid composition in the bioaccumulation of lipophilic pollutants in gill-breathing animals. *Meeresforschung* **29**: 69–79.

Subramanian, An., S. Tanabe, Y. Fujise and R. Tatsukawa. 1986. Organochlorine residues in Dall's and True's porpoises collected from northwestern Pacific and adjacent waters. *Mem. Natl Inst. Polar Res.* **44**: 167–173.

Subramanian, An., S. Tanabe, R. Tatsukawa, S. Saito and N. Miyazaki. 1987. Reduction in the testosterone levels by PCBs and DDE in Dall's porpoises of northwestern North Pacific. *Mar. Pollut. Bull.* **18**: 643–646.

Tanabe, S. 1985. Distribution, behavior and fate of PCBs in the marine environment. *J. Oceanogr. Soc. Japan* **41**: 358–370.

Tanabe, S., Y. Nakagawa and R. Tatsukawa. 1981. Absorption efficiency and biological half-life of individual chlorobiphenyls in rats treated with Kanechlor products. *Agric. Biol. Chem.* **45**: 717–726.

Tanabe, S., K. Maruyama and R. Tatsukawa. 1982. Absorption efficiency and biological half-life of individual chlorobiphenyls in carp (*Cyprinus carpio*) orally exposed to Kanechlor products. *Agric. Biol. Chem.* **46**: 891–898.

Tanabe, S., T. Mori, R. Tatsukawa and N. Miyazaki. 1983. Global pollution of marine mammals by PCBs, DDTs and HCHs (BHCs). *Chemosphere* **12**: 1269–1275.

Tanabe, S., T. Mori and R. Tatsukawa. 1984a. Bioaccumulation of DDTs and PCBs in the southern minke whale (*Balaenoptera acutorostrata*). *Mem. Natl Inst. Polar Res.* **32**: 140–150.

Tanabe, S., H. Tanaka and R. Tatsukawa. 1984b. Polychlorobiphenyls, ΣDDT, and hexachlorocyclohexane isomers in the western North Pacific ecosystem. *Arch. Environ. Contam. Toxicol.* **13**: 731–738.

Tanabe, S., S. Miura and R. Tatsukawa. 1986. Variations of organochlorine residues with age and sex in Antarctic minke whale. *Mem. Natl Inst. Polar Res.* **44**: 174–181.

Tanabe, S., B. G. Loganathan, A. N. Subramanian and R. Tatsukawa. 1987a. Organochlorine residues in short-finned pilot whale: possible use as tracers of biological parameters. *Mar. Pollut. Bull.* **18**: 561–563.

Tanabe, S., R. Tatsukawa and D. J. H. Phillips. 1987b. Mussels as bio-indicators of PCB pollution: a case study on uptake and release of PCB isomers and congeners in green-lipped mussels (*Perna viridis*) in Hong Kong waters. *Environ. Pollut.* **47**: 41–62.

Tanabe, S., S. Watanabe, H. Kan and R. Tatsukawa. 1988. Capacity and mode of PCB metabolism in small cetaceans. *Marine Mammal Sci.* **4**: 103–124.

Taruski, A. G., C. E. Olney and H. E. Winn. 1975. Chlorinated hydrocarbons in cetaceans. *J. Fish. Res. Board Can.* **32**: 2205–2209.

Wagemann, R. and D. G. C. Muir. 1984. Concentrations of heavy metals and organochlorines in marine mammals of northern waters: Overview and evaluation. In *Canadian Technical Report of Fisheries and Aquatic Science*, No. 1279. Western Region, Department of Fisheries and Oceans, Canada.

Walker, C. H. and G. C. Knight. 1981. The hepatic microsomal enzymes of sea birds and their interaction with liposoluble pollutants. *Aquat. Toxicol.* **1**: 343–354.

Walker, C. H., A. C. C. Craven and M. Kurukgy. 1975. The metabolism of organochlorine compounds by microsomal enzymes of the shag. *Environ. Physiol. Biochem.* **5**: 58–64.

Watanabe, S., T. Shimada, S. Nakamura, N. Nishiyama, N. Yamashita, S. Tanabe and R. Tatsukawa. 1989. Specific profile of liver microsomal cytochrome P–450 in dolphin and whales. *Mar. Environ. Res.* **27**: 51–65.

Watanabe, S., S. Tanabe and R. Tatsukawa. Unpublished data.

Wiemeyer, S. N., T. G. Lamont, C. M. Bunck, C. R. Sindelar, R. J. Gramlich, J. D. Fraser and M. A. Byrd. 1984. Organochlorine pesticide, polychlorobiphenyl, and mercury residues in bald eagle eggs — 1969–79 — and their relationships to shell thinning and reproduction. *Arch. Environ. Contam. Toxicol.* **13**: 529–549.

Woodwell, G. M., C. E. Wurster Jr. and P. A. Issacson. 1967. DDT residues in an east coast estuary: a case of biological concentration of a persistent insecticide. *Science* **15**: 821–824.

Part V
Fish-eating Birds

CHAPTER 8

Patterns and Trends of Organic Contaminants in Canadian Seabird Eggs, 1968–90

J. E. ELLIOTT*, D. G. NOBLE†, R. J. NORSTROM†, P. E. WHITEHEAD*, M. SIMON†, P. A. PEARCE‡ and D. B. PEAKALL†

** Canadian Wildlife Service, Environment Canada, Box 340, Delta, B.C. V4K 3Y3, Canada*

† Canadian Wildlife Service, Environment Canada, Ottawa, Ont. K1A 0H3, Canada

‡ Canadian Wildlife Service, Environment Canada, Box 400, Fredericton, N.B. E3B 4Z9, Canada

INTRODUCTION

There has been an effort over the past few decades at both the national and international level to regulate the release of chemicals to the oceans (UNEP, 1985). However, many areas of the marine environment, in particular the North Atlantic and North Pacific oceans, still receive large quantities of chemical wastes from land-based manufacturing and waste dumps, from marine transport accidents, ocean dumping and incineration. There is, therefore, a need to monitor chemical contamination of the marine environment. This paper discusses the use of seabirds for monitoring pollution of the marine environment by persistent organic chemicals in the context of examples drawn from a Canadian seabird monitoring project.

METHODS

The advantages of biological monitoring, and criteria for selecting seabird species for monitoring, can be found in Gilbertson *et al.* (1987). The species selected for the Canadian seabird monitoring project are outlined in Table 8.1. Further details on the relevant ecology of each

TABLE 8.1 *Seabird Species Used for Monitoring Organochlorines Around the Coasts of Canada*

	Atlantic	Arctic	Pacific
Inshore—omnivore			
Herring gull (*Larus argentatus*)	+		
Glaucous-winged gull (*Larus glaucescens*)			+
Ivory gull (*Pagophila eburnea*)		+	
Inshore—subsurface feeder			
Double-crested cormorant (*Phalacrocorax auritus*)	+		+
Continental shelf—subsurface feeder			
Atlantic puffin (*Fratercula arctica*)	+		
Rhinoceros auklet (*Cerorhinca monocerata*)			+
Thick-billed murre (*Uria lomvia*)		+	
Offshore—surface feeder			
Leach's storm-petrel (*Oceanodroma leucorhoa*)	+		+
Northern fulmar (*Fulmarus glacialis*)		+	

Some of the species breed in zones other than those indicated, but are not used for contaminants monitoring in those areas

species are available: for the Atlantic coast in Pearce *et al.* (1989), the Pacific coast in Elliott *et al.* (1989a) and the Arctic in Peakall and Nettleship (1987).

The samples consisted of fresh eggs taken from randomly selected nests early in the nesting season. Eggs were collected by hand and temporarily refrigerated. Egg contents were then removed and placed into chemically cleaned (acetone and hexane) glass jars with a foil-lined lid. Egg contents were frozen at −40°C until analyzed. Sample size was normally 5 eggs; however, beginning in 1990, 15 eggs were collected for each sample on the Pacific coast and analyzed as 5 pools of 3 eggs each in order to reduce the variance and increase the power to detect change, as recommended by Gilbertson *et al.* (1987).

Samples from 1971 to 1983 were analyzed for organochlorines by the Ontario Research Foundation following procedures described in Reynolds and Cooper (1975). Samples dating from 1984 and archived samples [e.g., Leach's storm-petrels (*Oceanodroma leucorhoa*) from Great Island, Newfoundland, 1968; double-crested cormorant (*Phalacrocorax auritus*) eggs from Mandarte Island, B.C., 1973] drawn from the Canadian Wildlife Service Specimen Bank (Elliott, 1985) were analyzed at the National Wildlife Research Centre using methods described in Peakall *et al.* (1986). Polychlorinated biphenyls (PCBs) were estimated as a 1:1 mixture of Aroclors 1254 and 1260, which was dictated by the procedures employed in earlier analyses. This number overestimates the true sum of PCB congeners, but is useful for comparing older packed column and newer capillary column analytical methods (Turle *et al.*, 1991). Polychlorinated dibenzo-*p*-dioxins (PCDDs) were determined in pooled samples of double-crested cormorant eggs by methods described in Norstrom *et al.* (1990).

Residue levels were calculated on a wet weight basis in mg/kg (parts per million) for the organochlorine pesticides and PCBs and in ng/kg (parts per trillion) for the PCDDs. Generally the values exhibited a log-normal distribution and data are given as geometric means with 95% confidence intervals.

RESULTS AND DISCUSSION

Offshore Environment

Figure 8.1 compares temporal trends of selected organochlorines in Leach's storm-petrels from a relatively remote location on the Atlantic coast of Newfoundland to those on the Pacific coast of Vancouver Island. Levels in storm-petrels reflect contamination of surficial food chains in pelagic regions of both oceans (Table 8.1). Since 1970, mean DDE levels declined by about one-third in eggs from Newfoundland and about one-half in eggs from Vancouver Island. None of the other organochlorines or PCBs in Leach's storm-petrel eggs show any significant trends with time (linear regression, $P < 0.001$), even 20 years after initial restrictions on usage in North America. Comparing PCB levels in petrel eggs to levels in either Atlantic puffins (*Fratercula arctica*) from the Atlantic coast or rhinoceros auklets (*Cerorhinca monocerata*) from the Pacific coast, it is apparent that while levels were similar or lower in the petrels at the earliest time period, mean levels are now about two times higher in petrels. This pattern also appears at other locations where we have both petrel and alcid data, such as the Bay of Fundy (Pearce *et al.*, 1989, unpublished data). The pattern might reflect dispersal of very persistent compounds such as the PCBs from highly contaminated inshore areas to the marine environment in general (Stout, 1986; Tanabe, 1988). It might also be due to atmospheric transport and deposition of PCBs (Murphy *et al.*, 1985).

PCB, dieldrin, oxychlordane and HCB levels have been consistently higher in storm-petrel eggs from the Atlantic than the Pacific coasts (Fig. 8.1). Tanabe and Tatsukawa (1986) also reported higher PCBs in fish and marine mammals from the Atlantic. In contrast, DDE and HCH levels are higher in Pacific than Atlantic storm-petrel eggs. These

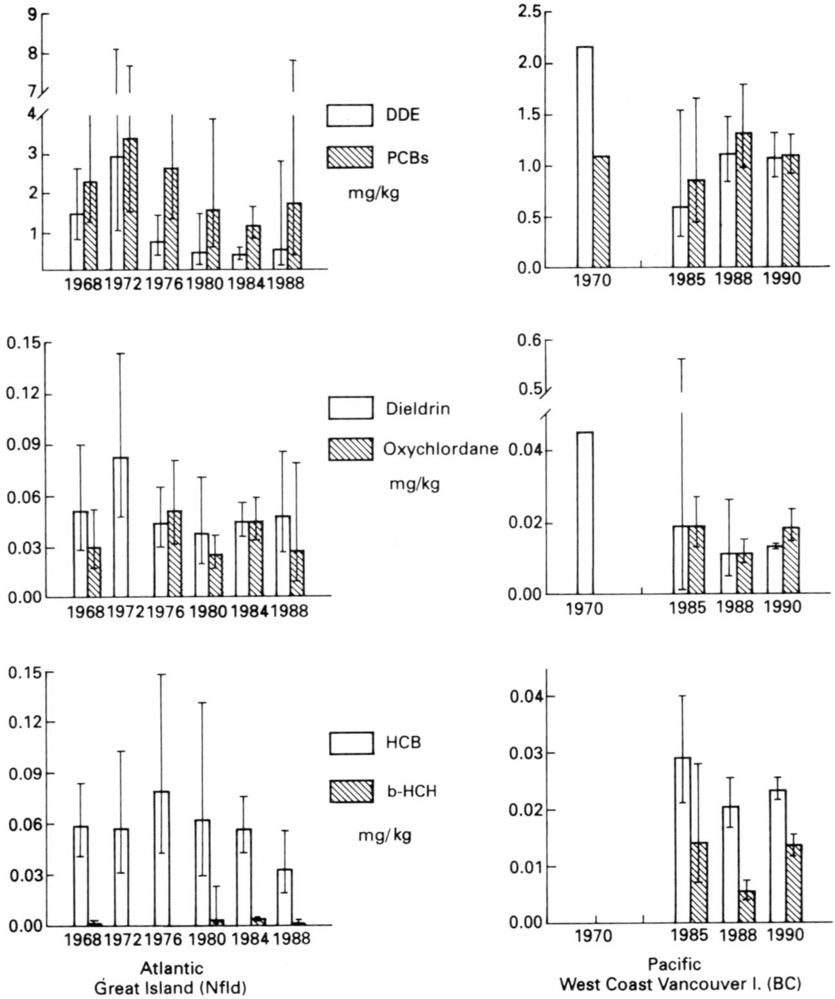

FIG. 8.1 Mean concentrations of selected organochlorines in eggs of Leach's storm-petrels collected between 1968 and 1990 (geometric mean and 95% confidence interval).

results are consistent with both published information on air and water samples and with relative sources and prevailing wind patterns as discussed previously by Elliott *et al.* (1989a).

Continental Shelf

Figure 8.2 compares trends for selected organochlorines in eggs of the Atlantic puffin from Machias Seal Island in the Bay of Fundy and

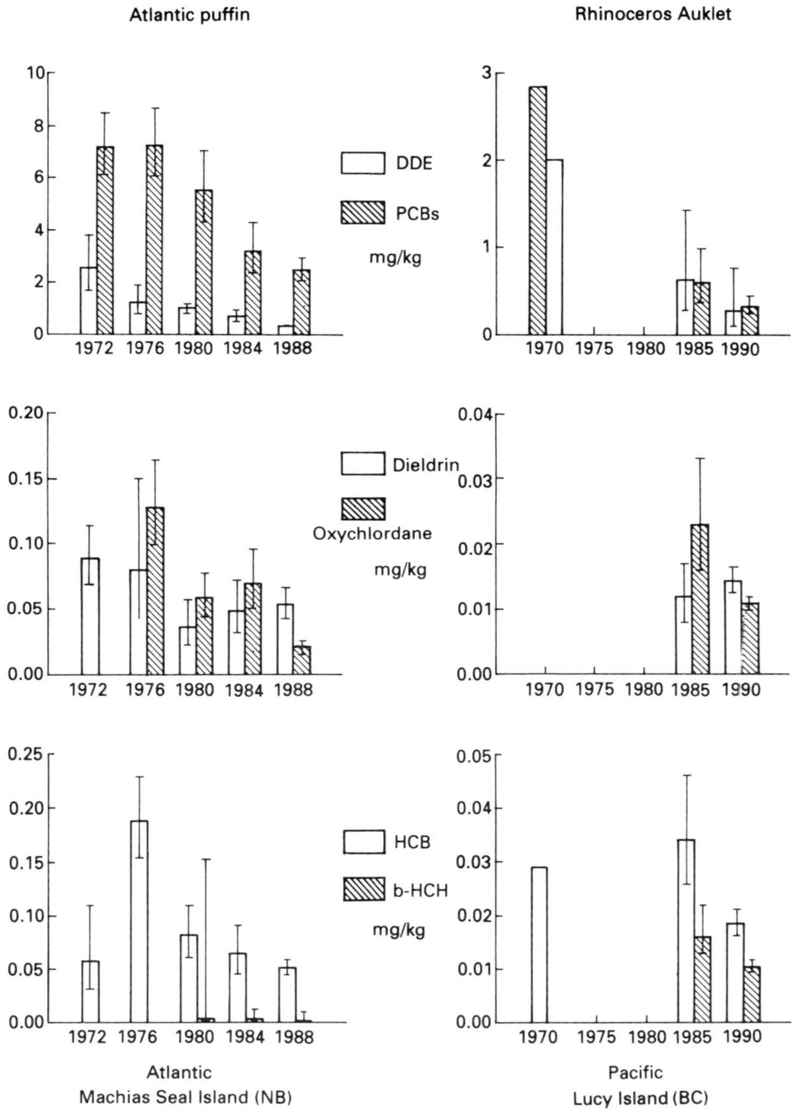

FIG. 8.2 Mean concentrations of selected organochlorines in eggs of Atlantic puffins from Machias Seal Island and rhinoceros auklets from Lucy Island, between 1970 and 1990 (geometric mean and 95% confidence interval).

the rhinoceros auklet from Lucy Island located near Prince Rupert in north-western British Columbia. Both species feed mainly on fish taken from deeper waters, but along the continental shelf during the breeding season (Table 8.1).

PCB production was voluntarily restricted in North America by

Monsanto Corporation in 1971. In 1977, use of PCBs in non-electrical equipment was curtailed by legislation in both the US and Canada. In the Bay of Fundy, initial restrictions did not result in any change in PCB levels between eggs sampled in 1972 and 1976; however, by 1980 PCB levels had begun to decline. Mean concentrations have subsequently decreased in a first-order fashion. PCB levels in guillemot (*Uria aalge*) eggs from the Baltic Sea also increased during the early 1970s, with decreasing mean concentrations later in the decade (Olsson and Reutergardh, 1986). PCB levels in salmon (*Oncorhynchus kisutch*) and herring gulls (*Larus argentatus*) from the Great Lakes (Norstrom *et al.*, 1985) showed initial sharp decreases during the late 1970s followed by relatively minor changes in concentration during the 1980s.

Agricultural use of DDT was restricted in Canada in 1970 and in the US in 1972. DDT was used extensively for spruce budworm control in the forests of New Brunswick; cessation of this use has resulted in an ongoing decline in mean DDE concentrations in the food chains of the Bay of Fundy. In the early 1970s, DDE levels were slightly higher in Pacific auklet compared with Atlantic puffin eggs (Fig. 8.2). Apparently the food chains of the Pacific continental shelf were as contaminated as the Atlantic, presumably as a result of DDT use in the forests of the north-western states and perhaps also from both agricultural and industrial inputs from California.

Mean concentrations of dieldrin in alcid eggs did not change significantly between 1970 and 1990. Although use of aldrin/dieldrin was severely restricted by legislation in North America in the early 1970s, sales had been declining, at least in Canada, since the early 1960s. In contrast, Canadian sales of technical chlordane increased during the 1970s until restricted in 1978. Increasing or stable levels of chlordane-related compounds during the 1970s and 1980s have been reported in biological samples from other locations (Zitko, 1978; Norstrom *et al.*, 1985; Pearce *et al.*, 1989; Norstrom *et al.*, 1988).

Inshore Environment

Figure 8.3 compares temporal trends of selected chlorinated hydrocarbons in double-crested cormorants from Manawagonish Island in the Bay of Fundy and Mandarte Island in the Strait of Georgia. This species feeds mainly on fish caught in the inshore zone.

PCB concentrations were remarkably similar in eggs from both coasts. They also share a similar temporal pattern of initial sharp declines followed by relatively stable levels. DDE levels were initially higher in the Bay of Fundy and also declined more dramatically there than in the Strait of Georgia. This is consistent with the heavy use of DDT in New Brunswick forests and is similar to the temporal trend reported

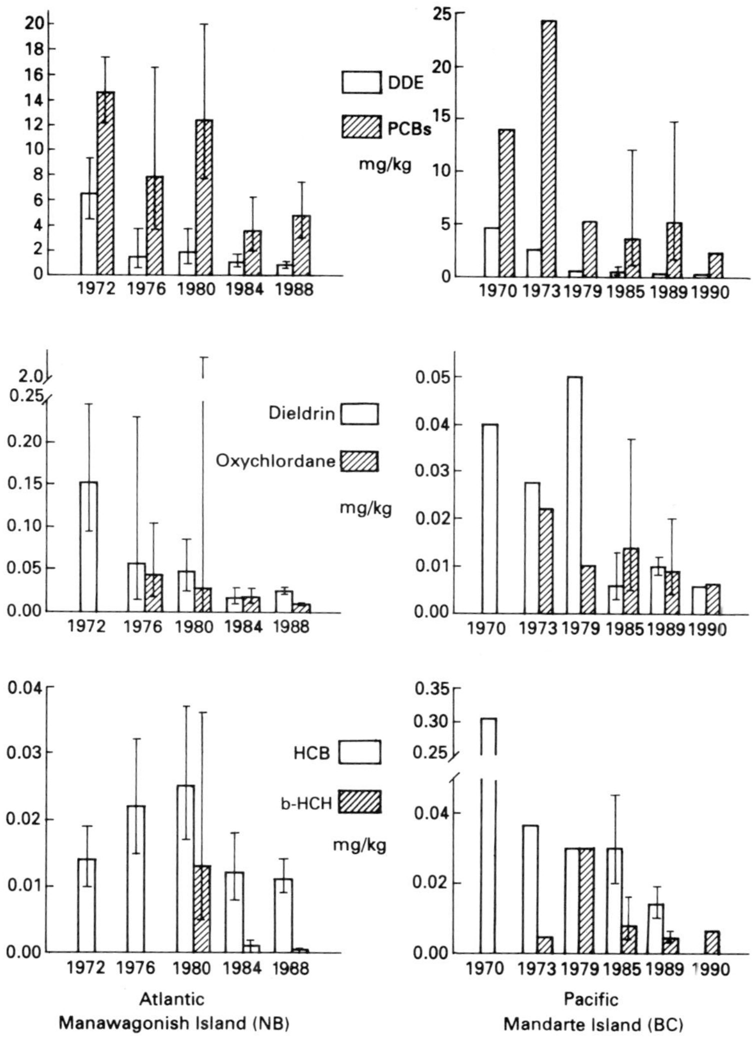

FIG. 8.3 Mean concentrations of selected organochlorines in eggs of double-crested cormorants collected between 1970 and 1990 (geometric mean and 95% confidence interval).

previously in gannets (*Sula bassanus*) from the Gulf of St. Lawrence (Elliott *et al.*, 1988).

Figure 8.4 compares mean levels of the major polychlorinated dibenzo-*p*-dioxin (PCDD) isomers in cormorant egg pools collected in 1979 from the two coasts. Levels of 2,3,7,8-TCDD were similar at both locations and are probably typical of background exposure. Mean levels in cormorant eggs from lakes in the Canadian prairies (Norstrom,

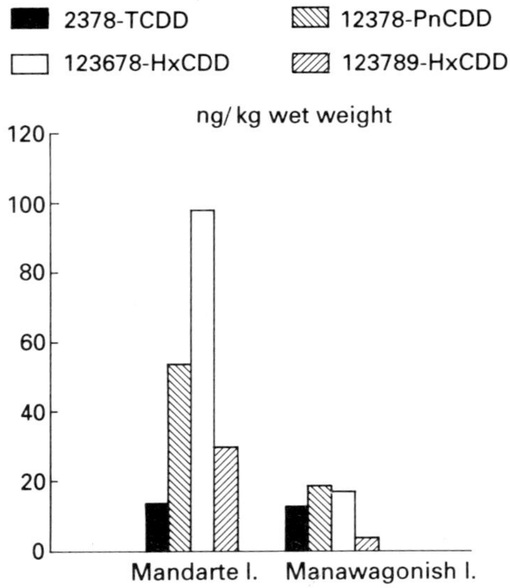

FIG. 8.4 Polychlorinated dibenzo-*p*-dioxin levels in eggs of double-crested cormorants collected in 1979 from Mandarte Island (Pacific coast) and Manawagonish Island (Atlantic coast) of Canada. Analyses were of pooled samples ($N = 10$).

unpublished data), which are remote from any immediate industrial sources, are about the same. Much higher levels of 2,3,7,8-TCDD were found in cormorant eggs from colonies in Lake Ontario and near kraft pulp mills in British Columbia (ibid.)

Concentrations of 1,2,3,7,8-PnCDD, 1,2,3,6,7,8-HxCDD and 1,2,3,7,8,9-HxCDD are 3 to 7 times higher in eggs from the British Columbia colony compared to the Atlantic coast. For several decades tetrachlorophenol and pentachlorophenol formulations have been used in large amounts in British Columbia to protect lumber for export and to preserve wood products, respectively. Tetra- and pentachlorophenols contain a variety of PCDDs and polychlorinated dibenzofurans (PCDFs), including those isomers found in cormorant eggs (Hagenmaier and Brunner, 1987). In addition, chlorophenol-contaminated woodchips have been used by pulp mills in British Columbia. Chlorophenols are therefore the most likely source of the higher chlorinated PCDDs in the Strait of Georgia. Contamination of the cormorant food chain at Mandarte Island, which is removed from any immediate industrial sources, indicates that pollution of the Strait of Georgia by these chemicals is pervasive. Elevated levels of PCDDs and PCDFs have also been reported recently in great blue herons (*Ardea herodias*) and in a number of species of diving ducks from around the Strait of Georgia (Elliott *et al.*, 1989b; Whitehead *et al.*, 1990).

Arctic Environment

Table 8.2 reports levels of chlorinated hydrocarbons in ivory gulls (*Pagophila eburnea*) from Seymour Island in the western Canadian Arctic. Total PCBs were the major contaminant in both sets of ivory gull eggs and mean levels increased by about one-half over the 11-year period. PCB levels were the highest of the chlorinated hydrocarbon groups in northern fulmar (*Fulmarus glacialis*), black-legged kittiwake (*Rissa tridactyla*) and thick-billed murre (*Uria lomvia*) tissues collected at Prince Leopold Island in the mid-1970s (Peakall and Nettleship, 1987), but PCB levels in those species declined between the 1970s and the 1980s (Canadian Wildlife Service, unpublished data). Polar bear (*Ursus maritimus*) tissues collected from various locations in the Canadian Arctic in the early 1980s also had high PCB levels (Norstrom *et al.*, 1988). Norstrom *et al.* (1988) reported that total PCBs were higher in adipose samples of polar bears collected in the 1980s compared to samples from 1969. This difference is now thought to be an age-related effect on body burden, the two samples having had different age distributions (Norstrom, unpublished data), rather than an actual increase in total PCBs in Arctic food chains over that time period. Data for female ringed seals (*Phoca hispida*) from the western arctic (Addison

TABLE 8.2 *Organochlorine Levels (mg/kg wet weight) in Eggs of Ivory Gulls from Seymour Island, Northwest Territories*

	1976 (pooled sample n = 10)	1987 (N = 9, mean ± S.D.)
% fat	8.6	9.2 ± 0.9
% H_2O	80.1	75.4 ± 1.1
Total PCBs	1.09	1.53 ± 0.48
1,2,4,5-TCB[a]	0.006	0.011 ± 0.005
PnCB	0.011	0.007 ± 0.004
HCB	0.098	0.092 ± 0.014
Octachlorostyrene	0.007	0.008 ± 0.002
Trans-nonachlor	0.028	0.052 ± 0.02
DDE	0.819	1.08 ± 0.51
DDT	0.025	0.017 ± 0.004
Photo-mirex	0.008	0.017 ± 0.006
Mirex	0.011	0.017 ± 0.006
α-HCH	0.003	0.005 ± 0.003
β-HCH	0.003	0.009 ± 0.004
Oxychlordane	0.082	0.162 ± 0.007
Cis-chlordane	0.003	0.003 ± 0.001
DDD	0.005	0.004 ± 0.003
Cis-nonachlor	0.005	0.009 ± 0.003
Heptachlor epoxide	0.028	0.038 ± 0.013
Dieldrin	0.043	0.038 ± 0.012
2,3,7,8-TCDD	5 ng/kg	8 ng/kg

[a] TCB, PnCB, HCB = tetrachlorobenzene, pentachlorobenzene, hexachlorobenzene.

and Zinck, 1986) show that total PCBs declined during the 1970s. Mean organochlorine levels were also generally lower between the mid-1970s and the mid-1980s in the fulmars, murres and kittiwakes discussed above (Canadian Wildlife Service, unpublished data).

Mean concentrations of total chlordane-related compounds increased by about a factor of 2 in ivory gull eggs between 1976 and 1987. These results are consistent with the pattern of increased use of chlordane during the 1970s. Levels of chlordane-related compounds also increased in polar bear fat between 1969 and the early 1980s, although this may be in part due to differences in age structures of the samples as discussed above for PCBs.

Qualitative Analysis

Figure 8.5 shows GC/MS chromatograms of pools of Leach's storm-petrel eggs, in this case from Great Island, 1984, and from the Storm Islands, 1987, which are located on the north-east coast of Vancouver Island. Only the second fractions are shown; the first fraction includes DDE, PCBs, chlorobenzenes and mirex; the third fraction includes polar organochlorines such as dieldrin and heptachlor epoxide. In both samples, but particularly in that from Great Island, many of the dominant peaks were toxaphene-related. Toxaphene is a broad spectrum insecticide which can contain up to 177 different chlorinated camphenes. A pooled sample of storm-petrel eggs from Great Island, 1968, had total toxaphene levels of 0.9–1.0 mg/kg. By 1984, levels had increased to 1.4–1.9 mg/kg, suggesting that toxaphene contamination of North Atlantic food chains increased during that interval. Levels in storm-petrel eggs on the west coast were much lower, 0.070 mg/kg.

Elevated concentrations of toxaphene in North Atlantic food chains are consistent with heavy use of this pesticide during the 1970s in the Mississippi drainage basin and recent use in the countries of Central America (Burton and Philogene, 1986). Bidleman and Olney (1975) reported toxaphene in air samples from the western North Atlantic, 1200 km from the nearest point sources. Zell and Ballschmiter (1980) found highest levels of toxaphene in a salmon caught off the west coast of Ireland, and attributed this to long-range transport from the cotton fields of the southern US via the gulf stream or prevailing westerly winds. Toxaphene has been reported previously in marine birds from Alaska (Ohlendorf *et al.*, 1982). In samples taken from beached cetaceans in eastern Canada in the early 1980s, toxaphene levels were the highest of all the measured organochlorines (Muir *et al.*, 1988). Toxaphene levels similar to total PCBs and greater than total DDT-related compounds were also reported in Atlantic cod (*Gadus morhua*)

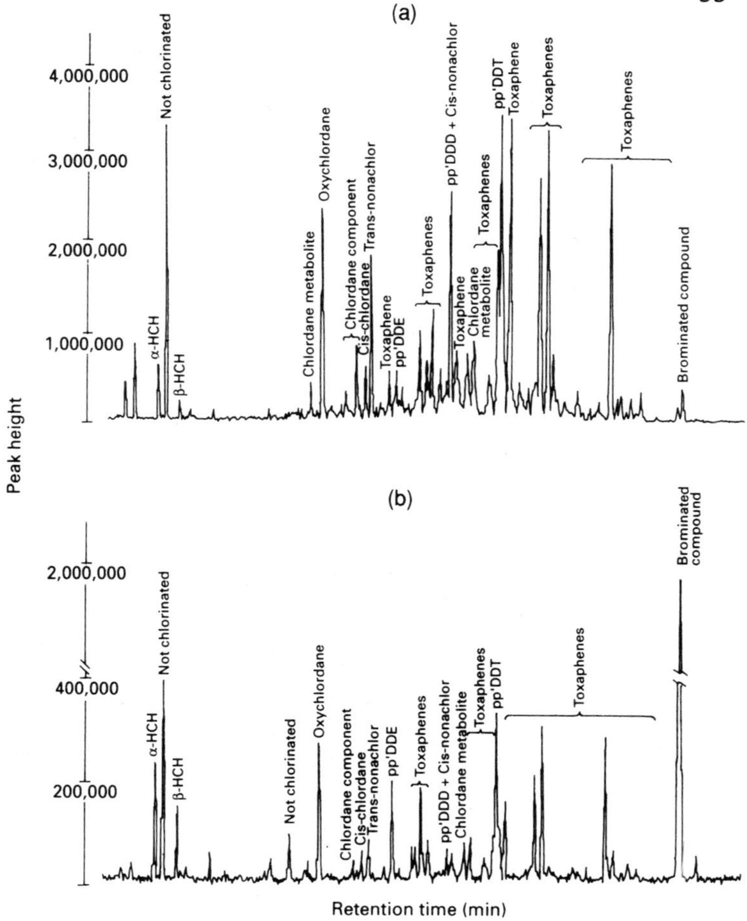

FIG. 8.5 GC/MS chromatograms (second fraction) of pools of Leach's storm-petrel eggs collected from (a) Great Island (Atlantic coast) in 1984 and (b) the Storm Islands (Pacific coast) in 1987.

liver and herring (*Clupea harengus*) muscle from eastern Canadian waters (Musial and Uthe, 1983; Freeman *et al.*, 1983).

A previously unidentified brominated aromatic compound was detected in Leach's storm-petrel eggs from both coasts, and is the major contaminant in the second fraction (at an estimated concentration of 80 μg/kg) in the Storm Island sample. This compound contains five bromines, and likely originates from polybrominated biphenyls.

SUMMARY

Since the end of the organochlorine era (early 1970s), mean residue levels of organochlorine pesticides and industrial compounds have declined in the food chains of inshore and continental shelf seabirds on both coasts of Canada. In contrast, mean residue levels of one industrial organochlorine compound, HCB, have not declined except in double-crested cormorants from the Strait of Georgia. Although data are scarce and levels are low, mean levels of β-HCH appear to be declining on the Atlantic but not on the Pacific coast. In open ocean food chains (Leach's storm-petrel), only DDE appears to be declining.

Estimates, based on full-scan GC/MS analysis of total toxaphene levels in Leach's storm-petrels, show that levels are much higher in Atlantic food chains than the Pacific. Toxaphene levels in the Atlantic samples also increased between 1968 and 1984 and are now the highest of all the groups of organochlorines found in storm-petrel eggs. Increasing toxaphene levels reflect increased use of that chemical as an insecticide in North and Central America. Leach's storm-petrel eggs also contain a previously unidentified brominated compound which is a major contaminant in storm-petrel eggs from the Pacific coast colony. GC/MS analysis of double-crested cormorant eggs collected in 1979 showed that levels of 2,3,7,8-substituted pentachloro- and hexachloro-dibenzodioxins were much higher in the Strait of Georgia, compared to the Bay of Fundy. Higher concentrations of those dioxins were related to forest industry use of chlorophenols.

Organochlorines in resident ivory gull eggs from the western Canadian Arctic showed that mean residue levels remained the same or increased for all compounds, except DDT, over the period 1976 to 1987. This is in sharp contrast to species which only breed in the Arctic for which organochlorine levels in eggs and other tissues generally declined over the same period.

Acknowledgments We thank all of the people who collected seabird samples over the years, in particular Neville Garrity, Stewart MacDonald and Anne Harfenist. Henry Won did most of the organochlorine analysis.

REFERENCES

Addison, R. F. and M. E. Zinck. 1986. PCBs decline more than DDT-group residues in Arctic ringed seals (*Phoca hispida*) between 1972 and 1981. *Environ. Sci. Technol.* **20**: 253–256.

Bidleman, T. F. and C. E. Olney. 1975. Long range transport of toxaphene insecticide in the atmosphere of the western North Atlantic. *Nature* **257**: 475–477.

Burton, D. K. and B. J. R. Philogene. 1986. An overview of pesticide usage in Latin America. Canadian Wildlife Service, Ottawa, unpublished report. 122 pp.

Elliott, J. E. 1985. Specimen banking in support of monitoring for toxic contaminants in Canadian wildlife. In S. A. Wise and R. Zeiser (eds.), *Intern. Rev. Environ. Specimen Banking*. NBS Spec. Publ. 706. US Dept. of Commerce, Washington DC, pp. 4–12.

Elliott, J. E., R. J. Norstrom and J. A. Keith. 1988. Organo-chlorines and eggshell thinning in northern gannets (*Sula bassanus*) from Eastern Canada, 1968–1984. *Environ. Pollut.* **52**: 81–102.

Elliott, J. E., D. G. Noble, R. J. Norstrom and P. E. Whitehead. 1989a. Organochlorine contaminants in seabird eggs from the Pacific coast of Canada, 1971–1986. *Environ. Monit. Assess.* **12**: 67–82.

Elliott, J. E., R. W. Butler, R. J. Norstrom and P. E. Whitehead. 1989b. Environmental contaminants and reproductive success of great blue herons (*Ardea herodias*) in British Columbia. *Environ. Pollut.* **59**: 91–114.

Freeman, H. C.. J. F. Uthe, and P. J. Silk. 1983. Polychlorinated biphenyls, organochlorine pesticides, and chlorobenzenes content of livers from Atlantic cod (*Gadus morhua*) caught off Halifax, Nova Scotia. *Environ. Monit. Assess.* **4**: 389–394.

Gilbertson, M., J. E. Elliott and D. B. Peakall. 1987. Seabirds as indicators of marine pollution. In A. W. Diamond and F. L. Fillion (eds), *The Value of Birds*, International Council for Bird Preservation, Cambridge, Tech. Pub. No. 6, pp. 231–48.

Hagenmaier, H. and H. Brunner. 1987. Isomer-specific analysis of pentachlorophenol and sodium pentachlorophenate for 2,3,7,8- substituted PCDD and PCDF at sub-ppb levels. *Chemosphere* **16**: 1759–1764.

Muir, D. C. G., R. Wagemann, N. P. Grift, R. J. Norstrom, M. Simon and J. Lien. 1988. Organochlorine chemical and heavy metal contaminants in white-beaked dolphins (*Lagenorhynchus labirostris*) and pilot whales (*Clobicephala melaena*) from the coast of Newfoundland, Canada. *Arch. Environ. Contam. Toxicol.* **17**: 613–629.

Murphy, T. J., L. J. Formanski, B. Brownawell and J. A. Myer. 1985. Polychlorinated biphenyl emissions to the atmosphere in the Great Lakes region. Municipal landfills and incinerators. *Environmental Science and Technology* **19**: 942–946.

Musial, C. J. and J. F. Uthe. 1983. Widespread occurrence of the pesticide Toxaphene in Canadian East Coast marine fish. *Internat. J. Environ. Anal. Chem.* **14**: 117–126.

Norstrom, R. J., T. P. Clark and D. V. Weseloh. 1985. Great Lakes monitoring using herring gulls. In T. C. Hutchinson and S. M. Evans (eds.) *Hazardous Contaminants in Ontario: Human and Environmental Effects*. Institute for Environmental Studies, Toronto, pp. 86–98.

Norstrom, R. J., M. Simon, D. C. G. Muir and R. E. Schweinsburg. 1988. Organochlorine contaminants in Arctic marine foodchains: identification, geographical distribution, and temporal trends in polar bears. *Environ. Sci. Technol.* **22**: 1063–1071.

Norstrom, R. J. and M. Simon. 1990. Determination of specific polychlorinated 1:4 dibenzodioxins and dibenzofurans in biological matrices by gel permeation-carbon chromatography and gas chromatography-mass spectrometry. In *Environmental Carcinogens—Methods of Analysis and Exposure Measurement*. Vol. 11—*Polychlorinated dibenzodioxins, dibenzofurans and biphenyls*. World Health Organization, International Association for Research on Cancer, Lyon, France.

Ohlendorf, H. M., J. C. Bartonek, G. J. Divoky, E. E. Klaas and A. J. Krynitsky. 1982. Organochlorine residues in eggs of Alaskan seabirds. *U.S. Fish* and *Wildlife Service, Special Scientific Report* No. 245, 41 pp.

Olsson, M. and L. Reutergardh. 1986. DDT and PCB pollution trends in the Swedish aquatic environment. *Ambio* **15**: 103–109.

Peakall, D. B., R. J. Norstrom, A. D. Rahimtula and R. D. Butler. 1986. Characterization of mixed-function oxidase systems of the nestling herring gull and its implications for bioeffects monitoring. *Environ. Toxicol. Chem.* **5**: 379–385.

Peakall, D. B. and D. N. Nettleship. 1987. Organochlorine residue levels in three high Arctic species of colonially-breeding seabirds from Prince Leopold Island. *Marine Pollution Bulletin* **18** (8): 434–438.

Pearce, P. A., J. E. Elliott, D. B. Peakall and R. J. Norstrom. 1989. Organochlorine contaminants in eggs of seabirds in the Northwest Atlantic, 1968–84. *Environ. Pollut.* **56**: 217-235.

Reynolds, L. M. and T. Cooper. 1975. Analysis of organochlorine residues in fish. Spec. Techn. Publ. 573, ASTM, pp. 196–205.

Stout, V. F. 1986. What is happening to PCBs? Elements of environmental monitoring as illustrated by an analysis of PCB trends in terrestrial and aquatic organisms. In J. S. Waid (ed.), *PCBs and the Environment*. CRC Press, Boca Raton, Florida, Vol. 1, pp. 163–206.

Tanabe, S. 1988. PCB problems in the future: foresight from current knowledge. *Environ. Pollut.* **50**: 5–28.

Tanabe, S. and R. Tatsukawa. 1986. Distribution, behaviour and load of PCBs in the oceans.

In J. S. Waid (ed.), *PCBs and the Environment*. CRC Press, Boca Raton, USA, pp. 143–162.

Turle, R., R. J. Norstrom and B. Collins. 1991. Comparison of PCB quantitation methods: re-analysis of archived specimens of herring gull eggs from the Great Lakes. *Chemosphere* **22**: 201–213.

UNEP (United Nations Environment Program). 1985. Protection of the marine environment against pollution from land-based sources. UNEP working group 120/3 (Part IV): Environment Policy and Law. Elsevier Science Publishers B.V., North Holland.

Whitehead, P. E., J. E. Elliott and R. J. Norstrom. 1990. PCDD and PCDF contamination of waterfowl in the Strait of Georgia, British Columbia, Canada, 1989–90. In O. Hutzinger (ed.), *Dioxin '90*, Ecoinforma Press, Bayreuth, FRG, Vol. 1, pp. 459–462.

Zell, M. and K. H. Ballschmiter. 1980. Baseline studies of the global pollution II. Global occurrence of hexachlorobenzene (HCB) and polychlorocamphenes (toxaphene) (PCC) in biological samples. *Fresenius Z. Anal. Chem.* **300**: 387–402.

Zitko, V. 1978. Nonachlor and chlordane in aquatic fauna. *Chemosphere* **1**: 3–7.

CHAPTER 9

Problems of Bioaccumulation in Fish-eating Birds from the Mediterranean

A. RENZONI

Department of Environmental Biology
University of Siena

INTRODUCTION

The well-known process of bioaccumulation of persistent contaminants has been demonstrated in representatives of many animal groups. Bioaccumulation has been more frequently reported in fishes, birds and mammals than in other groups, but this could merely reflect the large number of species in the first group and the degree of interest in the second and third group. Within birds, the bioaccumulation of persistent lipophilic pollutants in the trophic chain was detected after the discovery of declines of the populations of raptors and/or of fish-eating birds in the UK, in North America and in Sweden (Moore and Walker 1964; Borg *et al.*, 1969; Blus *et al.*, 1974; Fimreite *et al.*, 1970, 1971; Lindberg 1975).

Repeated observations during the 5-year-long FAO/UNEP project for the monitoring of the Mediterranean (1976–80) and subsequent observations for the Mediterranean Action Plan have shown that the residues of several contaminants (particularly mercury, organic and inorganic, and PCBs) are present in the eggs and in the tissues of the studied birds at higher levels than have been reported in similar material from other areas of the world. This trend is closely related to the relatively high levels occurring in fish and other representatives of the Mediterranean marine fauna (Baldi *et al.*, 1978; Renzoni *et al.*, 1978; Buffoni *et al.*, 1982); in some species of fish the levels reported have often been well above the so-called background levels (International Register Potentially Toxic Chemicals, 1978).

During the first stage of research, the monitoring of pollutants in the marine fauna was the primary aim of the project: during the second one, the levels of contaminants in fish-eating birds were studied in many resident as well as migratory species; in the third one (still in force), attention has been directed towards the study of possible strategies of detoxification by various enzyme systems.

COMPARISON OF ATLANTIC AND MEDITERRANEAN ECOSYSTEMS

The first series of studies have clearly shown the difference between Atlantic and Mediterranean pelagic fish species (anchovy *Engraulis encrasicolus*, sardine *Sardinella aurita* and tuna *Thunnus thynnus thynnus*) with regard to the mercury accumulation in the trophic chain and in relation to age. Residues of the metal are generally much higher in the Mediterranean than in the Atlantic.

To confirm these findings in pelagic fish, a migratory species feeding upon surface organisms of pelagic waters was analysed in great detail (Renzoni *et al.*, 1986). The species in question was Cory's shearwater (*Calonectris diomedea*) which winters in Central and in South Atlantic waters and breeds on Mediterranean and Atlantic islands. The study of the xenobiotic levels in this species was performed at one Atlantic and three Mediterranean island stations.

The results are reported in Table 9.1 and Fig. 9.2, and from that study the following conclusions could be drawn:

(a) The amount of mercury in the eggs collected in the Mediterranean

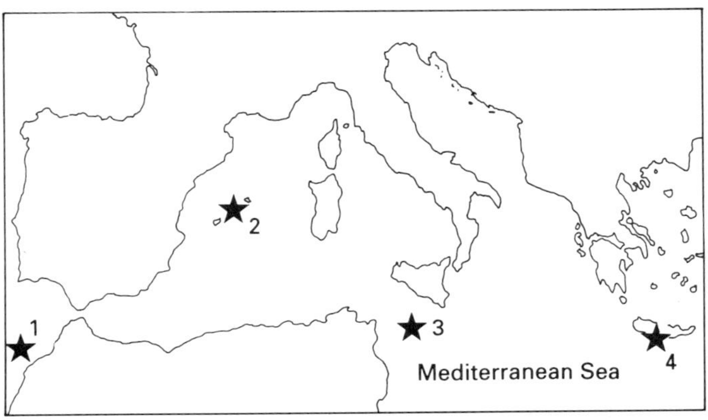

FIG. 9.1 Stations of collection of Cory's shearwater. 1: Selvagen Island; 2: Cabrera, a small Balearic Island; 3: Linosa, a Pelagian Island; 4: Dragonada, a small island near Crete.

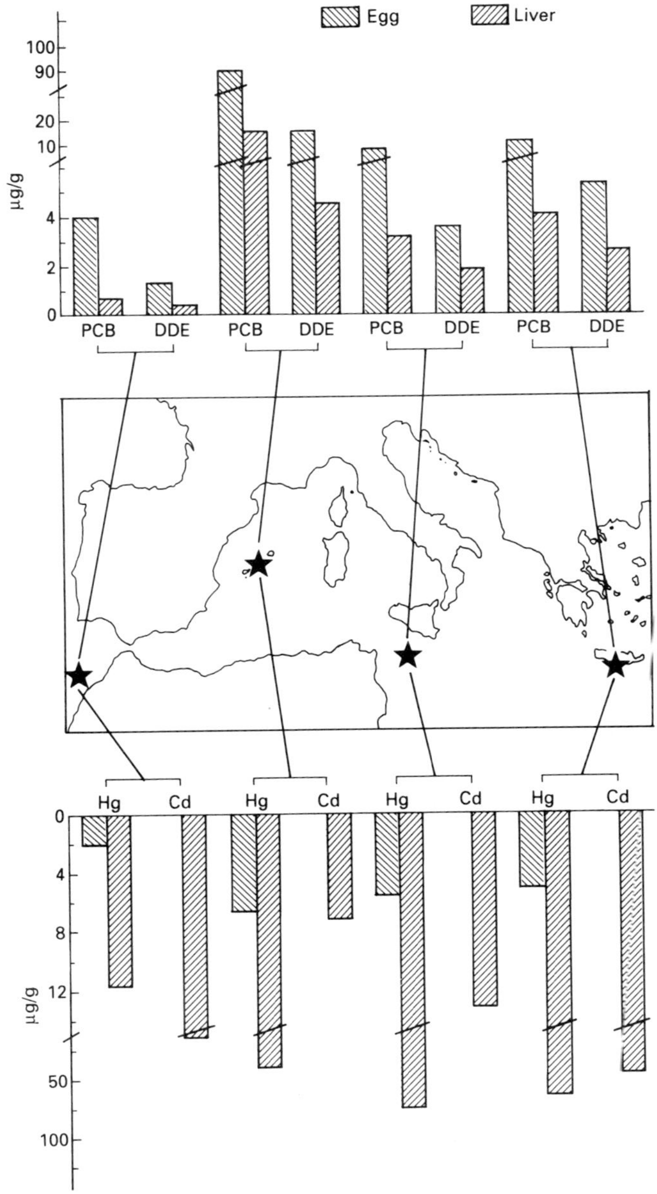

FIG. 9.2 Residue levels (μg/g dry weight) in eggs and liver of Cory's shearwater from the four stations (see Fig 9.1).

TABLE 9.1 *Three Elements in Tissues and Organs of Cory's Shearwater of the Four Stations of Fig. 9.1 ($\mu g\ g^{-1}$ dry weight)*

	Hg		Cd		Pb	
	$\bar{X}$	S.D.	$\bar{X}$	S.D.	$\bar{X}$	S.D.
Station 1 (n = 3)						
Brain	1.95	0.70	0.97	0.40	1.07	0.50
Muscle	2.56	0.75	2.07	1.10	0.84	0.71
Kidney	4.63	1.48	214.17	45.85	1.08	0.75
Liver	12.53	2.28	26.45	1.30	0.97	0.66
Station 2 (n = 5)						
Brain	5.45	1.79	3.00	0.73	1.56	1.15
Muscle	4.76	1.48	3.93	1.09	0.61	0.19
Kidney	14.96	4.51	42.82	5.88	0.48	0.39
Liver	49.69	23.08	7.52	2.38	0.60	0.38
Station 3 (n = 5)						
Brain	6.82	1.54	2.63	1.62	0.89	0.58
Muscle	7.20	1.55	2.97	0.46	0.56	0.31
Kidney	17.79	10.64	106.21	46.80	0.98	0.60
Liver	86.16	99.16	13.91	15.17	0.81	0.48
Station 4 (n = 5)						
Brain	3.43	1.26	1.06	1.73	0.95	0.72
Muscle	5.91	1.03	2.00	0.78	0.87	0.34
Kidney	18.89	5.98	187.50	62.62	0.51	0.28
Liver	74.85	50.94	55.92	39.76	0.76	0.47

were from 2.5 to 3.5 times higher than those collected in the Atlantic station. The values in the three Mediterranean stations follow a consistent pattern. In the analysed tissues, mercury increases progressively from the fat to the uropygial gland, brain, muscle, kidney and liver. As in the case of the eggs, in all tissues the mercury content was higher in the material collected in the Mediterranean stations than in that from the Atlantic one. In particular, the liver residues of birds from the Mediterranean are four to six times higher than those from the Atlantic.

(b) In the eggs, cadmium levels were always below the detection limits of the instrument (0.03 $\mu g\ g^{-1}$). In the kidney, cadmium levels reached their maximum, especially in stations 1 (214 $\mu g\ g^{-1}$) and 4 (187 $\mu g\ g^{-1}$). In the liver, cadmium levels were also relatively high (from 7 to 55 $\mu g\ g^{-1}$). In the fat, uropygial gland, muscle and brain, concentrations varied from 0.21 to 3.93 $\mu g\ g^{-1}$.

(c) The eggs from all the stations had concentrations of lead below the limit of detection (0.3 $\mu g\ g^{-1}$). In organs and tissues, levels never exceeded 1.6 $\mu g\ g^{-1}$.

(d) Atlantic eggs had the lowest concentrations of *pp'*DDE. Within the Mediterranean, the residues in Station 2 (18.7 $\mu g\ g^{-1}$) were much higher (5–6 times) than in Stations 3 and 4. Similar results were obtained for the tissues and organs: in Station 2 of the Mediterranean DDE (1,1-dichloro-2,2-bis(4 chlorophenyl) ethylene) values were particularly high (80.2 μg g^{-1} in the fat). In general, the trend of distribution in tissues and organs shows the increase of *pp'*DDE from the kidney to the brain, liver, muscle, uropygial gland and fat. If we normalize the concentrations on a lipid basis, the trend does not change, except for the uropygial gland and brain: the two last organs become the compartments with the lowest DDE residues.

(e) In the eggs of the Mediterranean stations, PCB (polychlorinated biphenyls) values were from 2 to 25 times higher than in the Atlantic. In Station 2 there was an exceptional level of PCB residues (as high as 95 μg g^{-1}). Residues in tissues and organs of adults also reached remarkable levels in Station 2; the levels were about 20 times higher than those in Station 1. The distribution of PCBs in the organs was similar to that of DDE.

Whilst organochlorine residues are clearly of anthropic origin, for mercury there is strong evidence that most of the element present in the Mediterranean waters, sediments, and biota is of natural origin; more than 50% of the world supply of this metal derives from the basin.

Pelagic fish species and their avian predators are very suitable indicator species for the level of pollutants in the Mediterranean as a whole; for a study of local impact of human activities, fish-eating birds feeding on coastal areas and/or brackish lagoons are evidently more appropriate.

BIRDS MIGRATING TO THE MEDITERRANEAN AREA FROM NORTHERN OR SOUTHERN REGIONS

During the last decade several species of migratory birds have been analysed. These are: wintering cormorants *Phalacrocorax carbo* and grebes *Podiceps nigricollis* (Cottiglia *et al.*, 1982), grebes (Leonzio *et al.*, 1986), breeding little terns *Sterna albifrons* (Focardi *et al.*, 1980) and common terns *Sterna hirundo* (Renzoni *et al.*, 1982) from various Italian lagoons. It is evident from the data collected during the last 15 years that the Mediterranean is an unusual basin, where, as a consequence of the natural weathering of rocks and of the many anthropic activities, the levels of some pollutants in the biota are higher than in the same or in closely similar species living in the Atlantic Ocean or in other large basins. The levels of contaminants here could be compared with those found 15–20 years ago in the North Sea and in the Great Lakes (Bourne and Bogan, 1976; Gilman *et al.*, 1977).

But within this comprehensive remark some comments are necessary. For most common pollutants in the Mediterranean the slow turnover of the water, shallowness, small number of rivers (flowing through highly industrialized areas), intense agricultural practices and many other human activities lead to a general diffusion and persistence of certain toxic substances. For mercury, with levels consistently much higher in the Mediterranean material (eggs, organs or tissues) than in that from Atlantic, the explanation could be found in the occurrence of the mercury ore cinnabar in several inland areas around this basin and the consequent release of substantial quantities of inorganic mercury following weathering.

However, for this last metal especially, the results of bioaccumulation are quite different between the Cory's shearwaters feeding in pelagic waters on the one hand and cormorants, grebes and terns feeding in coastal areas and/or in lagoons contaminated by human activity on the other. The mercury levels in the second group are higher, especially in the cormorant, the most specialized fish-eating bird in the group (Table 9.2).

TABLE 9.2 *Mercury Levels in Tissues and Organs of Two Fish-eating Birds Feeding in Mercury Contaminated Lagoons (µg g^{-1} dry weight)*

	Cormorant Hg (µg g^{-1}; n = 5)	Black-necked grebe Hg (µg g^{-1}; n = 9)
Brain	23.62 (11.75–36.50)	28.58 (10.50–57.00)
Muscle	33.89 (18.10–54.87)	29.88 (11.60–86.74)
Kidney	220.16 (42.84–399.84)	97.65 (34.75–300.92)
Liver	130.60 (38.00–284.55)	74.25 (26.00–150.00)

For cadmium, the high levels found in tissues and organs of the Cory's shearwater seem to be correlated with the natural content of cadmium in the pelagic organisms upon which the birds feed (Anderlini *et al.*, 1972; Hutton 1981) rather than the contamination of the Mediterranean basin.

Regarding the most common contaminants of the organochlorine group in the Mediterranean basin, there are no important differences between our findings and those of previous researchers (Bourne and Bogan, 1976, 1980; Bijleveld *et al.*, 1979; Viviani *et al.*, 1974; Mendola *et al.*, 1977; Vannucchi *et al.*, 1978; Focardi *et al.*, 1980; Fossi *et al.*, 1984). Within the Mediterranean, the material from the island of Mallorca indicates higher levels of *pp'*DDE and of PCBs than in all other Mediterranean areas. Since the industrial activities in the Balearic Archipelago are very limited and the agriculture is mainly old fashioned, the most likely explanation for such findings is the arrival of toxic substances from the Rhone and the coastal waters of

Marseille transported in a counter-clockwise stream in the western Mediterranean.

THE DELTA OF THE DANUBE

The Danube river crosses the land of several countries and collects its waters from highly industrialized and intensely cultivated areas (Germany, Romania, etc.). The delta is one of the best nesting areas in the world for a wide range of species of aquatic birds. The analysis of eggs of birds breeding in the delta, besides giving an opportunity to evaluate the level of contamination in that area (reflecting clearly the contamination of the entire river too), has clearly indicated the different residue levels in birds at different trophic levels. Examples have been studied of predominantly primary consumers, feeding mostly on plants (mallard *Anas platyrhynchos*); secondary consumers, feeding only on small aquatic organisms (glossy ibis *Plegadis falcinellus*) or on those and fish (grey heron *Ardea cinerea*, night heron *Nycticorax nycticorax*); and of tertiary consumers feeding exclusively on fish (common cormorant *Phalacrocorax carbo*, pygmy cormorant *Phalacrocorax pygmeus* and white pelican *Pelecanus oncrotalus*). Results are reported in Tables 9.3 and 9.4.

TABLE 9.3 *Mercury Levels in Eggs of Seven Species of Water-birds Collected in the Delta of Danube (μg g⁻¹ dry weight)*

Species	n	Hg		fw/dw	
		$\bar{X}$	S.D.	$\bar{X}$	S.D.
Mallard	4	0.27	0.06	5.01	0.88
Glossy ibis	2	0.52	–	6.07	–
Grey heron	3	0.55	0.03	4.71	1.33
Night heron	10	0.96	0.53	6.17	1.14
Pygmy cormorant	8	3.13	1.69	4.45	1.10
White pelican	12	3.55	1.94	4.89	0.99
Common cormorant	19	4.98	2.23	6.39	0.09

The striking differences in contamination between the species which we examined are related to their respective diets. In the case of fish-eaters, the type, age and size of prey was also related to levels of contamination. In fact the low levels of both metals and chlorinated hydrocarbons that were found in the eggs of the mallard clearly reflect the low levels in blood and organs of adults which are predominantly herbivores. The values obtained for the eggs of glossy ibis, night heron and grey heron were very similar (even though the ibis feed more extensively than herons on invertebrate primary consumers), but were higher than those for the mallard. The pygmy cormorant, white pelican

TABLE 9.4 *Chlorinated Hydrocarbons in the Same Material as Table 9.3 (μg/g dry weight)*

		Mallard $n = 4$	Glossy Ibis $n = 2$	Grey Heron $n = 3$	Night Heron $n = 13$	Pygmy Cormorant $n = 10$	White Pelican $n = 12$	Common Cormorant $n = 13$
EOM	$\bar{X}$	40.7	28.7	24.7	28.5	22. 8	23.9	24.5
(% d.w.)	S.D.	3.6	nc	1.5	3.3	2.2	1.7	7.6
HCB	$\bar{X}$	0.18	0.16	0.17	0.19	0.47	0.32	1.30
	S.D.	0.02	nc	0.03	0.05	0.25	0.08	0.82
α–HCH	$\bar{X}$	nd	nd	nd	nd	nd	0.20	0.44
	S.D.	nd	nd	nd	nd	nd	0.13	0.49
γ–HCH	$\bar{X}$	0.27	0.28	0.65	0.52	0.46	1.15	2.01
	S.D.	0.03	nc	0.04	0.21	0.11	0.42	0.98
pp′DDE	$\bar{X}$	1.19	3.50	6.57	5.34	18.55	15.38	57.20
	S.D.	0.18	nc	0.28	1.64	5.40	5.34	19.35
pp′DDD	$\bar{X}$	0.02	0.05	0.13	0.08	0.26	2.51	1.11
	S.D.	0.01	nc	0.01	0.02	0.23	0.90	0.37
pp′DDT	$\bar{X}$	0.06	0.45	0.65	0.83	0.50	0.86	1.63
	S.D.	0.01	nc	0.02	0.27	0.14	0.29	0.48
PCB	$\bar{X}$	0.98	2.40	2.04	2.33	14.95	5.58	23.59
	S.D.	0.09	nc	0.05	1.06	7.86	2.23	12.86
$\frac{PCB}{DDE}$	$\bar{X}$	0.83	0.68	0.31	0.46	0.72	0.39	0.41
	S.D.	0.22	nc	0.02	0.19	0.12	0.16	0.12

and common cormorant, which contained the highest residues, are all basically fish-eaters (Cramp, 1980), even though small crustaceans, small mammals and leeches are occasionally included in the diet of the pygmy cormorant. Observing carefully the data obtained, one may wonder why then do the eggs of those three species, gathered at the same nesting-sites, contain different amounts of contaminants? In particular, cormorants contained much higher levels of DDE and PCBs than either of the other two species. Possible explanations concern differences in feeding areas and in the size of the prey. According to A. Mihaileanu (personal communication), common cormorants and pelicans fish in different areas: the former in the sea off the Danube delta, and the latter in branches of the delta and in brackish-water lagoons south of the delta. The feeding area can also vary in depth. It is known that the common cormorant does its fishing by swimming underwater to a greater or lesser depth. In fact, Cramp (1980) reports that in Vorso the common eel (*Anguilla anguilla*), a typically benthic animal, accounted for 38% of its diet from April to October. From this and other evidence, Cramp concluded that the cormorant's prey consisted of adult and subadult organisms, themselves mostly predators, coming from the layer of water between the surface and a depth of 9 metres. This type of prey usually bears higher levels of contaminants than the young or adult organisms in the top metre, where the pelican catches most of its prey (Crivelli, personal communication).

Together with the previously reported data and with the many available in the literature, including those obtained with Dalmatian pelican along the Danube river (Mitchev and Dilchev, 1986) and in the south European Lake Mikri Prespa (Crivelli *et al.*, 1989), these findings are additional evidence of the process of bioaccumulation occurring in birds high in the trophic system.

CONSEQUENCES OF CONTAMINATION

From the study performed in the Mediterranean, there are no observations regarding the possible consequences towards the health of avian population and/or their reproductive success. One parameter repeatedly studied in the past concerns the reduced thickness of the eggshell in the birds with a high content of organochlorine residues in their tissues (Peakall and Peakall, 1973; Ohlendorf *et al.*, 1978; Fox and Donald, 1980; Blus *et al.*, 1979).

In tissues of Cory's shearwater sampled in the Mediterranean, residues of *pp*'DDE (a compound which can cause eggshell thinning) are higher than in samples from the Atlantic (see Fig. 9.2) and a slight thinning of eggshells from the Mediterranean was thought to occur (Renzoni *et al.*, 1986).

However, even though all the eggs of the Mediterranean colonies are thinner than the Atlantic ones, within the first group (i.e. those from the Mediterranean), the most polluted eggs are not the ones with the thinnest shells. More data are therefore necessary to confirm this finding and to evaluate the reduction on a statistical basis (Table 9.5).

TABLE 9.5 *Chlorinated Hydrocarbons (μg/g dry weight) in Eggs of Cory's Shearwater from the Four Stations of Fig. 9.1. The Thickness of the Shell is Reported*

	Station 1 $n = 20$		Station 2 $n = 10$		Station 3 $n = 11$		Station 4 $n = 2$	
	$\bar{X}$	S.D.	$\bar{X}$	S.D.	$\bar{X}$	S.D.	$\bar{X}$	S.D.
Lipids % dry weight	36.8	5.4	39.1	3.2	35.5	6.7	40.2	36.9
Hexachlorobenzene	0.04	0.01	0.03	0.02	0.07	0.02	0.03	0.05
*pp'*DDE	1.49	0.68	18.77	12.46	3.75	2.08	4.79	9.99
PCBs	4.19	1.70	95.08	56.72	10.59	4.18	7.96	20.81
PCBs/DDE	2.87	0.62	5.5	2.20	3.00	0.72	1.66	2.08
Thickness (μm)	385	18	330	24	308	16	324	318

As far as the hatchability index is concerned, not many precise observations have been made under natural conditions. The decreased density of some bird populations in Scandinavia and in North America in the late sixties was mostly attributed to the eggshell thinning, even though there were indications that reduced hatchability could have been responsible. In nature, total mercury levels ranging from 2 to 16 ppm in eggs did not appear to affect the hatching and fledging of herring gulls (*Larus argentatus*) in North America (Vermeer *et al.*, 1973); in laboratory experiments with pheasants, chicken and quail dosed with mercury and PCB (injected into egg or administered with the diet), it was found that in the whole egg the levels associated with a reduction in hatching capacity (Briggs and Harris, 1972; Britton and Huston, 1972) range from 0.5 to 2 ppm (mercury) and from 2.2 to 4.5 ppm (PCB), or over 6 ppm (PCB) (Borg *et al.*, 1969; Scott *et al.*, 1971; Keplinger *et al.*, 1971; Fimreite *et al.*, 1971; Britton and Huston, 1972; Scott *et al.*, 1975).

There is a difficulty in performing laboratory studies on the uptake and metabolic fate of PCBs which are relevant to the environmental situation. In particular, the routes of uptake in the field are apparently complex and improperly understood.

BIOMODIFICATION AND BIODETOXIFICATION STRATEGIES

Great emphasis has recently been placed on the modifications undergone by members of the PCB group once in the environment, and especially their metabolism within the animal body. Studies of the last

20 years (De Freitas and Norstrom, 1974; Burse *et al.*, 1976; Baker *et al.*, 1977; Matthews and Tuey, 1980) have demonstrated that many of the lower chlorinated di-, tri- and tetrachloro isomers and congeners, and selected higher chlorinated PCBs, were preferentially eliminated from body tissues and organs. Data of other authors confirm such findings in mammals and in birds (Safe, 1980), and are the basis for a more general subdivision into persistent and non-persistent PCB congeners (Bush *et al.*, 1974; Kuroki and Masuda, 1977; Wolfe *et al.*, 1982; Braune and Norstrom, 1989).

Partially modified or in their original form, most chemicals released by man to the environment reach living organisms directly or indirectly. Some of them are easily eliminated, some are accumulated in different proportions in the various structures of the body and some others are metabolized.

In the last two decades a monooxygenase system [also known as mixed function oxidase (MFO)] has been studied in various groups of animals, and it has been considered one of the most important substrate-inducible defensive systems employed in the biomodification and detoxification of lipophilic xenobiotics, such as aromatic and aliphatic hydrocarbons and polychlorinated biphenyls (Walker, 1980; Knight *et al.*, 1981; Knight and Walker, 1982; Safe *et al.*, 1985; Borlakoglu *et al.*, 1988).

A recent study concerning the levels of this complex in the liver of the black-headed gull shows a remarkable difference between a population feeding in a lagoon and a population feeding on a dump (Fossi *et al.*, 1986, 1988). The livers of the specimens feeding on the dump demonstrate higher MFO activity than those feeding in the lagoon

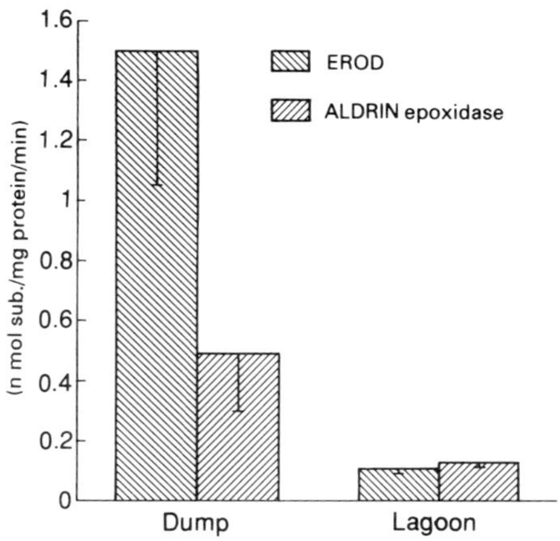

FIG. 9.3 Aldrin epoxidation and EROD activities in black-headed gull feeding in dump and lagoon sites (arithmetic mean and 95% confidence limit: Dump, $n = 12$; Lagoon, $n = 8$).

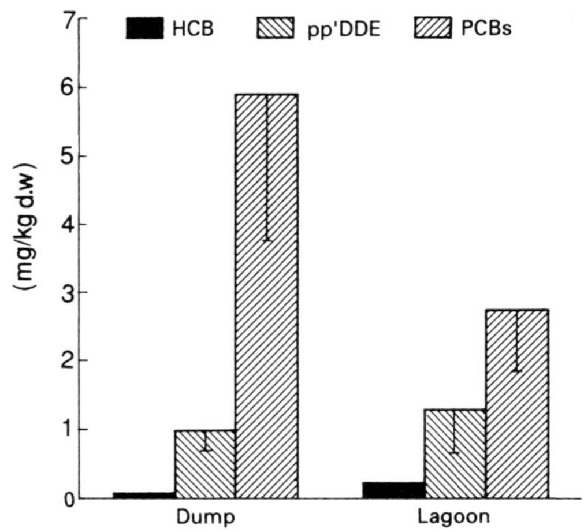

FIG. 9.4 HCB, *pp*'DDE, and PCBs concentration in liver of black-headed gull feeding in dump and lagoon sites (arithmetic mean and 95% confidence limit: Dump, n = 12; Lagoon, n=8).

(Fig. 9.3), suggesting that induction may have been caused by pollutants (Fig. 9.4) in the most contaminated area.

TRENDS IN ENVIRONMENTAL CONTAMINANT LEVELS

For persistent contaminants, organisms at the top of the food chain are the most appropriate for the evaluation of the contamination level of the area where they live. There is evidence that in several areas (once highly polluted) there is a trend toward an improvement of the environment. This is the case in the Great Lakes of North America and in the North Sea. In the Mediterranean area the limited data available over the last 5-year period do not allow definite conclusions to be drawn. The only comparative data comes from the analyses of the eggs of Cory's shearwater in Linosa, Central Mediterranean Island. The results of 1988 compared with those of 1982 seem to show a slight

TABLE 9.6 *Levels of Contaminants in Eggs of Cory's Shearwater Collected in 1982 and 1988 in Station 3 (see Fig. 9.1) ($\mu g\ g^{-1}$ dry weight)*

	Year	
	1982 (n = 11)	1988 (n = 12)
Mercury	5.90	7.50
pp'DDE	3.75	4.87
PCBs	10.59	22.08
HCB	0.07	0.05

increase in mercury and DDE levels and an evident increase of PCB (Table 9.6). However, the sizes of the sample and area were too small to permit the drawing of any definite conclusion.

CONCLUSIONS

From the available data on residue levels of persistent pollutants in fish-eating birds of the Mediterranean basin, a few comprehensive statements may be made:

1. Residue levels of mercury are higher than in similar material of several other areas of the world; the values could be considered close to those obtained in the Great Lakes 10–15 years ago.
2. Organochlorine contaminants are in the same range of values as those reported in the literature and obtained in several areas of the Northern Hemisphere; only in the material from Mallorca are values clearly higher than elsewhere.
3. No clear trend in the Hg, DDE and PCB levels in the eggs of Cory's shearwater collected in Linosa were found over the period 1982–8. However, the period from the first analysis (1982) to the second one (1988) is perhaps too short to show possible changes.

REFERENCES

Anderlini, V. C., P. G. Connors, R. W. Risebrough and J. H. Martin. 1972. Concentrations of heavy metals in some Antarctic and North America seabirds. *Proc. Colloq. Conserv. Probl. Antarctica*, B. C. Parker (ed.), pp. 49–62. Blacksburg, Virginia Polytechnic Institute and State University.

Baker, F. D., C. F. Tumasonis and F. C. Lo. 1977. Toxicity and persistence of low level PCB in adult Wistar rats, fetuses and young. *Arch. Environ. Contam. Toxicol.* **5**: 143–156.

Baldi, F., A. Renzoni and M. Bernhard. 1978. Mercury concentrations in pelagic fishes (anchovy, mackerel and sardine) from the Italian coast and Strait of Gibraltar. *Journées Etud. Pollutions, 4th, Antalya*, CIESM, pp. 251–254.

Bijleveld, M. F. I. J., P. Goeldlin and J. Mayol. 1979. Persistent pollutants in Audouin's gull (*Larus audounii*) in the Western Mediterranean: a case-study with wide implications? *Environ. Conserv.* **6**: 139–142.

Blus, L., B. S. Neely, A. A. Belisle and R. M. Pronty. 1974. Organochlorine residues in brown pelican eggs: relation to reproductive success. *Env. Poll.* **7**: 82–91.

Blus, L., T. G. Lamont and B. S. Neely Jr. 1979. Effects of organochlorine residues on eggshell thickness, reproduction, and population status of brown pelicans (*Pelecanus occidentalis*) in South Carolina and Florida, 1969–1976. *Pestic. Monit. J.* **12**: 172–184.

Borg, K., H. Wanertorp, K. Erne and E. Hanks. 1969. Alkyl mercury poisoning in terrestrial Swedish wildlife. *Viltrevy* **6**: 301–379.

Borlakoglu, J. T., J. P. G. Wilkins and C. H. Walker. 1988. PCBs in fish-eating birds—molecular features and metabolic interpretations. *Mar. Envir. Res.* **24**: 15–19.

Braune, B. M. and R. J. Norstrom. 1989. Dynamics of organochlorine compounds in herring gulls: III. Tissue distribution and bioaccumulation in Lake Ontario gulls. *Envir. Toxic. Chem.* **8**: 957–968.

Briggs, D. M. and J. R. Harris. 1972. Polychlorinated biphenyls influence on hatchability. *Poult. Sci.* **52**: 1291–1294.

Britton, W. M. and T. M. Huston. 1972. Influence of polychlorinated biphenyls in the laying hens. *Poult. Sci.* **52**: 1620–1624.

Bourne, W. R. P. and J. A. Bogan. 1976. Estimation of chlorinated hydrocarbons in marine birds. In K. Johnston (ed.), *Marine Pollution*, pp. 482–502. London, Academic Press.

Bourne, W. R. P. and J. A. Bogan. 1980. Estimation of chlorinated hydrocarbons in marine birds. *Environ. Conserv.* **7**: 277–278.

Buffoni, G., M. Bernhard and A. Renzoni. 1982. Mercury in the Mediterranean tuna? Why is their level higher than in Atlantic tuna? A model. *Thalassia Jugosl.* **18**: 231–243.

Burse, V. W., R. F. Moseman, G. W. Sovocol and E. C. Villanueva. 1976. PCB metabolism in rats following prolonged exposure to Aroclor 1242 and Aroclor 1016. *Bull. Environ. Contam. Toxicol.* **15**: 122–128.

Bush, B., C. F. Tumasonis and F. D. Baker. 1974. Toxicity and persistence of PCB homologs and isomers in the avian system. *Arch. Environ. Contam. Toxicol.* **2**: 195–209.

Cottiglia, M., S. Focardi, C. Leonzio, C. Mascia, A. Renzoni and C. Fossi. 1982. Contaminants in tissues of shore-birds from a polluted lagoon of the island of Sardinia. *Journées Etud. Pollutions, 6th, Cannes*, CIESM, pp. 293–298.

Cramp, S. 1980. *Handbook of the Birds of Europe, the Middle East and North Africa*. Vol. I. Oxford University Press.

Crivelli, A., S. Focardi, C. Fossi, C. Leonzio, A. Massi and A. Renzoni. 1989. Trace elements and chlorinated hydrocarbons in eggs of *Pelecanus crispus*, a world endangered bird species nesting at Lake Mikry Prespa, North Western Greece. *Environ. Poll.* **61**: 235–247.

De Freitas, A. S. and R. J. Norstrom. 1974. Turnover and metabolism of PCBs in relation to their chemical structure and the movement of lipids in the pigeon. *Can. J. Physiol.* **52**: 1080–1094.

Fimreite, M., R. W. Fyfe and J. A. Keitle. 1970. Mercury contamination of Canadian prairie seed eaters and their avian predators. *Canad. Fld. Nat.* **84**: 269–276.

Fimreite, N., W. N. Holsworth, J. A. Keith, P. A. Pearce and I. M. Gruchy. 1971. Mercury in fish and fish-eating birds near sites of industrial contamination in Canada. *Canad. Fld. Nat.* **85**: 211–220.

Fishbein, L. 1974. Toxicity of chlorinated biphenyls. *Ann. Rev. Pharmacol.* **14**: 139–154.

Focardi, S., C. Leonzio and A. Renzoni. 1980. Chlorinated hydrocarbons and trace metals in eggs of *Larus argentatus michaellis* Naumann and *Sterna albifrons albifrons* Pallas. *Atti 1° Congr. Soc. Ital. Ecol., Salsomaggiore Terme Parma* **11**: 345–350.

Fossi, C., S. Focardi, C. Leonzio and A. Renzoni. 1984. Trace metals and chlorinated hydrocarbons in birds eggs from the delta of Danube. *Envir. Conserv.* **11**: 345–350.

Fossi, C., C. Leonzio and S. Focardi. 1986. Mixed function oxidase activity and cytochrome P-450 forms in black-headed gulls feeding in different areas. *Mar. Poll. Bull.* **17**: 546–548.

Fossi, C., C. Leonzio, S. Focardi and A. Renzoni. 1988. The black-headed gull's adaptation to polluted environments: the role of the mixed-function oxidase detoxification system. *Env. Cons.* **15**: 221–224.

Fox, G. A. and T. Donald. 1980. Organochlorine pollutants, nest-defense behavior and reproductive success in merlins. *Condor*, **82**: 81–84.

Gilman, A. P., G. A. Fox, T. R. Peakall and G. T. Haymes. 1977. Reproductive parameters and contaminants levels of Great Lakes herring gulls. *Jour. Wildl. Management* **41** (3): 456–468.

Hutton, M. 1981. Accumulation of heavy metals and selenium in three seabird species from the United Kingdom. *Environ. Pollut. Ser. A* **26**: 129–145.

Keplinger, M. L. *et al.* 1972. Quoted by Fishbein, 1974.

Knight, G. C. and C. H. Walker. 1982. A study of the hepatic microsomal monooxygenase of seabirds and its relationship to organochlorine pollutants. *Comp. Biochem. Physiol.* **73C**: 211–221.

Knight, G. C., C. H. Walker, D. C. Cabot and M. P. Harris. 1981. The activity of two hepatic microsomal enzymes in seabirds. *Comp. Biochem. Physiol.* **68C**: 127–132.

Kuroki, H. and Y. Masuda. 1977. Structure and concentrations of the main components of polychlorinated biphenyls; correlations in marine mammals. *Nature, Lond.* **8**: 469–474.

Leonzio, C., C. Fossi and S. Focardi. 1986. Heavy metals and selenium variation in a migratory bird wintering in a mercury polluted lagoon. *Bull. Environ. Contam. Toxicol.* **37**: 219–225.

Lindberg, P. 1975. The peregrine falcon in Sweden. World Conference of Birds of Prey. *Proceedings*, Vienna, pp. 329–338.

Matthews, H. B. and D. B. Tuey. 1980. The effect of chlorine position on the distribution and the excretion of four hexachlorobiphenyl isomers. *Toxicol. Appl. Pharmacol.* **53**: 377–388.

Mendola, J. T., R. W. Risebrough and J. Blondel. 1977. Contamination de l'avifaune camarguaise par des residues organochlorés. *Environ. Poll.* **13**: 21–23.

Mitchev, T. and N. Dilchev. 1986. Chlororganic pesticides in eggs of some bird species from Biosphere Reserve 'Srebarna'. In *Proceedings of the National Conference on Ecological and Genetical Problems of the Pollution of the Environment*, June, pp. 235–238 (in Bulgarian).

Moore, N. W. and C. H. Walker. 1964. Organochlorine insecticide residues in wild birds. *Nature* **201**: 1072–1073.

Ohlendorf, H. M., E. E. Klaas and T. E. Kaiser. 1978. Environmental pollutants and eggshell thinning in the black-crowned night heron. In A. Sprunt, J. C. Ogden and S. Winckler (eds.), *Wading Birds.* National Audubon Society Report No. 7, pp. 63–82.

Peakall, D. B. and M. L. Peakall. 1973. Effect of polychlorinated biphenyl on the reproduction of artificially and naturally incubated dove eggs. *J. Appl. Ecol.* **10**: 863–868.

Renzoni, A., M. Bernhard, R. Sarà and M. Stoeppler. 1978. Comparison between the Hg body burden of *Thunnus thynnus* from the Mediterranean and the Atlantic. C.I.E.S.M. – Antalya, pp. 255–260.

Renzoni, A., S. Focardi, C. Leonzio, and A. Mocci Demartis. 1982. Contaminants in residents and migratory birds of the Mediterranean Sea. *Thalassia Jugosl.* **18**: 245–252.

Renzoni, A., S. Focardi, C. Fossi, C. Leonzio and J. Mayol. 1986. Comparison between concentration of mercury and other contaminants in eggs and tissues of adults of Cory's shearwater (*Calonectris diomedea*) collected on Atlantic and Mediterranean islands. *Environ. Poll. Series A* **40**: 17–35.

Safe, S. 1980. Metabolism, uptake, storage and bioaccumulation. In R. D. Kimbrough (ed.), *Halogenated Biphenyls, Terphenyls, Naphthalenes, Dibenzodioxins and Related Products*, pp. 81–107. Amsterdam, Elsevier/North Holland.

Safe, S., S. Bandiera, T. Sawyer, L. Robertson, L. Safe, T. Parkinson, O. E. Thomas, D. E. Ryan, L. M. Reik, W. Levin, M. A. Demonne and T. Fujita. 1985. PCBs: structure-function relationship and mechanism of action. *Environ. Health Perspect.* **60**: 47–56.

Scott, M. L. D., D. V. Vadehra, P. A. Mullenhoff, G. L. Rumsey and R. W. Rice. 1971. Results of experiments on the effects of PCBs on laying hen performance. *Proc. Cornell Nutr. Conf.* pp. 56–64.

Scott, M. L., J. R. Zimmerman, S. Marinsky, P. A. Mullenhoff, G. L. Rumsey and R. W. Rice. 1975. Effects of PCBs, DDT and mercury compounds upon egg production, hatchability and shell quality in chickens and Japanese quail. *Poul. Sci.* **54**: 350–368.

Vannucchi, C., S. Sivieri and M. Ceccanti. 1978. Residues of chlorinated naphthalenes, other hydrocarbons and toxic metals (Hg, Pb, Cd) in tissues of Mediterranean seagull. *Chemosphere* **6**: 483–490.

Vermeer, R., F. A. J. Armstrong and R. M. Hatch. 1973. Mercury in aquatic birds at Clay Lake, Western Ontario. *Jour. Wildlife Management* **37** (1): 58–61.

Viviani, R., G. Crisetig, P. Cortesi and E. Carpené. 1974. Résidues de polychlorobiphenils (PCB) et de pesticides chlorés dans les poissons et les oiseaux du delta du Po. *Rev. Int. Oceanogr. Méd.* **35/36**: 79–90.

Walker, C. H. 1980. Species variations in some hepatic microsomal enzymes that metabolize xenobiotics. *Prog. Drug. Metabol.* **5**: 113–164.

Wolfe, M. S., J. Thorton, A. Fishbein, R. Likis and I. J. Selikoff. 1982. Disposition of polychlorinated biphenyl congeners in occupationally exposed persons. *Toxicol. Appl. Pharmac.* **62**: 294–306.

CHAPTER 10

The Ecotoxicology of Persistent Pollutants in Marine Fish-eating Birds

C. H. WALKER

Department of Biochemistry and Physiology, School of Animal and Microbial Sciences, University of Reading, Whiteknights, P.O. Box 228, Reading RG6 2AJ

INTRODUCTION

This chapter will be largely concerned with fish-eating seabirds belonging to the orders Procellariformes, Pelecaniformes and Charadriformes which occupy positions at the top of marine food chains. The species of particular interest are specialized predators of the families Alcidae, Phalacrocoracidae, Procellaridae, Hydrobatidae and Salidae. Only limited attention will be given to omnivorous gulls which sometimes take fish, e.g. herring gull (*Larus argentatus*) and black-headed gull (*Larus ridibundus*).

Like other predators, these species tend to bioaccumulate relatively high levels of persistent lipophilic pollutants which undergo biomagnification during their passage through marine food chains. It has been frequently observed that the highest levels of pollutants such as dieldrin (HEOD, 1,2,3,4,10,10 hexachloro-1,3,3a,5,6,7,8,8a-octahydro-6,7-epoxy-exo-1,4-endo 5,8-dimethanonaphthalene), DDE (*pp'*dichlorodiphenyltrichloro-ethylene) and polychlorinated biphenyl (PCB) isomers are found in predators, rather than in organisms below them in the food chain (Moore and Walker, 1964; Robinson *et al.*, 1967; Walker, 1990).

Predatory birds often contain a very wide range of persistent compounds including polychlorinated dibenzo-*p*-dioxins (PCDDs), polychlorinated dibenzofurans (PCDFs) and organomercury compounds, in addition to the compounds mentioned above. This strong tendency towards bioaccumulation makes them valuable indicator species, which can provide an index of the levels of contamination in environmental monitoring. In this context the sampling of eggs is a particularly useful strategy—it is relatively easy to accomplish, can be carried out in a random fashion, ensuring representative sampling and, if carried out with care, need not have any harmful effects on the population. These species are seen to be particularly at risk because of the quantity and variety of the potentially toxic pollutants that they contain—but also because of their vulnerability to toxic compounds on account of their relatively small numbers and their slow reproductive rates.

The ensuing account will consider first the levels of some of the more important lipophilic organochlorine pollutants in seabirds, before dealing with questions of their metabolism and toxic effects.

LEVELS OF PERSISTENT ORGANOCHLORINE POLLUTANTS IN SEABIRDS

Levels of persistent organochlorine pollutants were measured in tissues such as liver, fat, muscle, brain and in eggs of seabirds in a number of studies conducted in different areas over the past 27 years. The levels found were related to species, geographical location and year (Walker, 1990). These studies have been very useful for the environmental monitoring of marine pollutants. Some of these data are summarized in Table 10.1. The pattern of these residues will first be discussed before going on to consider the explanation of these residues and their possible toxicological significance.

DDE, Dieldrin and Total PCBs

*pp'*DDE, a persistent metabolite of *pp'*DDT, was found in all the eggs and tissues analysed. With the shag (*Phalacrocorax aristotelis*) and/or cormorant (*Phalacrocorax carbo*), egg analyses were carried out for a number of years, from a number of different locations in British and Irish coastal sites, sampled between 1963 and 1983 and egg residues were below 4 μg/g in all cases. Considerably higher levels were reported from the Mediterranean (6.4–7.1 μg/g) and the Danube Delta (19–57 μg/g) for the shag and cormorant, respectively. Levels of 7.3–17 μg/g DDE were found in cormorant livers sampled in The Netherlands in 1970 (these birds contained very high levels of total PCB). In common tern (*Sterna hirundo*) eggs, higher DDE levels were found in samples from the Mediterranean taken in 1980–2, than in samples taken in Great Britain and The Netherlands in 1964–6. *pp'*DDE levels were also relatively high in the eggs of other species from the Mediterranean, viz. < 5.1 μg/g in little tern (*Sterna albifrons*) and < 8.8 μg/g in Cory's

TABLE 10.1 *Levels of Persistent Pollutants in Seabirds from Different Areas* (pp'*DDE, dieldrin and total PCBs*)

Species	Material	Location	Year	μg/g in term of wet weight *pp*'DDE	Dieldrin	PCBs
Shag (*Phalacrocorax aristotelis*)	Eggs	G.B. (E. Coast)[a]	1963–71	0.96–3.7	0.18–2.9	3
	Eggs	Ireland (S. Coast)[a]	1964	0.3–0.7	0.1–0.5	
	Eggs	Mediterranean[c]	–	6.4–7.1		24–26
	Liver	Mediterranean[c]		0.98–1.2		23–26
Cormorant (*Phalacrocorax carbo*)	Eggs	G.B. (E. Coast)[b]	1964	1.5–3.8	0.3–0.8	0.3–1.2
	Liver	G.B. (E. Coast)[b]	1965/6	4.1	0.19	0.9–2.8
	Eggs	Ireland (S. Coast)[f]	1983	0.080–0.088	0.024–0.031	
	Liver	Netherlands	1970	7.3–17	0.76–2.2	250–340
	Eggs	Romania[d] (Danube Delta)		19–57		13–24
	Liver	Mediterranean[c]	–	0.31–2.8		1.1–6.2
Razorbill (*Alca torda*)	Eggs	G.B. (E. Coast)[a]	1963–7	0.64–2.9	0.35–1.6	6
	Eggs	G.B. (W. Coast)[a]	1967	1.1	0.5	7
	Liver	G.B. (W. Coast)[c]	1969	8.4	< 1	26
Guillemot (*Uria aalge*)	Eggs	G.B. (E. Coast)[a]	1963–7	0.7–4.0	0.1–2.0	3
	Eggs	G.B. (W. Coast)[a]	1967	2.0	0.3	8
	Liver	Irish Sea[g]	1969	10	0.4	111
	Eggs	Ireland (S. Coast)[a]	1964	1.3-2.1	< 0.2	
Gannet (*Sula bassanus*)	Liver	G.B. (E. Coast)[b]	1965–6	1.4	0.35	
	Liver	G.B. (W. Coast)[e]	1972			200
	Eggs	Canada[h]	1968–84	1.1–3.4	0.15–0.17	9.5–41

TABLE 10.1 *Continued*

Species	Material	Location	Year	µg/g in term of wet weight *pp'*DDE	Dieldrin	PCBs
Puffin (*Fratercula arctica*)	Eggs	Canada[h]	1968–84	0.5–2.6	1.0–8.5	
Common tern (*Sterna hirundo*)	Eggs	G.B. (E. Coast)[a]	1964	0.2–0.7	0.1–0.5	
	Eggs	Netherlands[c]	1965–6	0.44–0.55	0.29–0.70	
	Eggs	Italy[c]	1980–2	0.80–6.0		9.7–25
Sandwich tern (*Sterna sandvicensis*)	Eggs	G.B. (E. Coast)	1963-5	0.24–1.5	0.15–1.5	
	Eggs	Netherlands[i]	1965–6	0.61	0.23–0.80	
	Eggs	Ireland (S. Coast)	1965	0.64	0.12	
Little tern (*Sterna albifrons*)	Eggs	Mediterranean[c]	1980–2	0.45–5.09		2.2–7.8
Cory's shearwater (*Calomectris diomedea*)	Eggs	Mediterranean[c]	1982–4	0.22–18.8		0.48–95.1

[a] Moore and Tatton (1965). [b] Robinson *et al.* (1967). [c] Lambertini and Leonzio (1986).
[d] Fossi *et al.* (1984). [e]Parslow *et al.* (1973). [f] Earley (1987 and personal communication).
[g] NERC (1971 and 1983). [h] Elliott *et al.* (1988a). [i] Koeman *et al.* (1967).
PCBs = polychlorinated biphenyls; DDE = *pp'*dichlorodiphenyltrichloro-ethylene.

shearwater (*Calomectris diomedea*). High levels of *pp*′DDE were reported in the eggs of gannet (*Sula bassanus*) sampled in Canada (ref. Chapter 8 of this book). In general, higher levels of DDE were reported from Mediterranean sites, the Danube Delta, and from one Canadian site than were found in Great Britain and Ireland.

Dieldrin residues were generally lower than *pp*′DDE residues. In several samples of cormorant and shag eggs from Britain and Ireland, residue levels were found not to exceed 3 ppm. However, this is not wholly reassuring, since dieldrin is a much more toxic compound than *pp*′DDE, tissue levels in excess of 10 μg/g being regarded as high enough to cause lethal poisoning (Koeman and van Genderen, 1972).

Total PCB residues in eggs and tissues were extremely variable. Particularly high levels were found in two cases where poisoning by these compounds was suspected in the field. Thus liver residues of 254–340 μg/g were detected in cormorants found dead in the Netherlands in 1970 and a mean level of 111 μg/g was reported for guillemots (*Uria aalge*) found dead in a seabird kill in the Irish Sea in 1969. These two incidents are discussed later.

Structural Features of Molecules Which Undergo Strong Bioaccumulation

PCB Congeners

Studies on herring gulls in the Great Lakes (Norstrom, 1988) reported the occurrence of a distinctive pattern of some twenty persistent PCB congeners, accounting for 80% or more of the total PCB residues (see Fig. 10.1). These congeners were found to be at relatively high levels compared to another group of isomers (Table 10.2). Compared to levels of congeners in fish (*Alosa pseudoharengus*, the alewife), upon which the birds feed, the first group were bioaccumulated to a greater extent than other PCB congeners. Borlakoglu *et al.* (1988) reported a similar picture for cormorant, shag, razorbill (*Alca torda*), guillemot (*Uria aalge*) and puffin (*Fratercula arctica*) from British and Irish coastal sites collected during 1978–80. An "enrichment index" was calculated which compared the abundance (% of total PCBs) of individual congeners in the avian fat sample with abundance in the Aroclor 1260 mixture, a product in which higher chlorinated compounds are well represented. Thus, figures above the value of 1 indicate relatively strong persistence. In both studies the group of congeners that bioaccumulated strongly were characterized by a particular structural feature, namely the absence of unsubstituted adjacent *meta* (*m*) and *para* (*p*) positions on the biphenyl ring. By contrast, the less persistent compounds

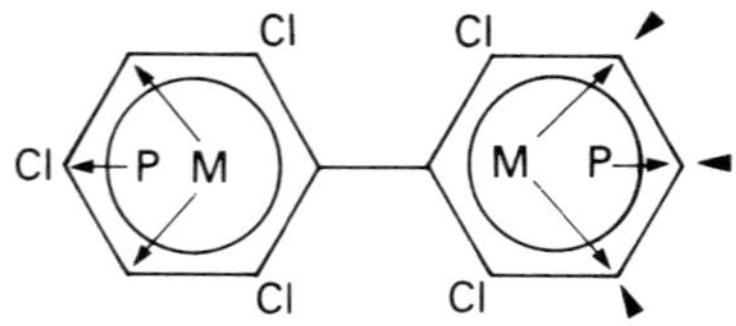

FIG. 10.1 Metabolism of PCB isomers by hepatic microsomal monoxygenase. Oxidative attack readily occurs in areas of biphenyl ring where there are unsubstituted adjacent *meta* (*m*) and *para* (*p*) positions. Thus, in the example given, 2,2,4,6,6-pentachlorobiphenyl can undergo attack in the *meta* and *para* positions of the right-hand ring indicated by arrows, to yield phenolic metabolites.

each possessed unsubstituted *m* and *p* positions, strongly suggesting that this structural feature is necessary for effective oxidative detoxication to occur.

In a study of PCB congeners in eggs of Cory's shearwater collected from the Mediterranean and Atlantic, Renzoni *et al.* (1986) reported a similar pattern of persistent congeners to those mentioned above.

Polychlorinated Dibenzo-p-dioxins (PCDDs) and Polychlorinated Dibenzofurans (PCDFs)

Livers from cormorants sampled in The Netherlands during 1980–82 were analysed for PCDDs and PCDFs (Van den Berg *et al.*, 1987). All the congeners found had chlorine substitution in the 2, 3, 7 and 8 positions (Fig. 10.2). The following residues were found: 2,3,4,7,8-pentachlorodibenzofuran (2,3,4,7,8-PnCDF), 1,2,3, 6,7,8-hexachlorodibenzo-*p*-dioxin (1,2,3,6,7, 8-HxCDD), 2,3,7,8-tetrachlorodibenzo-*p*-dioxin (2,3,7,8-TCDD) and 1,2,3,7,8-pentachlorodibenzo-*p*-dioxin (1,2,3,7,8-PnCDD). These were also present in the eels, which are the major food species for cormorants in the study area. Other higher chlorinated PCDDs and PCDFs were found in cormorants but not in the samples of eels that were analysed. By comparing residue levels in cormorant livers with those in eels, apparent bioaccumulation factors were estimated. These were 12, 340 and 30 for 2,3,7,8-TCDD,2,3,4,7,8-Pn CDF and 1,2,3,6,7,8-HxCDD, respectively. It appears that these compounds with chloro substitution in the 2,3,7 and 8 positions of the ring system are resistant to monooxygenase attack, as are the PCBs which lack free adjacent *m* and *p* positions on the biphenyl ring (see last section) and that cormorants metabolize them very slowly or not at all. The relatively high levels of these compounds in cormorants could not have occurred if *in vivo* metabolism had been rapid. Van den Berg *et al.* estimated a bioaccumulation factor for 2,3,4,7,8-Pn CDF, using a one-compartment model incorporating a

TABLE 10.2 *Bioaccumulation of PCB Isomers by Seabirds*

IUPAC No.	Isomer chlorine substitution Ring 1	Ring 2	Number of unsubstituted adjacent carbons *o'm*	*mp*	Relative[a] BF in herring gull	Enrichment[b] index seabirds
74	4	2,4,5	2	0	0.86	
66	2,4	3,4	2	0	0.43	
99[c]	2,4	2,4,5	1	0	0.74	
114	4	2,3,4,5	2	0	<0.1	
118[c]	3,4	2,4,5	1	0	0.83	23.6
146	2,3,5	2,4,5	0	0	1.00	
153[c]	2,4,5	2,4,5	0	0	1.00[c]	1.44
105[c]	3,4	2,3,4	2	0	0.57	
138[c]	2,3,4	2,4,	0	0	0.93	3.1
178	2,3,5	2,3,4,6,6	0	0	0.52	
187[c]	2,4,5	2,3,5,6	0	0	0.94	
128[c]	2,3,4	2,3,4	2	0	0.77	
177	2,3,4	2,3,5,6	1	0	0.59	
180[c]	2,4,5	2,3,4,5	0	0	1.10	1.28
170[c]	2,3,4	2,3,4,5	1	0	1.05	1.35
201	2,3,5,6	2,3,4,5	0	0	0.98	
196[c]	2,3,4,5	2,3,4,6	0	0	1.07	
194[c]	2,3,4,4	2,3,4,5	0	0	0.90	
202	2,3,5,6	2,3,5,6	0	0		3.52
52	2,5	2,5	0	2	<0.1	N.D.
70	2,5	3,4	1	1	0.02	
101	2,5	2,4,5	0	1	0.24	0.40
97	2,3	2,4,5	1	1	<0.1	0.20
87	2,5	2,3,4	1	1	0.12	
110	3,4	2,3,6	1	1	0.21	
151	2,5	2,3,5,6	0	1	0.02	0.10
149	2,3,6	2,4,5	0	1	0.18	0.17
141	2,5	2,3,4,5	0	1	0.19	
174	2,3,6	2,3,4,5	0	1	0.07	

[a] Bioaccumulation factor (BF) $\frac{\text{herring gull (\textit{Larus argentatus}) eggs}}{\text{Alewife (\textit{Alosa pseudoharengus}).}}$ relative to PCB No. 153, Norstrom (1988)

[b] "Enrichment index" is $\frac{\text{PCB as a \% of total PCB in seabird depot fat.}}{\text{PCB as a \% of total PCB in Aroclor 1260.}}$

[c] These were amongst the 14 PCB congeners found at the highest concentrations in eggs from the Mediterranean and Atlantic (Renzoni *et al.*, 1986). These authors also found Nos. 156, 172 and 183, the latter two of which were also reported by Borlakoglu *et al.* (1988).

half-life for another compound (2,3,7,8-TCDD) obtained for the rat. The estimated factor was reasonably close to that calculated from field data. A similar pattern of persistent PCDDs and PCDFs has been reported by Elliott *et al.* (1988b) for blue herons (*Ardea herodias*) in British Columbia.

Other Persistent Pollutants

Most other compounds shown to bioaccumulate to relatively high concentrations in birds at or near the top of the food chain are

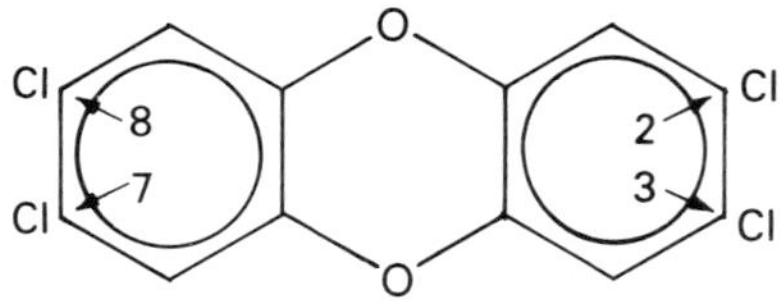

FIG. 10.2 Structure of 2,3,7,8-tetrachlorodibenzodioxin (TCDD).

also highly halogenated (in most cases chlorinated). In general, the substitution of chlorine for hydrogen in organic molecules creates an effective barrier to oxidative metabolism by MOs. The persistent pollutants dieldrin and *pp*-DDE which have been widely reported in fish-eating seabirds (see earlier discussion) have this feature. Other persistent pollutants which have been reported less often in the literature, mainly because they have not been looked for, also fall into this category. For example, the following organochlorine compounds have been detected in the eggs of seabirds: hexachlorobenzene, *cis*-chlordane, oxychlordane, mirex, photomirex and nonachlor isomers (Elliott *et al.*, 1988a and Chapter 8 of this book).

THE METABOLISM OF LIPOPHILIC XENOBIOTICS BY SEABIRDS

Fish-eating seabirds, like terrestrial mammals and birds, do not have an effective mechanism for eliminating absorbed lipophilic xenobiotics in their unchanged forms. In this respect they differ from aquatic organisms such as fish or molluscs, which are able to "lose" such substances to ambient water by passive diffusion (Walker, 1987; see also Chapter 1). There is, consequently, a risk that lipophilic xenobiotics present in the diet will be bioaccumulated to toxic levels over a period of time, unless there is some other mechanism for eliminating them. The effective elimination of such substances by terrestrial vertebrates is usually accomplished by metabolizing them to water-soluble products (especially conjugates) which are readily excreted in bile or urine. Thus, when considering questions of strong bioaccumulation, such as have been reported earlier in this chapter, there is a strong implication that metabolic detoxification has failed. This suggests a deficiency in an enzyme or enzymes which can metabolize the xenobiotics to water-soluble products. The organochlorine compounds mentioned here, which tend to undergo strong bioaccumulation as they pass from fish to fish-eating birds, are not readily metabolized.

The principal enzyme system which can metabolize them is microsomal monooxygenase, which is particularly well represented in liver. Even here metabolism of these highly halogenated compounds is often very slow. Since these compounds have only appeared in the environment relatively recently, it is not surprising that enzymes capable of efficiently metabolizing them have yet to evolve. In the case of dieldrin, *pp'*DDE and various PCB isomers, monooxygenase attack leads to the formation of hydroxylated metabolites, (alcohols or phenols, see Fig 10.1). There is no evidence, as yet, of any other enzyme systems which are of major importance in the metabolism of organochlorine compounds. However, there are reports that mice can convert 3,3′,4,4′-tetrachloro biphenyl to a methyl sulphoxyl derivative (Darnerud *et al.*, 1986). It is not yet known whether this enzymic conversion occurs in seabirds.

Hepatic Microsomal Monooxygenase Activities of Seabirds

The activity of hepatic microsomal monooxygenases (HMOs) has been measured, using different substrates in the following species—cormorant, double-crested cormorant (*Phalacrocorax auritus*), shag, guillemot, razorbill, Manx shearwater (*Puffinus puffinus*) and puffin (*Fratercula arctica*) (Walker, 1990).

The substrates aldrin, aminopyrine, HCE and aniline measured activities attributable to metabolism by a range of cytochrome P-450 forms. With the exception of the puffin, all species showed HMO activities substantially lower than those of mammals of similar body size. The activities were also low in comparison to those found for most omnivorous or herbivorous birds (Walker, 1980, 1990).

The differences in activity between species and groups may be compared by expressing all activities (relative to that of the male rat, which is ascribed an arbitrary value of 1, making a correction for differences between species in relative liver weight) (Fig. 10.3, Walker, 1980; Ronis and Walker, 1989). As noted elsewhere, mammals show a negative correlation between log relative activity and log body weight (Walker, 1980). The same correlation is found for fish-eating birds other than the puffin, except that the activities are only about 7% of those of mammals of the same body size. The activities for omnivorous and herbivorous birds fall between these two lines. It has been argued that the low activities of HMO in fish-eating birds and other specialized predators may reflect a limited requirement for detoxification of lipophilic xenobiotics (Walker, 1980). Why the puffin should, in this respect, differ from seven other species of fish-eating birds, including two species of auks (*Alcidae*), is not clear.

The relative HMO activities of razorbills, puffins and cormorants from different countries can be compared. Puffins had the highest activities of the species investigated both in Canada and in Britain and Ireland, although activities were somewhat higher in Britain and

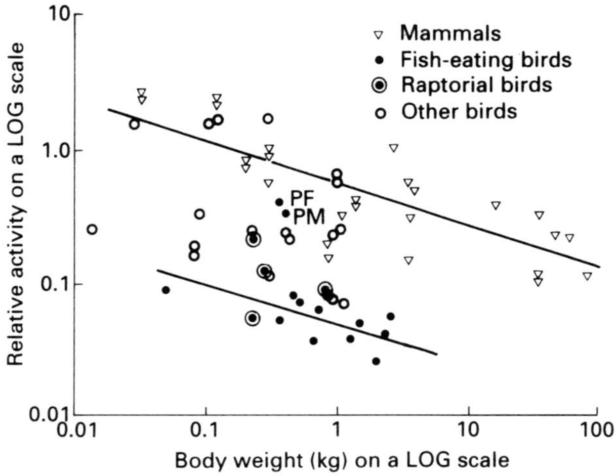

FIG. 10.3 Relationship between body weight and hepatic monooxygenase activity. Values for mammals show a regression coefficient of −0.378 and a correlation coefficient of −0.808. Values for fish-eating seabirds, with the exception of the puffin which has unusually high values, show a regression coefficient of −0.396 with a correlation coefficient of −0.76 ($p>99.8$) (after Ronis and Walker, 1989). PF = female puffin; PM = male puffin. Other seabirds represented: shag, cormorant, double-crested cormorant, Manx shearwater, Leach's petrel, razorbill and guillemot. Males and females plotted separately for all species except for Manx shearwater.

Ireland. Guillemots from Canada showed higher activity than did those from Britain and Ireland. Cormorants from Ireland and Italy (C. Fossi, personal communication) and double-crested cormorants from the USA all showed low activities, similar to those found in Britain.

There was considerable variation of activity within species. This was particularly the case with razorbills and puffins from Britain and Ireland which showed a difference of about 100-fold between the highest and lowest activities in the sample. When plotted on frequency diagrams, the activities for razorbills, guillemots, and puffins were resolved into two or more peaks (Fig. 10.4. Knight and Walker, 1982a). By contrast, shags showed a narrower range of activities which appeared to represent a single peak on a frequency diagram. In the case of razorbills the highest activities were found in females with large ovaries early in the breeding season, but otherwise no clear temporal variations were found. However, it should be emphasized that (in all of these studies) the sample sizes were fairly small, and were limited to the breeding season. Thus the opportunity for identifying temporal variation was limited, and any major trends outside of the breeding season would not have been measured. Possible reasons for the polymorphism, in regard to HMO activities in certain seabird populations, will be discussed later.

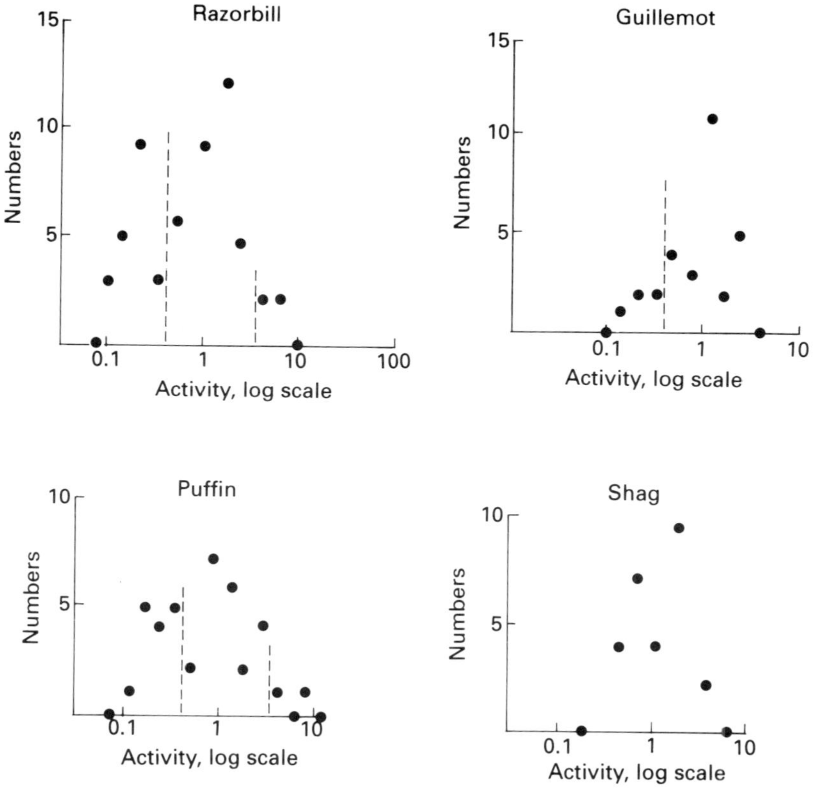

FIG. 10.4 Frequency diagrams showing distribution of HMO activities in populations of different species of fish-eating seabirds. Each of the points represents the number of individuals having HMO activities within a particular range. The activities were measured using the organochlorine substrates aldrin (1,2,3,4,10,10-hexachloro-1,4,4a,5,8,8a in hexahydro-exo-1,4-endo-5,8-dimenthanonaphthalene) and HCE (1,2,3,4,9,9-hexachloro-1,4,4a,5,6,7,8,8a-octahydro-exo-7,8-epoxy-1,4-methanonaphthalene), and were expressed on a relative scale (activity in male rat = 1). Males and females are represented in all species and no distinction is made between them. The range of activities is very wide for the razorbill, guillemot, and puffin, the highest values being 40–100 fold greater than the lowest values, and the distribution of them appears to be bimodal or trimodal. The values for the shag are less variable, highest value 10–fold greater than the lowest values, and appear to be arranged into a single peak. The dotted lines indicate boundaries between peaks (after Knight and Walker, 1982a).

The activity of microsomal epoxide hydrolase was also studied in the cormorant, shag, guillemot, razorbill, puffin and Manx shearwater from British and Irish coastal sites (Knight and Walker, 1982b). Activity towards the dieldrin analogue HEOM (HEOM = 1,2,3,4,9,9-hexachloro-1,4,4a,5,6,7,8,8a-octahydro-hydro-6,7-epoxy-1,4-methanonaphthalene) was found to be comparable to that of other birds, although lower than that of most mammals. In female puffins, there were indications that some of this activity occurs in mitochondria. This enzyme may be important in detoxifying carcinogenic epoxides generated by the microsomal monooxygenase system.

Induction of HMOs

Different isozymes of the haem protein cytochrome P-450 constitute the catalytic centres of hepatic microsomal monooxygenases. Certain of these cytochrome P-450s are inducible by lipophilic xenobiotics, including some pollutants. Cytochrome P-450s belonging to gene families P-450 IA1 and P-450 IA2 (according to the classification of Nebert and Gonzalez, 1987) have been found to be inducible by aromatic hydrocarbons such as 3-methyl cholanthrene, by coplanar PCBs and by dioxins in mammals, birds and fish (Livingstone and Stegeman, 1989). Cytochrome P-450s belonging to one IIB subfamily are inducible by phenobarbitone in mammalian species, but this induction has not been demonstrated in fish, and birds are less readily induced than mammals.

There have been a few studies of HMO induction in seabirds. In one study, immature puffins (approximately 40 days old) were dosed with DDE for 16–21 days before sacrifice (Bend *et al.*, 1977). Significant increases in aniline hydroxylase activity (*c*. 3.5-fold) and benzphetamine demethylase activity (*c* 5-fold) were observed, clearly indicating that induction had occurred. The birds were subject to prefledgling starvation and consequently mobilized stored DDE. The mean DDE content of livers was 2037 ppm, and toxic effects were seen.

In another study cormorants were dosed with phenobarbitone until anaesthesia. On studying their liver enzymes, no evidence was found of induction of the HMO system. There was no increase in either aldrin epoxidation or HCE hydroxylation (Ronis and Walker, 1989) or of the cytochrome P-450 content of microsomes. When solubilized microsomes were studied by Western immunoblotting, no cytochrome P-450s were found which cross-reacted with antibodies raised to phenobarbital-inducible cytochrome P-450s of the rat. Thus phenobarbitone did not appear to cause any induction of cytochrome P-450 in this species. This is in marked contrast to the situation in mammals, where phenobarbital readily induces upwards of four different cytochrome P-450 forms.

Liver microsomes of six species of fish-eating seabirds collected from British and Irish coastal sites were examined by Western blotting (Ronis *et al.*, 1989). The species studied were shag, cormorant, razorbill, guillemot, puffin and Manx shearwater. Certain individuals of each species were found to contain a cytochrome P-450 IA1; in other individuals this protein was not found. There was a general relationship between the levels of this form of cytochrome P-450 and the concentration of PCBs in the birds. This cytochrome P-450 may have been induced in certain individuals by marine pollutants.

Loss of PCBs by *In Vivo* Metabolism

Five cormorants captured in The Netherlands were maintained in captivity and dosed with the technical PCB mixture Clophen A60. The birds were dosed continuously with PCB mixture for periods of up to 124 days. All birds eventually died and the remaining PCB residues in brain, liver, fat and total body were measured. The total residues recovered ranged from 49 to 79% of the dose given, pointing to the remarkable persistence in this species of certain PCB congeners (Koeman *et al.*, 1973).

In a later study with cormorants the birds were given intraperitoneal injections of dieldrin (HEOD). Samples of blood and of whole carcasses were analysed at different times after dosing. The loss of half the adminstered dose occurred within 20–30 days. The initial rates of loss from blood or from whole carcasses were only 10–16% of those reported for male rats having similar tissue levels of the chemical (Walker *et al.*, 1984).

A novel approach to the study of *in vivo* metabolism in seabirds is described in an investigation of the fate of PCBs in wild puffins (Harris and Osborn, 1981). Birds were dosed in the field by implanting 30–35 mg of the PCB mixture Aroclor 1254 under the skin. The birds were released and some of them were subsequently recaptured and analysed. Mean residues (ppm/wet weight) in depot fat at different times after dosing were 408 (1–3.5 months), 342 (9–12 months), 202 (16 months) and 99 (34 months): 5–9 birds were represented in each group. Control birds samples at 34 months contained 47 ppm PCB in depot fat. In broad terms, this suggests a halving of the PCB concentration in fat over a period of about 14 months, and a further halving over the next 18 months. At 34 months the PCB levels in fat were about two times greater in treated birds than in controls. As with the study upon cormorants referred to earlier, there is evidence here of the remarkable persistence of certain congeners of PCB mixtures. Interestingly, there was also evidence of considerable interindividual variation in residues in the birds sampled at 2 and 16 months after dosing (a 4.3-fold difference between highest and lowest in fat, and a larger difference in liver). The substantial differences in monooxygenase activity between individual puffins may have contributed to this variability by causing differential rates of metabolism of biodegradable PCB congeners.

Metabolic Detoxification and Activation

Very low HMO activities relative to those in mammals were found in seven of the eight species of fish-eating seabirds represented in Fig. 10.1. A number of these species are known to strongly bioaccumulate persistent organochlorine pollutants as was discussed in the previous section. It seems clear that these high levels of bioaccumulation are a consequence of slow metabolism; slow metabolism is the expected consequence of the low HMO activities. This point is illustrated by a comparison of dieldrin metabolism *in vivo* in male rats with that in cormorants. Rats, which have much higher HMO than cormorants, lose dieldrin six times more rapidly than the latter species (Walker *et al.*, 1984).

There is evidence that predators, in general, tend to have low HMO activities, and this suggests that they, as a group, are liable to strongly bioaccumulate persistent lipophilic compounds which are difficult to metabolize (Walker *et al.*, 1984). The low levels of HMO activities in predators may be a reflection of a low requirement for the oxidative detoxication of lipophilic xenobiotics in their food in comparison to omnivores and herbivores (Walker, 1980; Ronis and Walker, 1989).

Slow metabolism can reflect the low level of activity of enzymes, or the resistance of certain chemical structures to metabolism, or both of these things. In general, highly halogenated compounds, which do not contain functional groups (e.g. ester, keto, nitro- or amino-), are metabolized only slowly by HMO, the principal enzyme system of vertebrates which can catalyse their biotransformation. [For further information on enzymic metabolism of functional groups see Jakoby (1980).] In the case of halogenated aromatic compounds, it is now becoming clear that the ease with which oxidative metabolism occurs follows some structural rules. As explained earlier, PCBs which have unsubstituted *m/p* positions tend to be metabolized relatively rapidly, and consequently do not readily bioaccumulate. On the other hand, PCBs which do not have such a vacant position are only metabolized slowly and tend to reach high concentrations in fish-eating seabirds (Borlakoglu *et al.*, 1988, 1990a, b, c; Norstrom, 1988). A similar case has been presented for selective bioaccumulation by fish (Chapter 6 and Boon *et al.*, 1987). With PCDDs and PCDFs, the chlorination of ring positions 2, 3, 7 and 8 is associated with marked persistence, which is presumably due to slow metabolism (Van den Berg *et al.*, 1987). Thus, a picture is emerging of the particular structures of organochlorine compounds which are resistant to oxidative metabolism.

Metabolism of organochlorine compounds by HMOs usually leads to detoxication, because the phenols and diols produced are readily excreted as conjugates or, perhaps, in their original forms. On the other hand, metabolism may generate active products, e.g. epoxides, which are able to interact with cellular nucleophiles. It is not yet clear whether PCB isomers are activated by this mechanism. It is interesting, in this context, that cytochrome P-450 IA1, which appears to be induced by pollutants in seabirds, has an important role in the activation of carcinogens such as benzo(a)pyrene (Ioannides *et al.*, 1984).

In a recent study, it has been shown that the 4-hydroxy metabolite of 3,4,3′,4′-tetrachlorobiphenyl is responsible for a disturbance of thyroid hormone and vitamin A metabolism (Brouwer *et al.*, 1990). This toxic effect is attributed to the binding of the metabolite to transthyretin. This binding is in competition with the binding of L-thyroxin. This toxic mechanism, which has been demonstrated in the common seal (*Phoca vitilina*), is currently being investigated in the cormorant (A. Brouwer, personal communication). The conversion of 3,4,3′,4′-tetrachlorobiphenyl to its hydroxy metabolite may be accomplished by cytochrome P-450 IA1. This cytochrome P450 is inducible by 3,4,3′,4′-tetrachlorobiphenyl.

EFFECTS OF POLLUTANTS UPON SEABIRDS IN THE ENVIRONMENT

The presence of a wide range of organochlorine compounds in seabirds, sometimes at substantial levels, raises questions about their effects. Such questions are often difficult to answer because of the range of compounds present and the possibility that there may be interactive effects. In the following account lethal and sublethal effects will be considered separately.

Lethal Effects

During 1964 and 1965 a number of Sandwich terns died with typical symptoms of organochlorine poisoning in the area of the Dutch Wadden Sea (Koeman *et al.*, 1967). Upon analysis, certain birds were found to contain sufficiently high levels of telodrin and/or dieldrin and/or endrin to have had lethal effects (Koeman and van Genderen, 1972). The principal source of these residues was identified as a chemical plant at Rotterdam. Subsequently, measures were taken by the company concerned to control the release of these chemicals, and environmental residues fell.

The poisoning of birds was observed to have occurred mainly at two stages of development: first, in chicks shortly after hatching, second in juvenile birds when they started flying. Birds that died showed spasms and convulsions, typical symptoms of poisoning by these chlorinated cyclodiene insecticides. In chicks, it appeared that lethal effects followed the mobilization of the persistent insecticides from the lipids of the yolk sac. The juvenile birds which died had very low lipid levels in relation to control birds which were shot. This suggested that mobilization of insecticides from fat depots may have made an important contribution to the toxic effect.

In other incidents in 1970 and 1971 cormorants were found dead in several areas of The Netherlands. Some birds were analysed and found to contain high levels of PCBs comparable to those found in the livers of birds which had been poisoned experimentally (210–285 ppm; Koeman *et al.*, 1973). It was probable that all or most of these birds died from PCB poisoning.

In the autumn of 1969 there was a large seabird wreck in the Irish Sea (Natural Environment Research Council Report, 1971). Large numbers of birds, estimated to be in excess of 17,000, were washed ashore on the coastline bordering the Irish Sea. This was regarded as a substantial underestimate of the actual numbers involved, because many would not have been recovered in this survey due to the bodies sinking or drifting into the Atlantic. Most of the birds were guillemots (estimated to be nearly 17,000), but there were also significant numbers of razorbills (142), herring gulls (88) and gannets (35). Upon analysis, guillemots were found to contain PCB residues of 1–880 ppm in liver. Many birds contained residue levels of PCBs comparable to those found in the livers of cormorants which had been lethally poisoned with PCBs (Koeman *et al.*, 1972, see above). It seems very probable that the death of many of the birds was caused by the toxic action of PCB residues, although other factors (e.g. starvation, moult) were also implicated (Walker, 1990).

In the years following this wreck, there was evidence of marked declines in the number of guillemots and razorbills at breeding colonies in Wales and Co. Wexford, Ireland (Knight and Walker, 1982a). In a study of these and other species of fish-eating seabirds from Co. Wexford, it was found that there was marked polymorphism with regard to the activity of the HMO system towards organochlorine substrates. Frequency diagrams for HMO activity towards organochlorine compounds were biphasic or triphasic for guillemots, razorbills and puffins (Fig. 10.4). Since this enzyme system is the principal one concerned with metabolism of PCBs and other organochlorine pollutants, it may be that selection in favour of individuals with high HMO activity has contributed to this polymorphism (i.e. that there has been selection in favour of a common gene which controls the synthesis of cytochrome P-450 that can catalyse the metabolism of these substrates). There is no evidence, so far, that this polymorphism is the consequence of induction, although there is some evidence for the induction of an orthologue of cytochrome P-450 IA1 in certain individuals (Ronis *et al.*, 1989). The latter enzyme, however, does not appear to be involved in epoxidation of aldrin—one of the assays referred to above.

Sublethal Effects

A number of studies have been concerned with the effects of organochlorine pollutants on reproduction in seabirds. Broadly speaking, two major types of effects have been considered—eggshell thinning and direct embryo toxicity.

There is much evidence associating thin eggshells with residues of *pp′*DDE (Cooke, 1973, 1979). Other organochlorine residues do not appear to have an important role. Biochemical studies suggest that the inhibition of Ca^{2+} ATP-ase by *pp′*DDE may be the critical mechanism involved, although there is some evidence that inhibition of carbonic anhydrase may also be important (Lundholm, 1987). The following cases of eggshell thinning associated with DDE residues have been reported: Gannets in Great Britain (Parslow *et al.*, 1973) and Canada (Elliott *et al.*, 1988a), cormorants in The Netherlands (Koeman *et al.*, 1973) and Great Britain (Cooke, 1979), and Sandwich terns in The Netherlands (Koeman and van Genderen, 1972).

In the Canadian study on the gannet it is suggested that DDE-induced eggshell thinning was responsible for reduced reproductive success and consequent population decline in the 1960s. Following pollution control, residue levels declined and the population recovered. These effects were associated with DDE levels of 20–30 ppm in the eggs. Apart from effects upon shell thinning, there was also evidence that organochlorine residues may have caused embryo toxicity in seabirds. Elliott *et al.* (1988a) report higher levels of dieldrin, heptachlor epoxide, βHCH, oxychlordane, *trans* chlordane, *cis* chlordane, *cis* nonachlor, and PCBs in gannet eggs which failed to hatch than in fresh eggs. A similar observation was made with common tern eggs sampled in Massachusetts (Nisbet and Reynolds, 1984). Embryotoxicity of organochlorine compounds in different species of birds is discussed by Cooke (1973) and Peakall and Fox (1987).

The induction of enzymes is another sublethal effect that may be caused by lipophilic pollutants, although it is not, in most cases, a harmful one. There are, however, some situations where induction may promote the activation of pollutants (see next section).

Many lipophilic pollutants such as DDE, PCBs and aromatic hydrocarbons are known to be inducers of hepatic microsomal cytochrome P-450 in vertebrates, including birds (Ronis and Walker, 1989). The induction of cytochrome P-450 forms can be used as an indicator of exposure to pollutants (Stegeman and Kloepper-Sams, 1987).

CONCLUSIONS

The high levels of DDE, dieldrin and PCBs in fish-eating birds such as the cormorant and shag suggest strong bioaccumulation in relation to the levels present in the fish upon which they feed. The low levels of HMO in these birds, and correspondingly slow rates of metabolism will allow bioaccumulation.

It is thought that sophisticated enzymic detoxication systems have not evolved in avian species which are specialized predators, because there is no requirement for them. By contrast, many omnivores and herbivores necessarily have such systems to detoxify lipophilic xenobiotics in their food (Walker, 1980, 1981). Specialized predators, e.g. birds which eat mainly fish, or birds which mainly eat other birds, ingest food which is of similar composition to their own bodies (i.e. lipophilic xenobiotics are relatively few). On the other hand, omnivores and herbivores ingest food which contains a wide range of lipophilic compounds which are normal constituents of plants, but are not produced by animals (Scheline, 1978). Thus, in considering the problem of bioaccumulation in marine ecosystems, particular attention should be paid to top predators which are likely to be deficient in detoxifying enzymes. There is a case for developing predictive models for bioaccumulation using data from *in vitro* studies in species like this (Walker, 1987).

The presence of such a diversity of lipophilic pollutants in seabirds raises questions about interactive effects (Malins and Collier, 1981). Where compounds are related in structures and have the same site of action, the likelihood exists of additive effects. This was suggested in the case of terns which were poisoned in The Netherlands by combinations of dieldrin, endrin and telodrin (Koeman and van Genderen, 1972). It is also possible that there may be effects that are greater than additive. Synergism may be the consequence of the inhibition of a detoxifying enzyme by one pollutant leading to the stabilization of another pollutant, an effect that can be produced when both compounds are taken up simultaneously. Longer-term potentiation of the toxicity of a pollutant may be the consequence of induction of an enzyme caused by earlier exposure to other pollutants. In the latter case, it is interesting to refer again to the evidence of induction of a cytochrome P-450 form (orthologue of cytochrome P-450 (IA1) in seabirds from British and Irish coastal sites (Ronis *et al.*, 1989). Elevated levels of this cytochrome were associated with relatively high levels of PCBs (see earlier discussion). It seems likely that birds with high levels of this cytochrome P-450 may be particularly vulnerable to certain aromatic hydrocarbon pollutants, when there are severe oil spills at sea, because of enhanced activation to carcinogenic metabolites (Walker and Johnston, 1989).

The effect that persistent pollutants may have on seabird populations remains a contentious issue. There are, however, a few studies which suggest that such effects have occurred. As discussed earlier, there are strong indications that PCBs were an important factor in the Irish Sea bird wreck, and there was also evidence of a decline in guillemots and razorbills in that area following the incident. The decline of the gannet population at a Canadian site in the late 1960s was clearly associated with levels of DDE; eggshell thinning was evidently the main reason for decline. Apart from these, there have been other declines of seabirds, e.g. of puffins in several northern European sites (Harris, 1984), which are not obviously related to pollutants, but where effects of pollutants cannot be definitely ruled out. Seabird wrecks and sharp population declines were reported long before persistent pollutants such as PCBs could have been present in the environment (Natural Environment Research Council, 1971). In general, it is difficult to identify the causes of population declines of seabirds, and there may have been effects of pollutants on populations which have passed unnoticed.

Because of their marked ability to bioaccumulate lipophilic pollutants, fish-eating seabirds are suitable for monitoring persistent residues in marine ecosystems. Migratory species like the auks and shearwaters acquire these residue burdens over large areas, whereas "sedentary" species like the shag are useful indicators of more local pollution problems. The sampling of eggs provides a useful means of systematically monitoring levels in seabird populations, and has been widely employed. Very little used, but potentially of value, is the practice of taking blood samples from species that can readily be caught in the field.

Whereas most monitoring to date has been concerned with the measurement of residues, in future greater use may be made of the measurement of characteristic biochemical responses to pollutants. In this context, a number of techniques are potentially useful, viz. the measurement of blood enzymes whose levels may be modulated by the effects of pollutants (e.g. B-esterases), the use of marker enzymes for tissue damage, the use of tests to establish enzyme induction and the measurement of adduct formation with cellular macromolecules (Walker, 1988). The use of diagnostic kits which rely on immunochemical interactions is likely to be of growing importance. The development of techniques for measuring induction in fish is already underway (Stegeman and Kloepper-Sams, 1987).

Apart from their value in indicating exposure to pollutants, some of these biochemical techniques have an additional advantage—they can indicate sublethal toxic effects, an approach that has already been followed with fish-eating birds; for example, recent work on disturbances of vitamin A metabolism by PCBs in gulls has pointed the way to developing biochemical assays of serum that will give useful measurements of exposure and toxic effect (Brouwer *et al.*, 1989). It is interesting to note that earlier work by Jefferies and Parslow (1976) gave evidence of changes in the weight and structures of the thyroid gland of guillemots dosed with PCBs. The development of this approach has considerable promise, in that it provides evidence which can establish a link between the presence of a pollutant and a toxic effect, which in turn may have an effect on the abundance or genetic constitution of a population.

REFERENCES

Bend, J. R., D. S. Miller, W. B. Kinter and D. B. Peakall. 1977. DDE-induced microsomal mixed-function oxidases in the puffin. *Biochem. Pharmac.* **26**: 1000–1001.

Boon, J. P., P. J. H. Reijnders, J. Dols, P. Wensvoort and M. T. J. Hillebrand. 1987. The kinetics of individual PCB congeners in female harbour seals with evidence for structure-related metabolism. *Aquat. Toxicol.* **10**: 307–324.

Borlakoglu, J. T., J. P. G. Wilkins and C. H. Walker. 1988. Polychlorinated biphenyls in fish-eating seabirds—Molecular features and metabolic interpretations. *Marine Environ. Res.* **24**: 15–19.

Borlakoglu, J. T., J. P. G. Wilkins, C. H. Walker and R. R. Dils. 1990a b c. Polychlorinated biphenyls in fish-eating sea birds. I. Molecular features of PCB isomers and congeners in adipose tissue of male and female razorbills from British and Irish coastal waters/II. Molecular features of PCB isomers and congeners in adipose tissue of male and female puffins, guillemots, shags and cormorants of British and Irish coastal waters/III. Molecular features and metabolic interpretation of PCB isomers and congeners in adipose tissues. *Comp. Biochem. Physiol.* **97C**: 151–160/161–171/173–177.

Brouwer, A., A. J. Murk and J. H. Koeman. 1990. Biochemical and physiological approaches in ecotoxicology. *Functional Ecology* **4**: 275–282.

Cooke, A. S. 1973. Shell thinning in avian eggs by environmental pollutants. *Environ. Pollut.* **4**: 85–150.

Cooke, A. S. 1979. Eggshell characteristics of gannets, shags and great black backed gulls exposed to DDE and other environmental pollutants. *Environ. Pollut.* **19**: 47–65.

Darnerud, P. O., I. Brandt, E. Klasson-Wehler, A. Bergman, R. D'Argy, L. Dencker and G. O. Sperber. 1986. 3,3′,4,4′Tetrachloro[^{14}C]biphenyl in pregnant mice: enrichment of phenol and methyl sulphone metabolites in late gestational features. *Xenobiotica* **16**: 296–306.

Earley, J. J. 1987. Pesticides and PCBs in seabirds and coastal pollution. In D. H. S. Richardson (ed.), *Biological Indicators of Pollution*. Royal Irish Academy, Ireland, pp. 193–200.

Elliott, J. E., R. J. Norstrom and J. H. A. Keith. 1988a. Organochlorines and eggshell thinning in northern gannets from Eastern Canada 1968–1984. *Environ. Pollut.* **52**: 81–102.

Elliott, J. E., R. W. Butler, R. J. Norstrom and P. E. Whitehead. 1988b. Levels of polychlorinated dibenzodioxins and polychlorinated dibenzofurans in eggs of Great Blue herons in British Columbia 1983–87: possible impacts on reproductive success. *Canadian Wildlife Progess Notes* No. 176.

Fossi, C., S. Focardi, C. Leonzio and A. Renzoni. 1984. Trace metals and chlorinated hydrocarbons in birds eggs from the delta of Danube. *Environ. Conservat.* **11** (4): 345–350.

Harris, M. P. 1984. *The Puffin*. T. and A. D. Poyser-Calton.

Harris, M. P. and D. Osborn. 1981. Effect of a polychlorinated biphenyl on the survival and breeding of puffins. *J. Appl. Ecol.* **18**: 471–479.

Ioannides, C., P. Y. Lum and D. V. Parke. 1984. Cytochrome P-448 and the activation of toxic chemicals and carcinogens. *Xenobiotica* **14**: 119–137.

Jakoby, W. B. (ed.) 1980. *Enzymatic Basis of Detoxication*, Vols. I and II. Academic Press.

Jefferies, D. J. and J. L. F. Parslow. 1976. Thyroid changes in PCB-dosed guillemots and their indication of one of the mechanisms of action of these materials. *Environ. Pollut.* **10**: 293–311.

Knight, G. C. and C. H. Walker. 1982a. A study of the hepatic microsomal monooxygenase of sea birds and its relationship to organochlorine pollutants. *Comp. Biochem. Physiol.* **73C**: 211–221.

Knight, G. C. and C. H. Walker. 1982b. A study of the microsomal epoxide hydrolase of sea birds. *Comp. Biochem. Physiol.* **73C**: 463–467.

Koeman, J. H. and H. van Genderen. 1972. Tissue levels in animals and effects caused by chlorinated hydrocarbons, insecticides, chlorinated biphenyls and mercury in the marine environment, along the Netherlands coast in marine pollution and sealife. *Fishing News Books*, West Byfleet, pp. 1–8.

Koeman, J. H., A. A. G. Oskamp, J. Veen, E. Brouwer, J. Rooth, P. Zwart, E. v.d. Broek and H. van Genderent. 1967. Insecticides as a factor in the mortality of the Sandwich tern—a preliminary communication. *Medel. Rijksfacult. Landbouwwetenschap. Gent.* **32**: 841–854.

Koeman, J. H., H. C. W. van Velzen-Blad, R. de Vries and J. G. Vos. 1973. Effects of PCB and DDE in cormorants and evaluation of PCB residues from an experimental study. *J. Reprod. Fert. Suppl.* **19**: 353–364.

Lambertini, G. and C. Leonzio. 1986. Pollutant levels and their effects on Mediterranean sea birds. In Medmaravis and X. Monballiu (eds.), *Mediterranean Marine Avifauna*. NATO Series Vol. G12, pp. 359–378.

Livingstone, D. and J. J. Stegeman, (eds.) 1989. Cytochrome P-450 forms and functions in animals. The fate of foreign compounds in biological systems. *Xenobiotica* **19**: 1037–1200.

Lundholm, E. 1987. Thinning of eggshells of birds by DDE; mode of action on the eggshell gland. *Comp. Biochem. Physiol.* **88C**: 1–22.

Malins, D. C. and T. K. Collier. 1981. Xenobiotic interactions in marine organisms; effects on biological systems. *Aquat. Toxicol.* **1**: 257–268.

Moore, N. W. and J. O. 'G. Tatton. 1965. Organochlorine insecticide residues in the eggs of sea birds. *Nature* **207**: 42–43.

Moore, N. W. and C. H. Walker. 1964. Organochlorine insecticide residues in British birds. *Nature* **201**: 1072–1073.

Natural Environment Research Council. 1971. *The Sea Bird Wreck in the Irish Sea, Autumn 1969*. NERC Publications Series C, No. 4.

Natural Environment Research Council. 1983. *Contaminants in Top Predators*. NERC Publications Series C, No. 23.

Nebert, D. W. and F. J. Gonzalez. 1987. P450 genes. Structure, evolution and regulation. *Ann. Rev. Biochem.* **56**: 945–993.

Nisbet, I. C. T. and L. M. Reynolds. 1984. Organochlorine residues in common terns and associated estuarine organisms. *Marine Environ. Res.* **11**: 33–66.

Norstrom, R. J. 1988. Patterns and trends of PCB contamination in Canadian wildlife. *Proceedings of Conference on PCBs and Replacement Fluids, Montreal 1986*. IEEE Cat. No. THO, 151–1.

Parslow, J. L. F., D. J. Jefferies and J. M. Hanson. 1973. Gannet mortality incidents in 1972. *Mar. Pollut. Bull.* **4**: 41–44.

Peakall, D. B. and G. H. Fox. 1987. Toxicological investigations of pollutant-related effects in Great Lakes gulls. *Environ. Health Perspect.* **71**: 187–193.

Renzoni, A., S. Focardi, C. Fossi, C. Leonzio and J. Mayol. 1986. Comparison between concentrations of mercury and other contaminants in eggs and tissues of Cory's shearwater collected on Atlantic and Mediterranean Islands. *Environ. Pollut.* **40A**: 17–35.

Robinson, J., A. Richardson, A. N. Crabtree, J. C. Coulson and G. R. Potts. 1967. Organochlorine residues in marine organisms. *Nature, London* **214**: 1307–1311.

Ronis, M. J. and C. H. Walker. 1989. The monooxygenases of birds. *Rev. Biochem. Toxicol.* **10**: 301–384.

Ronis, M. J. J., T. Hansson, J. T. Borlakoglu and C. H. Walker. 1989. Cytochrome P-450s of sea birds: Cross reactivity studies with purified rat cytochromes. *Xenobiotica* **19**: 1167–1173.

Scheline, R. R. 1978. *Mammalian Metabolism of Plant Xenobiotics.* Academic Press, London.

Stegeman, J. J. and P. J. Kloepper-Sams. 1987. Cytochrome P-450 isozymes and monooxygenase activity in aquatic animals. *Environ. Health Perspect.* **71**: 87–95.

Van den Berg, M., F. Blank, C. Heeremans, H. Wagenaar and K. Olie. 1987. Presence of polychlorinated dibenzo-*p*-dioxins and polychlorinated dibenzofurans in fish-eating birds and fish from the Netherlands. *Arch. Environ. Contam. Toxicol.* **16**: 149–158.

Walker, C. H. 1980. Species variation in some hepatic microsomal drug-metabolism enzymes. *Prog. Drug. Metab.* **5**: 113–164.

Walker, C. H. 1981. The correlation between *in vivo* and *in vitro* metabolism of pesticides in vertebrates. *Prog. Pesticide Biochem.* **1**: 247–285.

Walker, C. H. 1987. Kinetic models for predicting bioaccumulation of pollutants in ecosystems. *Environ. Pollut.* **44**: 227–240.

Walker, C. H. 1988. Some biochemical responses of vertebrates to sublethal doses of pesticides. *Aspects Appl. Biol.* **17**: 257–263.

Walker, C. H. and G. O. Johnston. 1989. Interactive effects of pollutants at the toxicokinetic level—implications for the marine environment. *Marine Environ. Res.* **28**: 521–525.

Walker, C. H., G. C. Knight, J. K. Chipman and M. J. J. Ronis. 1984. Hepatic microsomal monooxygenases of sea birds. *Marine Environ. Res.* **14**: 416–419.

Walker, C. H. 1990. Persistent pollutants in fish-eating sea birds—bioaccumulation, metabolism and effects. *Aquatic Toxicol.* **17**: 293–324.

Part VI

Overview

CHAPTER 11

Pollutants in Marine Ecosystems: an Overview

DAVID R. LIVINGSTONE*, PETER DONKIN* and COLIN H. WALKER†

* NERC Plymouth Marine Laboratory, Citadel Hill, Plymouth, Devon PL1 2PB, UK

† Department of Physiology and Biochemistry, University of Reading, Whiteknights, PO Box 228, Reading RG6 2AJ, UK

INTRODUCTION

Persistent pollutants enter marine ecosystems from many sources and via many routes. Within the ecosystem, interactions occur between the various abiotic and biotic components, a significant part of the flux of the chemicals being determined by the natural progression along food chains. The lifetime of a chemical in an animal is determined by the balance of the processes of uptake, storage, metabolism and elimination. This can result in the bioaccumulation and persistence of some chemicals in animals lower down but not higher up the food chain, e.g. marked accumulation of polynuclear aromatic hydrocarbons (PAHs) occurs in invertebrates but not in fish (Stegeman, 1981; Van der Oost *et al.*, 1991), largely due to the higher rates of biotransformation in the latter. This is the opposite of what is observed for chemicals which are metabolized very slowly or not at all, such as particular polychlorinated biphenyls (PCBs) and certain other organochlorines, which generally bioaccumulate to highest concentrations in top predators. Such considerations, and other factors, affect the choice of which species to use to monitor which pollutant. Differences in these processes, and in particular in exposure levels, result in spatial and temporal differences in persistence, including "hot spots" of pollution. Persistence is a cause for concern because of the toxicity of the parent compound and/or metabolites, manifest at various levels of biological organization from molecular through to organismic, with consequences for the health and fitness of individuals and populations. Deriving mathematical models to predict the fate and toxicity of chemicals in ecosystems is an important aim, and investigatory tool, of understanding the processes which determine persistence.

The final chapter addresses these considerations by drawing on the findings and conclusions of the previous chapters, which variously have considered aspects of persistence from the water column and sediment through to the top predators. The discussion is concerned mainly with organochlorine compounds, particularly PCBs, but also includes other contaminants such as PAHs.

TRANSFER PROCESSES AND ROUTES OF UPTAKE

The foregoing chapters reported on the concentrations of persistent organochlorine and other pollutants at different trophic levels of marine ecosystems, progressing from phytoplankton and zooplankton, through invertebrates and fish, to vertebrate predators which exist at the top of the food chain. Data are presented from the Atlantic and Pacific Oceans, the Canadian Arctic and the Mediterranean Sea, thus covering a wide geographical and climatic range. Of particular importance are the major routes of entry of the pollutants into the marine ecosystems, and the critical routes of transfer within the ecosystems.

Considering first the routes of entry, much attention is given to the important industrial pollutants, the PCBs, and the polychlorinated dibenzodioxins (PCDDs) and dibenzofurans (PCDFs) which are associated with them (all chapters). These may arise from diverse sources including insulator fluids from transformers, hydraulic fluids, and copying paper. PCDDs and PCDFs are generated when PCBs are incompletely burned in combination furnaces. PCDDs also occur as impurities in herbicides such as 2,4,5-trichlorophenoxyacetic acid and in chlorophenols used in the wood and pulp industry. It seems clear that substantial quantities of these compounds reach seas and oceans by aerial transport (Chapters 7 and 8). However, industrial liquid effluent is the main source of these compounds in particular coastal areas, as is well illustrated in studies carried out on the coasts of the USA (Chapters 1, 3, 4 and 5). Distance from the source of pollution can be an important qualitative factor in exposure: thus, PCB homolog distribution in water, sediment and organisms in New Bedford Harbor, USA shifted toward the higher chlorinated forms the further away the location was from the source of the chemicals, consistent with the greater volatilization of the lower homologs (Chapter 4). Other persistent residues, such as dieldrin and 1,1-dichloro-2,2-*bis*(*p*-chlorophenyl)ethylene (*pp*′DDE), arise from the use of organochlorine insecticides (Chapters 5, 8 and 9). Like the foregoing other compounds, they are subject to aerial transport, on a global scale, but their occurrence at particular coastal sites may follow different trends from PCBs because of the difference in location of their original release. The very high levels of *pp*′DDE in birds in the Danube Delta, Romania illustrate this point (Chapter 9).

The organochlorine compounds under consideration have very low water solubilities and, consequently, tend to be strongly adsorbed by particulate matter of sediments, the concentrations dissolved in water being very low. The distribution of individual compounds between water and sediment is related to polarity (Chapter 4). Thus, the most hydrophobic organochlorines tend to be at relatively low levels in water compared to sediment, whereas the opposite is the case for the more polar ones. The same is observed for PAHs, 4 to 7 ring compounds being more tightly bound to sediment particles than 1 to 3 ring ones (Chapter 5). Invertebrates which live in sediment, e.g. certain polychaetes and molluscs, may obtain these pollutants by ingestion with food particles, but also by direct transfer from water or the surface of mineral particles (Chapters 1 and 2). The same is also true of fish (Chapter 5; Gobas *et al.*, 1986; Hawker and Connell, 1988; Clark *et al.*, 1990) and it can be difficult to establish the relative importance of these two routes in the field situation (Chapter 4). That both are important has been shown in, for example, studies on the polychaete *Capitella capitata* and the mullet *Mugil cephalus*, in which bioaccumulation of PCBs was highly correlated with the product of two factors representative of partition behaviour in water (octanol–water partition coefficient) and adsorption characteristics on to surfaces (chromatographic elution time on carbon) (Shaw and Connell, 1984). In field studies of fish from a large number of urban sites around the coasts of USA, correlations were found between levels of PAHs, PCBs and DDTs in sediments and the levels in liver or bile, indicating that the contaminants were bioavailable to the fish and/or their food sources (Chapter 5).

It is clear that where organochlorine compounds are present at measurable concentrations in water, marine invertebrates and fish can usually readily take them up by passive diffusion across gill and other surfaces. There is evidence, however, for PCBs, that optimal bioaccumulation occurs with trichloro-isomers, whereas larger molecules with higher levels of chlorination are taken up more slowly. This decrease in apparent efficiency of uptake (assimilation efficiency) with increasing levels of chlorination and/or molecular size was seen in species of polychaete, mussel, hard clam, crab, lobster and flounder (Chapter 4). This raises the possibility that, with larger molecules, passage is slower across permeable membranes (Barron, 1990). Restricted movement of particular chemicals across membranes has also been indicated from a number of other studies, on both model systems and organisms. Penetration of a model membrane system (mercury adsorbed dioleoylphosphatidylcholine monolayers) by individual PCBs decreased with increasing globularity (decreasing planarity) of the congeners, the latter being dependent on both the positioning and degree of chlorination, the extreme of this effect being most evident in the highly substituted bulky congeners (Nelson *et al.*, 1990). Similar results were obtained for PAHs with a molecular size cut-off for higher membered molecules (benzo[a]pyrene) (Nelson *et al.*, 1990); for polychlorinated naphthalenes with no measurable uptake of hepta- and octachloronaphthalenes by fish being detected (Opperhuizen *et al.*, 1985); and for high molecular weight organochlorines taken up by rainbow trout *Salmo gairdnerii*, the bioconcentration factors (BCFs) of which did not correlate with octanol/water partition coefficienct (K_{ow}) (Oliver and Nliml, 1985). Bioconcentration is in fact observed to be reduced for superlipophilic compounds ($K_{ow} > 10^6$) as a result of the increased time taken to establish effective equilibrium with the tissue compartment, viz. a minimum period of 0.5 year for $K_{ow} > 10^6$, increasing to 10 years for $K_{ow} = 10^6$, with no significant bioaccumulation being observed for compounds of $K_{ow} > 10^{13}$ (Hawker and Connell, 1985, 1988). Similarly, the uptake efficiency of superlipophilic chemicals by fish gills was observed to be reduced compared to compounds of lower K_{ow} values (McKim *et al.*, 1985). The mechanistic basis of this phenomenon presumably includes molecular size and the water solubility of the compound (determines the maximum exposure concentration.)

Another factor which affects the selective uptake of PCB congeners and other pollutants is the presence of particulate matter. The adsorption of PCBs on to a surface is similarly influenced, principally, by congener stereochemistry, adsorption decreasing with decreasing planarity of the molecule: thus the particle-mediated component of the bioaccumulation of PCBs in *C. capita* and *M. cephalus* decreased with decreasing planarity (increasing chlorination and/or changing structure) of the PCB congeners (Shaw and Connell, 1984). For high numbers of particles, and extremely hydrophobic PCBs, the rates of dissolution or desorption of the congener from the particles determined the rates of uptake of the chemical by the fish (Opperhuizen and Stokkel, 1988). Another dimension to particle-containment interactions is the possibility of the presence of food organisms facilitating contaminant desorption from the sediment particles (Chapter 5).

At the base of the food chain, uptake of contaminants appears to be equally likely from either the water or the food, e.g. for filter-feeding bivalves which pass large quantities of water through their system (Chapters 1 and 2). Uptake of pollutants by both water and dietary routes (Van Veld, 1990) is also well known for the fish (Chapter 5). However, above the lower trophic levels, there is good evidence that uptake from ingested prey becomes increasingly important, as was found in the study of PCB accumulation in the ecosystem of New Bedford Harbor (Chapter 4). This was also argued for the uptake of dioxins by the blue crab *Callinectes sapidus* (Chapter 3) and, more generally, of organochlorines by fish-eating seabirds (Chapter 8) and cetaceans (Chapter 6). Thus, above the lower trophic levels, movement of organochlorine compounds appears to occur largely through the food chain itself (mainly from prey to predator), with a decreasing or no contribution from direct transfer from the water to the organism.

The result of transfer of non-metabolizable compounds along the food chain is biomagnification, i.e. tissue concentrations are highest in the top predators. This was very evident for PCBs, (Chapter 4) and other persistent chemicals such as 1,1-*bis*(4-chlorophenyl)-2,2,,2-trichloroethane (DDT), and can be well illustrated with data from other chapters, viz. ranges for body burdens of total PCBs, in μg per gram net weight, were 0.002–13.9 (molluscs), 0.03–7.7 (crustaceans) (Chapter 1); 2–37 (fish) (Chapter 5); 1–800 (sea mammals) (Chapter 6); and 0.4–340 (seabirds) (Chapter 8). Many other examples of biomagnification of persistent compounds are known, e.g. in various fish species (Gobas *et al.*, 1988; Clark *et al.*, 1990).

The upper limit of bioaccumulation of persistent compounds in any organism is presumably set by lipid levels in the tissues (Chapters 1, 2, 5 and 6), the time taken for very lipophilic compounds to reach tissue equilibrium (see before) and the (lethal) toxic effects of the pollutant (see later). Lipid levels also mainly determine the tissue distribution of pollutants within an organism, e.g. the digestive gland/gonads (Chapter 1) and blubber (Chapters 5 and 6) are major lipid and pollutant deposition sites in, respectively, molluscs and whales. Exceptions to this exist, however, such as the selective bioaccumulation of α-hexachlorocyclohexane in brain of whales (Chapter 6).

Once contaminants have reached the adult predators at the top of food chains, further transfer can still occur. Persistent lipophilic compounds can be passed on to the next generation by transport into the eggs of birds (Chapters 7 and 8), and either directly, or via milk, into the progeny of cetaceans and other sea mammals (Chapters 5 and 6). These processes appear to be largely or entirely the consequence of the compounds travelling with "exported" lipoproteins, in which they are dissolved. Transport and other lipoproteins also occur in tissues and eggs of marine invertebrates (Lee, 1991), and presumably a similar potential for contaminant transfer therefore exists in these animals. Such transport also raises important questions about toxic effects upon the next generation.

The transfer of persistent contaminants by spawning, placental transfer or lactation also represents a means of reducing the body burden of contamination of the parent animal, as does growth of an animal. Depuration of chemicals in response to cessation of loading occurs for many organisms (Chapters 1 and 2) and was indicated to occur in New Bedford Harbor (Chapter 4). However, even with cessation of loading, such elimination processes are far from complete, and chemical structure clearly plays a part in decreased elimination, as it does for decreased uptake (see above). Increased persistence of PAHs with increased exposure times was very apparent for molluscs (Chapter 1). Species differences are also important, e.g. pentachlorophenol was not released at all to the water by the polychaete *Lanice conchilega* (Chapter 2). Elimination rate constants for very hydrophobic chemicals ($K_{ow} > 10^4$) are inversely proportional to hydrophobicity (Gobas *et al.*, 1986): and thus, decreasing excretion of PCB congeners with increasing degree of chlorination was observed in the New Bedford Harbor study (Chapter 4). The balance of the different means of reducing body burdens of persistent pollutants therefore changes with the structure of the chemical and other factors. Thus, excretion can be the main mechanism for lowering tissue levels of unchanged trichloro- and tetrachloro-PCBs, whereas body growth as a sink, and excretion, are comparable means for reduction of pentachloro- and hexachloro-PCB concentrations (Chapter 4). In cetaceans, body burdens of PCBs were similar in males and females from highly contaminated areas, but were lower in females than males from less contaminated areas due to the relatively greater impact of lactation and parturition on PCB loss from the tissues (Chapter 6). Spawning had a similar impact in female fish in reducing contaminant body burdens (Chapter 5).

In addition to the movement or cycling of persistent contaminants by physical and geochemical processes, they can also be moved considerable distances with the migration of the organisms in which they are stored. Fish, cetaceans and seabirds can travel over very large distances, so this can be an effective means of dispersal (Chapters 5, 6, 7 and 8). It can also raise uncertainties about the place of origin of residues in migrating animals such as cetaceans (see later). Dramatic demonstrations of the magnitude of movements of persistent chemicals include the presence of PCBs in fish from the deep sea (Stegeman *et al.*, 1986), in cetaceans inhabiting waters well away from industrialized coastal regions (Chapter 7) and in polar bears (Chapter 6) and seabirds (Chapter 8) in the Arctic.

SAMPLING STRATEGIES AND GEOGRAPHICAL DISTRIBUTION

A number of species have served as indicators of levels of pollution in particular marine ecosystems. These have ranged from static or relatively static sentinels such as bivalve and other invertebrates, to vertebrates with restricted movement patterns, to highly mobile vertebrates. The choice of particular species in monitoring design can be highly informative, e.g. use of invertebrates for monitoring a wide range of pollutants (see below); species-dependent spatial differences in PCB body burden were seen in polychaetes and flounder in New Bedford Harbor, consistent with increased exposure via the sediment (Chapter 4); and levels of a number of organochlorines in eggs either remained constant or increased for a resident Arctic bird species compared to declines over the same period for bird species which only breed there, consistent with the continued persistence of these chemicals in the Arctic (Chapter 8). The advantages of sampling birds' eggs as a pollution monitoring strategy have been discussed (Chapter 9). In the foregoing chapters, tissue body burden data of chemical contamination have been presented for a range of chemicals, species and locations, thereby presenting an opportunity for geographical and other comparisons.

Marine invertebrates have been extensively used for monitoring purposes because they bioaccumulate a wider range of organic pollutants than do vertebrates, largely because of their lower rates of xenobiotic metabolism (Chapters 1 and 2). Thus, for compounds such as PAHs, in addition to measurements of the sediment and/or water column, chemical pollution gradients have often been established by analyses of invertebrate tissues (Klungsøyr *et al.*, 1988). Amongst invertebrates, bivalves such as mussels have been extensively employed in so-called Mussel Watch monitoring programmes, producing chemical contamination data for many parts of the world including the east and west coasts of USA, the Gulf coast, Australia, Hong Kong waters, Brazil, Norway and the Mediterranean Sea (Chapter 1). In a review of the US Mussel Watch data of 1976–78, for 62 sampling locations, it was observed that PCB hydrocarbon tissue concentrations were consistently an order of magnitude or more higher near the larger urban areas (Farrington *et al.*, 1978), well illustrating the strength of invertebrate body burden tissue analysis in providing detailed information on localized gradients of pollution. Such monitoring studies have been supported by many experimental investigations, both in the laboratory and the field, examining the reliability of particular species for monitoring particular chemicals (Chapters 1 and 2) and the use of such information for other purposes, e.g. modelling and prediction (Chapter 4). Even factors such as the taxonomy of closely related species are now being considered in the design of biomonitoring programmes (Lobel *et al.*, 1990).

Assessing the exposure of vertebrates to compounds such as PAHs, which are readily metabolized, is obviously less straightforward than for invertebrates. In addition to the use of chemical analyses of sediment, water column and (invertebrate) food sources, and the responses of biotransformation enzymes (see next section), other approaches that have been successfully applied in fish are measurement of fluorescent aromatic compounds and specific metabolites in bile, and co-occurring PCBs and DNA adducts in liver (Chapter 5). As with invertebrates, extensive experimental studies have been carried out for a range of persistent and not so persistent compounds in order to interpret such field observations (Chapters 5, 6, 7 and 10; Zaroogian *et al.*, 1985).

Examining the vertebrate data, it is possible to make some observations of geographical significance. With seabirds, in the period 1969–73, very high levels of PCBs, associated with toxic effects, were reported in certain areas, viz. levels in excess of 100 ppm in liver of a number of razorbills, which died in the Irish Sea in 1969, and in cormorants found dead in The Netherlands in 1970 (Chapter 10). The source of these residues is still not clearly defined, although there is evidence that dumping of sewage wastes made a significant contribution to PCB levels in the Irish Sea at that time (Holden, 1970). Organochlorine pesticides and industrial compounds (except hexachlorobenzene) have declined in the food chains of inshore and continental shelf seabirds of both coasts of Canada: in contrast, in open ocean food chains (Leach's storm petrel) only *pp*′DDE appears to be declining (Chapter 8).

Residue studies upon eggs of seabirds indicate that the Mediterranean Sea is an area of relatively high pollution. Apart from the high levels of mercury—which mainly reflect the abundance of mercury in geological deposits in the area—there is evidence of relatively high levels of PCBs and *pp*′DDE. In Cory's shearwater, for example, levels of these compounds were higher for Mediterranean sites than Atlantic sites in 1982–4, with particularly high levels reported from the area of Majorca (Chapter 9). It appears likely that the latter is the consequence of pollution originating from the French coast, e.g. Rhone Delta, since there is very little industrial activity in the Balearic Islands. The Danube Delta was found to be an area with particularly high levels of pollution by *pp*′DDE. In studies of North America, particularly high levels of *pp*′DDE were found in the Gulf of St. Lawrence (Elliot *et al.*, 1988). Levels of PCDDs were higher in areas of the west than the east coast of Canada, probably related to the forest industry's use of chlorophenols: in contrast, levels of the insecticide toxaphene were higher in food chains of the Atlantic than the Pacific Ocean (Chapter 8).

Turning to cetaceans, the picture is complicated by the very extensive migration of certain species. Some of the smaller coastal species, e.g. harbour porpoise and common dolphin, do not migrate over large areas, and the high residues of organochlorine compounds in some of them bear testimony to high levels of pollution in areas such as the Baltic Sea, the Gulf of Bothnia, Canadian east coast, US north-east coast, and California coast (Chapter 7). Similarly, harbour seals from relatively pristine Arctic waters showed low levels of chlorinated hydrocarbons, whereas in contrast, highly migratory Stellar's sealions caught in the same area had very high contaminant levels, probably due to feeding in California waters (Chapter 5). With larger species of cetaceans which inhabit the open ocean, e.g. minke whale, sperm whale and sei whale, interesting comparisons were possible between large areas of water. Higher levels of organochlorine contaminants were found in animals from the northern Pacific than in those from the southern Pacific Ocean. In general, the higher levels were found in areas where the greatest pollution was anticipated as a consequence of human activity, consistent with the findings of invertebrate studies. However, in studying these data for sea mammals and other vertebrates, the long half-lives of the residues, and the long life expectancy of the animals, need to be taken into consideration. The levels found may thus be representative of an exposure to contaminants which has occurred over a period of years, and in a number of different locations.

Overall, the above evidence of invertebrate and vertebrate studies point clearly to certain coastal sites which are "hot spots", in regard to pollution by organochlorine compounds. In a number of cases these are related to high levels of estuarine contamination, e.g. the Gulf of St. Lawrence, the Danube Delta and New Bedford Harbor.

COMPARATIVE ENZYMOLOGY AND OTHER FACTORS AFFECTING PERSISTENCE OF CHEMICALS IN SPECIES

By definition, metabolism of a pollutant reduces the body burden of that chemical. Elimination of the metabolites from the animal completes the process and is critically important if the metabolites are toxic. Persistence can therefore be regarded in terms of the continuing presence in an animal of the unchanged parent compound alone, or of both this and any residues derived from it, such as metabolites and adducts (Chapters 1, 2 and 5). If metabolic capabilities and elimination rates can differ between species, and also be dependent on physiological (e.g. seasonality/reproduction, nutrition) and environmental (e.g. temperature, salinity) factors (Chapter 6), then it has been argued that such variables should also be specified in definitions of persistence (Chapter 2).

Phase I biotransformation enzymes introduce functional groups into compounds, so allowing the attachment of polar moieties by Phase II enzymes, and the production of products that are more water-soluble than the parent compound. In some cases the metabolism is predominantly phase I only, e.g. the biotransformation of *pp*′DDT to *pp*′DDE and 1,1-dichloro-2,2-*bis*(*p*-chlorophenyl/ethane (*pp*′DDD). Of central importance in phase I metabolism is the cytochrome P450 monooxygenase or mixed function oxygenase (MFO) system which through its many different P450 forms can metabolize numerous, structurally diverse chemicals, thus providing a basis for the detoxication of countless substances in the environment and the diet (Gonzalez and Nebert, 1990). The MFO system appears to be present in all animals (Chapters 1, 6, 7 and 9), and the levels have some relationship to overall metabolic activities of the organism (Walker and Ronis, 1989; Livingstone, 1991a). Not surprisingly, therefore, MFO capacity is considerably higher in marine vertebrates than marine invertebrates, in the form of higher cytochrome P450 specific contents and MFO activities, higher microsomal protein yields and much greater inducibility of the MFO system (Table 11.1). The levels of total cytochrome P450s and MFO activities such as benzo[a]pyrene hydroxylase (BPH) and aldrin epoxidase are, however, lower in sea mammals and seabirds than in terrestrial mammals such as rodents (Chapters 7 and 9). Although there is debate about the inducibility of the MFO system in marine invertebrates (Livingstone, 1991a, b), a significant role was indicated (for some species, at least) from the enhanced metabolism of PCBs by nereid polychaetes at polluted compared to clean field sites (Chapter 2).

From comparison of rates of metabolism of "functional group compounds" and hydrocarbons, the MFO system was indicated to be possibly rate-limiting in the metabolism of hydrocarbons in molluscs and crustaceans (Chapter 1). Higher rates of metabolism of both groups of compounds were observed in crustaceans than molluscs (Chapter 1), consistent with the former's higher cytochrome P450 levels (Table 11.1). For hydrocarbons in marine invertebrates, rates of uptake of PAHs generally exceeded rates of metabolism by an order of magnitude, so accounting for the marked bioaccumulation of these chemicals in these organisms (Chapter 1). As the uptake of organic pollutants from water by invertebrates and fish is largely passive and likely to occur at similar rates (Zaroogian *et al.*, 1985), then the much higher MFO activities in the fish (one or more orders of magnitude higher, particularly in induced [exposed] animals) readily explains the much lower levels of hydrocarbons found in their tissues in the field (Chapter 5). Amongst birds, the lowest MFO activities are seen in specialized predators (fish-eaters and raptors) which therefore may be more susceptible to pollutants than other bird species (Chapter 9). Amongst cetaceans, MFO activities towards unspecific substrates (aldrin, benzo[a]pyrene and aniline, i.e. substrates not metabolized by single P450 isozymes [see below]) were

TABLE 11.1 *Characteristics of the Microsomal Mixed-function Oxygenase System in Different Animal Groups*[a]

Animal group	Cytochrome P450 (pmol/mg)[b]	Microsomal protein (mg/g wet wt)	BPH (pmol/min/mg)[b]	EROD (pmol/min/mg)[b]	Number of isozymes[c]	Inducibility[d]	References
Whales/dolphins	230 ± 41 (4)	22 ± 1 (3)	395 ± 167 (4)	66 ± 52 (4)	–	–	See Chapters 6 and 7
Seals	177 ± 48 (3)	–	–	–	–	–	Chapter 6
Birds	318 ± 48 (15)	13 ± 3 (5)	150 (1)	50	> 1	+++ (3MC,PB)	Dalvi *et al.* (1987); Sijm and Opperhuizen (1989); Walker and Ronis (1989)
Fish	322 ± 35 (38)	14 ± 4 (4)	307 ± 149 (29)	486 ± 208 (9)	5	+++ (3MC)	Stegeman *et al.* (1979); Addison and Edwards (1988); Perdu-Durand and Craven (1989); Sijm and Opperhuizen (1989)
Echinoderms	63 ± 12 (2)	5–10	8.4 ± 5.5 (2)	–	–	+ (PCB)	Chapter 1; Den Besten (1991)
Crustaceans	328 ± 70 (14)	5–10	40 ± 31 (10)	29 (1)	4	+ (3MC, PCB)	Livingstone (1990, 1991b)
Polychaetes	82 (1)	–	8.5 ± 6.8 (2)	–	–	+ (3MC,PCB)	Chapter 1; Livingstone (1991b)
Molluscs	73 ± 10 (9)	5–10	21.2 ± 3.4 (5)	3 ± (2)	2	+ (3MC, PCB)	Livingstone (1990)

[a] Individual means were calculated for each species (generally "control" animals) and the data presented are the means ± SEM of these (number of species in parenthesis). The data are generally for the tissues with the highest P450 specific content, namely liver (vertebrates), pyloric caeca (echinoderms), hepatopancreas (crustaceans), digestive gland (molluscs) and intestines (polychaetes). BPH, benzo[a]pyrene hydroxylase; EROD, 7-ethoxyresorufin *O*-deethylase; – no information.

[b] mg microsomal protein.

[c] Number isolated or indicated.

[d] +++, markedly inducible (BPH or EROD increased by up to × 10 or more); +, limited inducibility (BPH generally either not increased or increased by much less than × 10). 3MC, 3-methylcholanthrene-type inducers; PB, phenobarbital-type inducers; PCB, polychlorobiphenyl mixtures.

very low in three predatory species, in comparison to activities in male rats (Chapter 7).

Qualitative as well quantitative differences in the MFO system are also critically important in the metabolic fate of pollutants, particularly with respect to particular PCB congeners which can be substrates, inducers and/or inhibitors of different P450 forms (Chapter 6). To date the gene superfamily of cytochrome P450 has been found to be composed of more than a hundred different forms, falling into 20 gene families: an "explosion" of new genes has occurred in the last 400 million years, particularly in the II family (for details of nomenclature of cytochrome P450 genes and gene products see Nebert *et al.*, 1987). The hydrocarbon-metabolizing P450IA family (requires planar substrates) is present in all marine vertebrate groups (Chapters 5, 6, 7 and 9) and has been shown to be highly inducible in fish and birds. In contrast, the P450IIB family (planarity not necessary) appears underdeveloped in all these groups: some induction is seen in birds, but none in fish.

The study of PCB residue profiles in a range of marine vertebrates has suggested differential metabolism of two groups of congeners, possibly reflecting differences in the cytochrome P450 forms present in the different animals (Chapters 6, 7 and 9). The two groups of congeners are: (1) planar PCBs, with vicinal H atoms in *o,m*-positions with no more than one Cl atom in an *o*-position and (2) PCBs (often non-planar) with vicinal H atoms in *m,p*-positions, and variable substitution of Cl atoms in *o*-positions. Groups (1) and (2) are thought to be metabolized predominantly by P450s of, respectively, subfamilies IA and IIB. The following patterns of metabolism were indicated. Seabirds and the polar bear metabolized PCBs of group (2), but the facility was reduced in sea mammals and particularly low in cetaceans. Cetaceans, the harbour seal and the polar bear metabolized PCBs of group (1). Seabirds metabolized PCBs of group (1) less effectively than PCBs of group (2). Whilst these observations raise interesting possibilities about the representation of cytochrome P450s of the subfamilies IA and IIB in the different vertebrate groups (viz. IA in all groups, IIB low in cetaceans, IIB higher than IA in birds), firm conclusions must await the detailed characterization of these enzymes, as for example has been achieved for the demonstration of a IA representative in teleost fish (Stegeman, 1989). It is dangerous to rely too much on making analogies with the cytochrome P450s of rodents which are relatively well studied. In the longer term, the development of a phylogenetic classification for all the major enzymes concerned with the metabolism of pollutants by marine organisms is to be hoped for. Such a classification should give insight into the evolution of the enzymes in response to the selective pressure exerted by naturally occurring xenobiotics such as plant metabolites.

Unique patterns of PCB congener metabolism were indicated for polar bear (metabolism of congeners without any vicinal H atoms; Chapter 6) and the polychaete *Nereis diversicolor* (congeners highly chlorinated in one phenyl only; Chapter 1). Information on the P450 forms present, or likely to be involved in PCB metabolism, in marine invertebrates is virtually non-existent, but metabolism of both planar congeners (2,5,4′- in *Nereis virens*; 2,5,3′,4′- in *Asterias rubens*) and non-planar congeners (2,2′- and 2,4,6,2′,4′- in *N. virens*) has been indicated or observed (Chapters 1 and 2).

The most significant fact for the metabolism of PCB congeners and other persistent pollutants by all the animal groups, of course, is that many are either not metabolized at all by the P450s (some PCBs inactivate or inhibit the P450s), or are metabolized at a much lower rate than for most xenobiotics, e.g. metabolism of PCBs by fish was minimal (Chapter 5). The limited impact of metabolism on PCB body burden was reflected in the fact that PCB congener profiles in crustaceans, fish (Chapter 4) and sea mammals (Chapter 6) were similar in different tissue types, despite the MFO system being localized in particular tissues. However, the greater metabolism in sea mammals and birds was also reflected in greater differences in PCB profiles between the animal and the source of the PCBs (water, sediment or food) (Chapter 7).

Other, perhaps more fundamental, differences in MFO systems are indicated in marine invertebrates, which could affect the nature of metabolic products formed, e.g. a less integrated functioning of cytochromes P450 and P450 reductase in crustaceans and molluscs, and one-electron oxidation as a prominent P450 catalytic mechanism in molluscs (Chapter 1). The latter is speculated to be a mechanism for macromolecular adduct formation, possibly contributing to the increased persistence of PAHs in molluscs with increased periods of exposure to these pollutants. It could also be involved where observed BCFs are greater than those predicted on the basis of K_{ow}, e.g. *bis*(tributyltin) oxide in *M. edulis* (Chapter 1).

Macromolecular adduct formation of pollutants to DNA, RNA or protein is clearly a widespread phenomenon in animals (Chapters 1 and 5), and will contribute to the persistence of xenobiotics and/or their derivatives. In fish, DNA-xenobiotic adducts were found to be relatively more persistent than either parent PAHs or unbound metabolites (Chapter 5). Also, there was some indication that BaP formed macromolecular adducts more readily than PCBs, possibly due to different mechanisms of metabolism.

The studies on invertebrates highlighted the slow release of metabolites, compared to parent compound, from these organisms (Chapters 1 and 2). This situation was observed for many different types of chemicals (PAHs and other hydrocarbons, PCBs and other organochlorines, phenols, nitroaromatics, others) and identified the requirement for an effective excretion system, as well as the necessary biotransformation enzymes, for the elimination of xenobiotics by this means. The paradoxical situation therefore exists for marine invertebrates that with transfer to clean water, or cessation of the input of pollutant, biotransformation is likely to lead to persistence of derivatives of the chemical. In contrast, in sea mammals and birds, the lack of gills or other permeable interfaces with ambient waters mean that metabolism and excretion of polar metabolites are the sole major mechanism of xenobiotic elimination for most tissues (Chapters 6, 7 and 9).

Future studies must focus on identifying the quantitative and qualitative differences in biotransformation enzymes, including properties and regulation, between different species and animal groups. Elaboration of the nature of the P450s, and the use of such knowledge in pollution monitoring (e.g. P450 IA; see later),. will be particularly important. Improved knowledge of quantitative aspects of *in vivo* metabolism and the relationships with *in vitro* enzyme activities also offer great potential for predictive application.

MODELLING THE FATE AND EFFECTS OF POLLUTANTS

Models have been developed which can describe and predict the fate of pollutants in the marine environment. Firstly, there has been some success in the development of models to predict the gross movement and distribution of chemicals over large areas (Jorgensen, 1990). The use of the concept of fugacity has proved valuable in the construction of such predictive models for the environmental fate of lipophilic organic chemicals in the aquatic environment (Vighi and Calamari, 1990).

This type of modelling has not been dealt with in the present book, where the emphasis has been upon the use of models to predict the behaviour of chemicals in ecosystems. The simplest example of this type of model relates to the uptake of lipophilic compounds by molluscs and other marine organisms, where the degree of bioconcentration has been closely related to partition coefficients (e.g. K_{ow}) or water solubilities. Here uptake and loss are largely determined by the process of passive diffusion (Chapters 1 and 2; Walker, 1987). Uptake is very largely direct from ambient water, and loss is predominantly due to diffusion into ambient water. Metabolism and elimination of metabolites are too slow to contribute significantly to loss (Chapter 1), and prediction of BCFs becomes possible when the organisms are in the steady state. If metabolic fate of a pollutant in a marine invertebrate is of interest, however, then the relationship identified in Chapter 1 indicates that predictions of flux are possible, based on tissue body burdens. With more study this could be extended to fate as well, e.g. it may be possible to predict what fraction of a pollutant ends up as a macromolecular adduct, or to what extent macromolecular adduct formation is a sink for pollutants in species, populations or communities.

More complex models than this are needed when other factors become important in determining rates of uptake and loss. Thus, with marine organisms, ingestion of pollutants with food is often a major route of uptake. Similarly, metabolism and the growth of an organism may contribute to reduction of body burdens (Chapter 4; Walker, 1987, 1990). In this more complex situation, the simple model for bioaccumulation using K_{ow} or water solubility is usually inadequate. More complex models, such as the one used in the study of PCB accumulation in New Bedford Harbor, are appropriate (Chapter 4). More complex models are also being developed for the fate of pollutants in individual species, e.g. multicompartment models, incorporating physiological variables such as blood flow to gills (Barron *et al.*, 1990; Erickson and McKim, 1990). Such models should presumably focus on tissues that are major deposition sites or important target organs. In the case of cetaceans, the best model for chemicals such as PCBs was uptake through food, blood as a central transporter and lipid-based partitioning into the different tissues (Chapter 6).

In the case of predators at the top of food chains, e.g. cetaceans and seabirds, other considerations apply (Walker, 1987, 1990). Uptake of lipophilic pollutants is very largely from food and loss is very largely by metabolism. There is hardly any direct uptake from water and there is very little loss of unchanged pollutant either by diffusion or by export with lipoproteins. Female cetaceans lose a small amount of their body burden of pollutants by excretion into milk (Chapters 6 and 7); female seabirds lose a minute portion by transfer into eggs (Chapters 8 to 10). These losses are, at most, relatively small, and of course do not occur in the males of the species. A steady-state model has been proposed, which bases the prediction of BAFs on *in vitro* metabolic data (Walker, 1987). With lipophilic compounds whose primary metabolism is carried out by enzymes located in the endoplasmic reticulum, kinetic plots may be obtained which relate concentration in microsomes to rate of metabolism. These plots can then be used to predict the concentration of a pollutant needed in the endoplasmic reticulum to give metabolic rates which will balance rates of uptake *in vivo*. If successful, such a model would enable prediction of BAFs of pollutants for a wide range of species based upon laboratory data that can be obtained rapidly and cheaply. Most importantly, it could be used to make predictions for rare and protected species, most of which, for practical and/or ethical reasons, will never be available for long-term *in vivo* studies of bioaccumulation.

Speaking generally, modelling has been used to study the movement, distribution and metabolism of lipophilic pollutants with a certain degree of success. This approach has been useful experimentally to improve understanding of the basic processes involved, e.g. in identifying the routes of uptake of pollutants (Chapter 4), or the mechanisms by which organisms eliminate pollutants (Chapter 1). Apart from this, there is a growing interest in the development of predictive models which can be used to facilitate pollution control. The risk assessment of industrial chemicals or pesticides will become easier if better models are available. Such models must of necessity embrace both fluxes and fates of xenobiotics, and their toxic effects. Approaches towards the latter include the elucidation of quantitative structure–activity relationships (QSARs) (see next section). Whilst encouraging success has been made in the types of modelling just described, far less progress has been achieved in modelling for effects on individuals or populations (Donkin and Widdows, 1990; Kooijman, 1991; Van der Hoeven, 1991). This is understandable, because of the complexity of ecosystems and the large number of variables which determine biological effects. Effects at the level of population or community are particularly hard to model (Somerville and Walker, 1990). It is also regrettable, because the ultimate concern of ecologists and conservationists is with harmful effects of chemicals upon ecosystems. The development of models predicting effects of chemicals upon marine ecosystems remains a challenge for the future.

TOXIC EFFECTS

Certain pollutants cause concern because they may have effects in ecosystems. Persistent pollutants cause particular concern because of their long biological half-lives, their transfer between animals, and the possibility of being recycled back to the base of the food chain. Testimony to the persistence and stability of, for example, PCBs is their global distribution, despite their production being banned or severely limited in most countries since the 1970s. Although the initial interactions of any toxicant is at the molecular level, the consequences of such interactions can be manifest at any level of biological organization, right through to the levels of organism, population and community. A myriad of potential sites of toxic action exist for any chemical, but some will be more important than others, whereas others may be compensated for and have little or no consequence. A number have been highlighted in the foregoing chapters.

Four potential sources of toxic molecular species can be identified, derived either directly or indirectly from the pollutant, viz. the parent compound itself, metabolites and free radical derivatives of the compound, and enhanced production of reactive oxygen species or oxyradicals (O_2^-, H_2O_2, $\cdot OH$) (Livingstone, 1991a). These in turn will generate other reactive species by transfer and other free radical reactions with endogenous molecules (see Livingstone, 1991a). Oxidative damage to lipid and DNA has been observed in marine invertebrates and vertebrates (Di Giulio *et al.*, 1989; Livingstone *et al.*, 1990; Winston, 1991). A recent example was the detection of the guanine-derived lesion 2,6-diamino-4-hydroxy-5-formamido-pyrimidine (FapyGua; thought to arise from the attack of hydroxyl radical at C-8 position of guanine) in DNA from neoplastic hepatic tissues of English sole *Parophrys vetulus* from a polluted environment (Malins *et al.*, 1990). Genotoxic effects have been observed in both marine invertebrates and vertebrates (Chapter 5; Livingstone, 1991a; Chipman and Marsh, 1991), in some cases possibly organic radical-mediated, e.g. fish (Malins *et al.*, 1983) and mussels (Bihari *et al.*, 1990). In the case of two amphipod species, enhanced macromolecular adduct formation correlated with greater sensitivity to contaminant exposure (Chapter 5). Many other effects on other aspects of biochemistry have been recorded, e.g. on enzyme activities and levels of adenosine phosphates (Livingstone, 1985; Holwerda and Opperhuizen, 1991). Most biotransformation enzymes also have endogenous substrates and functions e.g. P450s and steroid metabolism, and deleterious effects of pollutants could be mediated by the induction or inhibition of such enzymes, e.g. on reproduction (see below).

PCBs and the related PCDDs and PCDFs have been extensively studied in mammals, particularly rodents and man (Safe, 1990; Silberhorn *et al.*, 1990). PCBs induce preneoplastic lesions, neoplastic nodules and hepatocellular carcinomas and can therefore be considered complete carcinogens. Females are usually more sensitive than males to the tumorigenic effects of PCBs. PCB mixtures and individual congeners are effective tumour promoters in liver: this ability is correlated with their ability to induce cytochromes P450IA and IIB (the mechanism of stimulating cell proliferation is likely different for the two types of PCB inducers). PCBs can also enhance or inhibit the process of carcinogenesis by altering, through induction of P450s, the metabolism of other initiating chemicals. Such phenomena have also been observed in fish (Bailey *et al.*, 1987) and birds (Brunström, 1991). PCDDs and PCDFs can cause similar effects through a common receptor-mediated mechanism (induction of P450 and other enzymes) of toxic action (see below). Other possible molecular mechanisms of PCB-mediated carcinogenicity include lipid peroxidation and free radical formation, depression of protective mechanisms such as selenium-dependent glutathione peroxidase (EC 1.11.1.9) and interference with vitamin A metabolism (involved in cellular growth and differentiation).

Sex differences were apparent for the effects of PCBs (Clophen A50) on reproduction in the starfish *Asterias rubens*. PCBs impaired the reproductive development of female but not male starfish (Den Besten *et al.*, 1990). Similarly, whereas normal fertilization could be achieved with spermatozoa from male PCB-exposed starfish, the early development of embryos was disturbed and the percentage success of normal larval development (bipinnaria larvae; 1 week) was reduced (Den Besten *et al.*, 1989). In pyloric caeca microsomes of PCB-exposed animals, cytochrome P450 content was reduced but BPH activity was increased: perhaps significantly, the increase in BPH activity was greater in female than male starfish (Den Besten, 1991). Reproductive and other biological effects, which could be cytochrome P450-mediated, are apparent in marine invertebrates. Developmental and other changes (molting, growth, limb regeneration), possibly due to the presence of PCDDs and PCDFs, were seen in juvenile blue crabs *Callinectes sapidus* exposed to field contaminants (Chapter 3). The development of imposex is promoted in the dog-whelk *Nucella lapillus* by organotin compounds (Bryan *et al.*, 1988).

PCBs (and related PCDFs and PCDDs) and DDTs are variously implicated in reproductive, teratogenic and immunological dysfunction in fish, sea mammals and seabirds (Chapters 5–10). Such chemicals have also been considered to cause the death of organisms, e.g. seabirds exposed to PCBs (Chapter 9). The mechanistic understanding of such processes are limited, but in marine mammals it is thought likely that PCBs potentially exert their toxicity in at least three different forms, viz. as parent compounds and as the hydroxylated and sulphonated metabolites (Chapter 6). Hydroxylated metabolites of particular congeners are known to reduce plasma levels of vitamin A and thyroxine in rodents, and the same situation has been implicated in seals reproductively impaired by PCB exposure. Correlations between blubber levels of PCBs and *pp'*DDE and reduced blood testosterone levels were seen in Dall's porpoise (Chapter 7). Induction of P450s of the IA family could also be an important toxic mechanism in fish (Chapter 5) and sea mammals (Chapter 6), exacerbating the effects of other contaminants such as PAHs. Correlations between P450IA levels (7-ethoxyresorufin *O*-deethylase [EROD] activity) and PCB body burdens have been observed in livers of short-finned pilot whales (Chapter 7). Equally well, it has been argued that cetaceans could be particularly susceptible to reproductive impairment by organochlorines because of the deficiency of P450IIB enzymes and the lower capacity of P450IA (BPH) ones to metabolize the contaminants (Chapter 7). Organochlorine toxicity in seabirds has been primarily associated with eggshell thinning and direct embryo toxicity (Chapters 9 and 10). Eggshell thinning is particularly associated with the action of *pp'*DDE.

There is evidence that the effect of *pp*′DDE is caused by inhibition of Ca^{2+} transport (mediated by Ca^{2+} ATPase) in the shell gland (Lundholm, 1987).

In terms of understanding and predicting the molecular toxicological effects of persistent pollutants, the various observations highlight two particularly important questions, viz. (a) if several types of potentially toxic molecular species exist, including the parent compound, which of them is likely to be toxicologically the most important, and (b) what influence do interactive effects at the toxicokinetic level have on the concentrations and effects of the pollutants (Chapter 10)? One approach to answering such questions is to carry out *in vivo* toxicity experiments in the presence or absence of known inhibitors or inducers of important metabolic pathways. For example, inhibition of phase 2 metabolism with salicylamide increased the toxicity of 3-trifluoromethyl-4-nitrophenol to rainbow trout (*Salmo gairdnerii*) by more than 60% (Lech, 1974). Using an alternative approach, Hermens *et al*., (1990) showd that pre-induction of the MFO system of rainbow trout with PCBs had no effect on the 96 h LC_{50} of 4-chloroaniline. However, the toxicity of certain PAHs to cultured bluegill sunfish cells was enhanced by as much as 80% by the addition of a rat-derived S-9 (9,000 g supernatant) fraction, presumably as a result of increased toxic metabolite production (Babich and Borenfreund, 1987).

These observations show that modulation of metabolism can increase, decrease or have no effect on toxicity. They also provide some indication of the nature of the toxic species. However, a more systematic way of establishing the molecular characteristics required to elicit toxicological responses is to establish quantitative relationships between molecular descriptors and toxicological response for a series of compounds.

Such QSARs have been established for the effects on aquatic organisms of a large number of industrial chemicals of differing degrees of persistence (Hermens, 1986; Donkin and Widdows, 1990). They commonly point to the importance of a non-specific narcotic mode of action, dominated in aquatic organisms by the influence of molecular hydrophobicity on bioaccumulation. Similar relationships are observed for lethal and a variety of sublethal endpoints and for organisms as diverse as fish and bacteria, with very different metabolic capabilities. These results strongly suggest that the primary mode of action of many pollutants is determined by the parent compound, e.g. by interaction with functional membrane proteins (Ahlers *et al*., 1991).

However, many of these aquatic QSAR studies are dominated by compounds of relatively low hydrophobicity ($K_{ow} < 10^5$), but when compounds of greater hydrophobicity are included, the slopes of the QSAR lines are often reduced indicating declining toxicity (Hermens, 1986; Donkin and Widdows, 1990). This could be indicative of a greater role for metabolism in the elimination of compounds with long depuration half-lives, but more common explanations relate to reduced diffusion rates and to the decreasing lipid solubility of very hydrophobic compounds (Donkin and Widdows, 1990; Connell and Hawker, 1988). Some QSARs show that toxicity of certain organic contaminants, expressed in terms of aqueous concentration, greatly exceeds that of non-specific narcotics of comparable hydrophobicity. Specific modes of action attributed to the parent compounds, such as neurotoxicity of organochlorine pesticides (Hermens, 1986; Bradbury *et al.*, 1991), and respiratory uncoupling by ionizable phenols (Schulz *et al.*, 1986), are generally invoked to explain such data.

In QSAR studies of PCBs, PCDDs and polybrominated dibenzo-*p*-dioxins in rats, it was shown that body weight loss and thymic atrophy were linearly correlated with induction of BPH and EROD activities (Safe, 1990). The compounds with less toxicity than predicted from the QSARs were those which were readily metabolized and excreted. Although the toxicity of PCBs and related compounds is thought to be mediated through binding to the *Ah* receptor (Roberts *et al.*, 1985), and this has been used to develop "toxic equivalency factors" (Safe, 1990), binding assays do not lead to such precise QSARs as the MFO activity induction determinations. Indeed McKinney *et al.*, (1985) suggested that the receptor mediating the toxic response is distinct from the *Ah* receptor involved in induction. Whatever the detailed mechanism, binding of parent compound is of paramount importance. The extent to which such mechanisms of toxic action apply to marine species is uncertain at present; in fish, similarities and differences from the rodent model are apparent (Hahn *et al.*, 1989).

In most instances, QSAR studies of persistent pollutants point to the importance of parent molecules in producing toxicity; metabolism is considered important when it reduces the tissue levels of these molecules and hence their toxicity. However, QSARs often combine data for compounds with toxicities varying over several orders of magnitude, so can be insensitive to small changes. Furthermore, the duration of toxicity determinations rarely exceeds 14 days, and is frequently shorter, so such experiments are a poor model for chronic environmental contamination which may give rise to significant metabolically mediated oxidative and genetic damage over longer timescales. There is a need for more closely coupled QSAR and metabolic experiments.

The study of mechanisms of toxicity in marine organisms at the molecular level is opening the way for the development of biomarkers (McCarthy and Shugart, 1990; Peakall, 1991). The measurement of characteristic biochemical changes (protective responses, alterations in steady-state concentrations of key molecules, end-points of damage) caused by pollutants, either singly or in combination, can give evidence of exposure and of early toxic action. One such example is the well-tried and almost routine use, at least in fish, of EROD activity as an indicator of cytochrome P450IA induction and impact by organic pollution (PAHs, PCBs, PCDDs, PCDFs, others). There is thus considerable interest in the development of simple, sensitive and selective assays which can be automated or incorporated into diagnostic kits for use in the field by workers without specialized biochemical knowledge. Again progress is being made with the measurement of cytochrome P450IA induction, viz. use of ELISA (Goksøyr, 1991) and microplate (EROD) (Grzebyk and Galgani, 1991) techniques. The use of such biomarkers should lead to rapid progess in the fields of monitoring and environmental impact assessment for marine pollutants. It is an approach which can contribute to establishing causal relationships between the presence of the pollutants and the toxic effects that they have at the levels of individual organism, population and community.

An important consideration of toxic effects on individuals is the extent to which decreases in animal fitness and performance, and/or increased mortality, are translated through to effects on populations and communities (Bayne *et al.*, 1988; Underwood and Peterson, 1988). A key factor in this is presumably the magnitude of the pollutant-mediated events compared to the natural ones. Various sublethal effects on individuals have been identified which could have consequences for the population, e.g. alterations in moulting and growth in crabs (Chapter 3) and aspects of reproduction in fish (Chapter 5), sea mammals (Chapters 6 and 7) and birds (Chapter 10). Pollutant impact is indicated to be possibly greater at certain life stages of an organism, e.g. organochlorine poisoning of birds occurred mainly at two stages of development—in chicks shortly after hatching, and in juveniles when they started flying (Chapter 10). Impact will obviously also depend on the effectiveness and flexibility of an organism's protective systems, e.g. cetaceans were considered to be likely susceptible to reproductive impairment by persistent organochlorines because of the deficiency of inducible cytochromes P450IIB (Chapter 7). Induction of detoxication systems and changes in other biochemical and physiological processes may lead to increased tolerance and adaptation (see respectively Chapters 2 and 3). Despite all these and other complexities, extrapolation of events through the biological hierarchical levels has obvious predictive and environmental management potential and continues to be a concern of ecotoxicology (De Kruijf, 1991; Seitz and Ratte, 1991).

EVOLUTIONARY CONSIDERATIONS

The interaction of xenobiotics with organisms is significant from an evolutionary point of view. Apart from the intrinsic interest, a knowledge of the evolution of enzymes and other biological components in response to the selective pressure of natural xenobiotics can provide a conceptual framework within which the effects of man-made contaminants upon organisms and ecosystems can be better understood. For example, the development of phylogenetic classifications for enzymes that metabolize xenobiotics, e.g. MFO system and esterases, can give useful guidance as to the ability of a wide range of different animal groups to perform particular types of biotransformations. This will aid in the identification of sensitive and/or bioindicator species, as well as facilitating risk assessment. The emergence of resistance mechanisms against man-made toxicants may be seen in relation to the longer-term evolution of protective systems against natural xenobiotics, e.g. enzymic detoxication of pesticides in insects and rodents compared to that of plant metabolites in herbivores. Such an evolutionary approach leads to a consideration of the energy costs associated with the detoxication of xenobiotics, and of the ability of organisms to adapt to contaminated (hostile) environments. An aspect of such adaptation is the largely unexplored area of the development of targets insensitive to the action of the toxicant, such as new forms of cholinesterase or sodium channels in insects resistant to insecticides.

A putative evolutionary development is most obvious for the MFO system, the key enzyme in the metabolism and recalcitrance of many pollutants, particularly those lacking any functional groups (Chapters 1, 5, 6, 7 and 9). As noted earlier, quantitative and qualitative increases in MFO capacity are observed during the course of evolution of animals, as indicated by larger quantities of endoplasmic reticulum containing higher levels of MFO components and activities, increased numbers of cytochrome P450 genes and isozymes and more responsive (inducible) systems of regulation. Lower down the phylogenetic tree, a less inducible and possibly more primitive MFO system is indicated (see below).

"Plant-animal warfare" i.e. combating natural xenobiotics, is proposed to have been a major selective pressure in driving the evolution of cytochrome P450s, viz. the great increase in the number of members of the IIB family coincides with the move to land and a more varied diet (Gonzalez and Nebert, 1990). The relationship with diet appears discernible within phyla, in the form of both increased total MFO activity and numbers of isozymes. Thus, aldrin epoxidase activity (catalysed mainly by P450IIBs in mammals) was lower in fish-eating birds than species with a more varied diet. The same has been observed for monophagous compared to polyphagous insects (Krieger *et al.*, 1971). The cytochrome P450IIB family was also indicated to be under-represented in marine mammals compared to terrestrial mammals, the former most likely having a more restricted diet. An alternative or additional argument put forward to explain the increased presence and inducibility of P450 forms of the II family in terrestrial animals (mammals, birds, insects) compared to aquatic ones (fish and perhaps marine mammals) is the presence of a limitless medium into which the latter can discharge unchanged xenobiotics (Chapter 7). In contrast, the only major elimination route for terrestrial organisms is biotransformation and the excretion of polar metabolites. A single member of the P450IA subfamily (P450IA1) was indicated in fish, with a possible radiation in two forms (P450IA1 and P450IA2) in sea mammals (Chapter 6). Recent evidence suggests that a similar situation with respect to diet exists for serum carboxylesterases in birds (Thompson *et al.*, 1991). Higher activities and numbers of isozymes were found in herbivores and/or omnivores than in specialized predators.

Although the mechanistic understanding of marine invertebrate MFO systems is limited, there is some evidence to suggest a less integrated functioning of cytochrome P450 reductase and cytochrome P450, with possibly a more prominent role for peroxidatic reactions (Livingstone, 1990). Given (1) the early need to utilize oxygen, yet closely control its toxicity by antioxidant systems and carrier molecules (Scott, 1991), (2) the likely early toxic effects of oxygen on organic molecules, including the production of lipid peroxides (Hassan and Schiavone, 1991); and (3) the assumption that cytochromes P450 and P450 reductase originally functioned as independent enzymes: a possible scenario in the primitive cell, or organism, is that oxidative detoxication first occurred mainly on the basis of cytochrome P450-mediated peroxidation utilizing endogenous hydroperoxides rather than MFO system-mediated monooxygenation utilizing dioxygen and reducing equivalents. By a similar analogy, the widespread, non-inducible flavoprotein monooxygenase (EC 1.14.13.8), which is a single protein, might be considered a more ancient detoxication enzyme than the multicomponent MFO system. Also, the former's capacity for *S*-oxidation may be no coincidence, given the sulphur-reducing conditions of early earth.

The marked accumulation of a much wider range of xenobiotics in marine invertebrates than vertebrates was attributed mainly to their lower levels of biotransformation enzymes, and possibly also to the inefficient excretion of metabolites. If a similar situation pertained for the uptake of natural xenobiotics from the environment and food, then it could be speculated that having to tolerate the presence of such high levels of foreign chemicals in the tissues, particularly say in membrane systems, may have been a major contributory factor in restricting their development and evolution. Alternatively, rates of uptake could be lower, and rates of metabolism higher, for natural xenobiotics than for the various chemicals examined. Other possible reasons as to why marine invertebrates have not evolved higher levels of biotransformation enzymes, particularly the MFO system, include (1) it is energetically too costly—there is very little, if any, information on this aspect (Calow, 1991), but if the detoxication systems (enzymes, scavengers, excretion systems) represented say 10% of the energy budget, it would be very difficult to achieve the 10-fold increase in effectiveness necessary to cope with the rate of xenobiotic uptake (see Chapter 1); (2) most biotransformation enzymes also have endogenous functions and to increase detoxication capacity without affecting the latter would require increased separation of the two functions and a more sophisticated system of control, i.e. more genes and more regulation; and (3) there is some evidence to suggest that levels of free radical scavengers, e.g. carotenoids, are particularly high in invertebrates (Livingstone, 1991a), and it may be that these are the main line of defence against xenobiotics, i.e. the main toxic threat in these organisms is from free radical reactions of the parent compound.

As well as possibly imposing restrictions on their evolution, the inadequacy of the biotransformation and excretion system of marine invertebrates will presumably also have given a selective boost to other mechanisms of dealing with xenobiotics, such as extracellular or intracellular compartmentation, e.g. binding to proteins and lipoproteins, and sequestration into lysosomes (Chapter 1). Aspects of macromolecular adduct formation could in fact be a detoxication mechanism if the macromolecules involved are designed for the purpose. Given that the metabolism of lipophilic compounds to more polar products restricts their movement across membranes, the apparent inability of marine invertebrates to excrete such metabolites raises the question as to the main function (original or present) of the xenobiotic-metabolizing capability of biotransformation enzymes. Possibilities include (1) the main threat of organic toxicants is penetration and disruption of membrane function, and conversion to polar metabolites prevents this; (2) in combination with hydrolytic enzymes (glucuronidases, sulphatases etc.), it allows control of intracellular (or extracellular) movement of the xenobiotic, viz. metabolism prevents random movement of xenobiotic across membranes, and hydrolysis at the required site releases parent compound (analogous to, or possibly the rudiments of, phase III metabolism [Livingstone, 1991a] of vertebrates); and (3) it is involved in resource acquisition, i.e. xenobiotics are trapped in the cell for possible exploitation.

A final important consideration is the extent to which man-made xenobiotics impose a toxic threat in relation to that already present in the diet from nature's pesticides (protective chemicals in plants) (Ames, 1989). The latter are present in high numbers and concentrations, and several have already been shown to bind to the *Ah* receptor and mimic the action of well-known pollutants such as TCDDs. Thus, it has been argued that in man a portion of broccoli (active ingredient: derivatives of indole carbinol) poses about 2×10^7 times the threat of the daily acceptable dose of TCDD. The counter to this, of course, is that detoxication enzymes have evolved largely to protect against natural xenobiotics, whereas anthropogenic pollutants are of relatively recent origin and may not be metabolized effectively by such enzymes. Such is the case with many of the highly chlorinated compounds, including PCBs, PCDDs, PCDFs and organochlorine pesticides, which consequently undergo bioaccumulation in predators at the top of food chains to an extent which is probably not encountered with natural xenobiotics.

The consequence of this evolutionary war between animals and plants is maintained to be a multilayered defence system, almost all of it inducible, which effectively buffers against toxicity of low doses of both man-made and natural chemicals (Ames, 1989). Clearly, elements of this will hold for any animal, but other considerations, obvious from the observed effects of pollutants on marine organisms, will also pertain, viz. (1) the degree of protection may vary with different animals, (2) dose levels may be high, particularly for persistent chemicals, (3) many of the defence systems seem geared towards DNA rather than immediate protection of functional protein molecules, and (4) very specific effects with very serious consequences can occur, e.g. *bis*(tributyl) tin and imposex in whelks.

REFERENCES

Ahlers, J., I. Cascorbi, M. Forêt, A. Gies, M. Köhler, W. Pauli and E. Rösick. 1991. Interaction with functional membrane proteins—a common mechanism of toxicity for lipophilic environmental chemicals? *Comp. Biochem. Physiol.* **100C**: 111–113.

Ames, B. N. 1989. Mutagenesis and carcinogenesis: endogenous and exogenous factors. *Environ. Molec. Mutagen.* **14**: supplement 16; 66–77.

Babich, H. and E. Borenfreund. 1987. Polycyclic aromatic hydrocarbon *in vitro* cytotoxicity to bluegill BF-2 cells: mediation by S-9 microsomal fraction and temperature. *Toxicol. Letts.* **36**: 107–116.

Bailey, G. S., D. Selivonchick and J. Hendricks. 1987. Initiation, promotion and inhibition of carcinogenesis in rainbow trout. *Environ. Health Perspect.* **71**: 147–153.

Barron, M. G., G. R. Stehly and W. L. Hayton. 1990. Pharmacokinetic modeling in aquatic animals. I. Models and concepts—review. *Aquat. Toxicol.* **18**: 61–85.

Bayne, B. L., R. F. Addison, J. M. Capuzzo, K. R. Clarke, J. S. Gray, M. N. Moore and R. M. Warwick. 1988. An overview of the GEEP workshop. *Mar. Ecol. Prog. Ser.* **46**: 235–243.

Bihari, N., R. Batel and R. K. Zahn. 1990. DNA damage determination by the alkaline elution technique in the haemolymph of mussel *Mytilus galloprovincialis* treated with benzo[a]pyrene and 4-nitroquinoline-*N*-oxide. *Aquat. Toxicol.* **18**: 13–22.

Bradbury, S. P., R. W. Carlson, G. J. Niemi and T. R. Henry. 1991. Use of respiratory-cardiovascular responses of rainbow trout (*Oncorhynchus mykiss*) in identifying acute toxicity syndromes in fish: seizure agents. *Environ. Toxicol. Chem.* **10**: 115–131.

Brunström, B. 1991. Toxicity and EROD-inducing potency of polychlorinated biphenyls (PCBs) and polycyclic aromatic hydrocarbons (PAHs) in avian embryos. *Comp. Biochem. Physiol.* **100C**: 241–243.

Bryan, G. W., P. E. Gibbs and G. R. Burt. 1988. A comparison of the effectiveness of tri-*n*-butyltin chloride and five other organotin compounds in promoting the development of imposex in the dog-whelk, *Nucella lapillus*. *J. Mar. Biol. Ass. U.K.* **68**; 733–744.

Calow, P. 1991. Physiological costs of combating chemical toxicants: ecological implications. *Comp. Biochem. Physiol.* **100C**: 3–6.

Chipman, J. K. and J. W. Marsh. 1991. Bio-techniques for the detection of genetic toxicity in the aquatic environment. *J. Biotechnol.* **17**: 199–208.

Clark, K. E., F. A. P. C. Gobas and D. Mackay. 1990. Model of organic chemical uptake and clearance by fish from food and water. *Environ. Sci. Technol.* **24**: 1203–1213.

Connell, D. W. and D. W. Hawker. 1988. Use of polynomial expressions to describe the bioconcentration of hydrophobic chemicals by fish. *Ecotoxicol. Environ. Safety* **16**: 242–257.

Dalvi, R. R., V. A. Nunn and J. Juskevich. 1987. Studies on comparative drug metabolism by hepatic cytochrome P–450-containing microsomal enzymes in quail, ducks, geese, chickens, turkeys and rats. *Comp. Biochem. Physiol.* **87C**: 421–424.

De Kruijf, H. A. M. 1991. Extrapolation through hierarchical levels. *Comp. Biochem. Physiol.* **100C**: 291–299.

Den Besten, P. J. 1991. Effects of cadmium and PCBs on reproduction of the sea star *Asterias rubens*. Ph.D. thesis, University of Utrecht, The Netherlands.

Den Besten, P. J., H. J. Herwig, A. C. Smaal, D. I. Zandee and P. A. Voogt. 1990. Interference of polychlorinated biphenyls (Clophen A50) with gametogenesis in the sea star, *Asterias rubens*. *Aquat. Toxicol.* **18**: 231–246.

Den Besten, P. J., H. J. Herwig, D. I. Zandee and P. A. Voogt. 1989. Effects of cadmium and PCBs on reproduction of the sea star *Asterias rubens*: aberrations in the early development. *Ecotoxicol. Environ. Safety* **18**: 173–180.

Di Giulio, R. T., P. C. Washburn, R. J. Wenning, G. W. Winston and C. S. Jewell. 1989. Biochemical responses in aquatic animals: a review of determinants of oxidative stress. *Environ. Toxicol. Chem.* **8**: 1103–1123.

Donkin, P. and J. Widdows. 1990. Quantitative structure–activity relationships in aquatic invertebrate toxicology. *Rev. Aquat. Sci.* **2**: 375–398.

Elliott, J. E., R. J. Norstrom and J. A. K. Keith. 1988. Organochlorines and eggshell thinning in Northern Gannets (*Sula bassanus*) from Eastern Canada 1968–1984. *Environ. Pollut.* **52**: 81–102.

Erickson, R. J. and J. M. McKim. 1990. A model for exchange of organic chemicals at fish gills: flow and diffusion. *Aquat. Toxicol.* **18**: 175–197.

Farrington, J. W., E. D. Goldberg, R. W. Risebrough, J. H. Martin and V. T. Bowen. 1983. US "Mussel Watch" 1976–1978. An overview of trace-metal, DDE, PCB, hydrocarbons, and artificial radionuclide data. *Environ. Sci. Technol.* **17**: 490–496.

Gobas, F. A. P. C., D. C. G. Muir and D. Mackay. 1988. Dynamics of dietary bioaccumulation and faecal elimination of hydrophobic organic chemicals in fish. *Chemosphere* **17**: 943–962.

Gobas, F. A. P. C., A. Opperhuizen and O. Hutzinger. 1986. Bioconcentration of hydrophobic chemicals in fish: relationship with membrane permeation. *Environ. Toxicol. Chem.* **5**: 637–646.

Goksøyr, A. 1991. A semi-quantitative cytochrome P450IA1 ELISA: a simple method for studying the monooxygenase induction response in environmental monitoring and ecotoxicological testing of fish. *Sci. Total Environ.* **101**: 255–262.

Gonzalez, F. J. and D. W. Nebert. 1990. Evolution of the P450 gene superfamily: animal-plant "warfare", molecular drive and human genetic differences in drug oxidation. *Trends in Genetics* **6**: 182–186.

Grzebyk, D. and F. Galgani. 1991. Measurement of the effect of organic pollution on marine organisms: rapid determination of EROD induction using plate readers. *Aquat. Living Resour.* **4**: 53–59.

Hahn, M. E., B. R. Woodin and J. J. Stegeman. 1989. Induction of cytochrome P–450E (P–450IA1) by 2,3,7,8-tetrachlorodibenzofuran (2,3,7,8–TCDF) in the marine fish scup (*Stenotomus chrysops*). *Mar. Env. Res.* **28**: 61–65.

Hassan, H. M. and J. R. Schiavone. 1991. The role of oxygen free radicals in biology and evolution. In C. Bryant (ed.), *Metazoan Life Without Oxygen*. Chapman and Hall, London, pp. 19–37.

Hawker, D. W. and D. W. Connell. 1985. Relationships between partition coefficient, uptake rate constant, clearance rate constant and time to equilibrium for bioaccumulation. *Chemosphere* **14**: 1205–1219.

Hawker, D. W. and D. W. Connell. 1988. Influence of partition coefficient of lipophilic compounds on bioconcentration kinetics in fish. *Wat. Res.* **22**: 701–707.

Hermens, J. L. M. 1986. Quantitative structure-activity relationships in aquatic toxicology. *Pestic. Sci.* **17**: 287–296.

Hermens, J. L. M., S. P. Bradbury and S. J. Broderius. 1990. Influence of cytochrome P450 mixed-function oxidase induction on the acute toxicity to rainbow trout (*Salmo gairdnerii*) of primary aromatic amines. *Ecotoxicol. Environ. Safety* **20**: 156–166.

Holden, A. V. 1970. Sources of PCB contamination in the marine environment. *Nature* **228**: 1220–1221.

Holwerda, D. A. and A. Opperhuizen (eds). 1991. Physiological and biochemical approaches to the toxicological assessment of environmental pollution. *Comp. Biochem. Physiol.* **100C**: Numbers 1/2, pp. 310.

Jorgensen. 1990. Modelling in ecotoxicology. *Developments in Environmental Modelling,* Vol. 18. Elsevier, Amsterdam.

Klungsøyr, J., S. Wilhelmsen, K. Westrheim, E. Saetvedt and K. H. Palmork. 1988. The GEEP workshop: organic chemical analyses. *Mar. Ecol. Prog. Ser.* **46**: 19–26.

Kooijman, S. A. L. M. 1991. Effects of feeding conditions on toxicity for the purpose of extrapolation. *Comp. Biochem. Physiol.* **100C**: 305–310.

Krieger, R. I., P. P. Feeny and C. F. Wilkinson. 1971. Detoxication in the guts of caterpillars: an evolutionary answer to plant defenses. *Science* **172**: 579–581.

Lech, J. J. 1974. Glucuronide formation in rainbow trout—effect of salicyclamide on the acute toxicity, conjugation and excretion of trifluoromethyl-4-nitrophenol. *J. Biochem. Pharmacol.* **23**: 2403–2410.

Lee, R. F. 1991. Lipoproteins from the hemolymph and ovaries of marine invertebrates. *Adv. Comp. Environ. Physiol.* **7**: 187–207.

Livingstone, D. R. 1985. Biochemical measurements. In *The Effects of Stress and Pollution on Marine Animals*, Praeger Scientific, New York, pp. 81–132.

Livingstone, D. R. 1990. Cytochrome *P*–450 and oxidative metabolism in invertebrates. *Biochem. Soc. Trans.* **19**: 15–19.

Livingstone, D. R. 1991a. Organic xenobiotic metabolism in marine invertebrates. *Adv. Comp. Environ. Physiol.* **7**: 45–185.

Livingstone, D. R. 1991b. Towards a specific index of impact by organic pollution for marine invertebrates. *Comp. Biochem. Physiol. Part C* **100C**: 151–155.

Livingstone, D. R., P. Garcia Martinez, X. Michel, J. F. Narbonne, S. O'Hara, D. Ribera and G. W. Winston. 1990. Oxyradical production as a pollution-mediated mechanism of toxicity in the common mussel, *Mytilus edulis* L., and other molluscs. *Funct. Ecol.* **4**: 415–424.

Lobel, P. B., S. P. Belkhode, S. E. Jackson and H. P. Longerich. 1990. Recent taxonomic discoveries concerning the mussel *Mytilus*: implications for biomonitoring. *Arch. Environ. Contam. Toxicol.* **19**: 508–512.

Lundholm, E. 1987. Thinning of eggshells of birds by DDE; mode of action on eggshell gland. *Comp. Biochem. Physiol.* **88C**: 1–22.

Malins, D. C., G. K. Ostrander, R. Haimanot and P. Williams. 1990. A novel DNA lesion in neoplastic livers of feral fish: 2,6-diamino-4-hydroxy-5-formamidopyrimidine. *Carcinogensis* **II**: 1045–1047.

McCarthy, J. F. and L. Shugart (eds.) 1990. *Biological Markers of Environmental Contaminants*. Chelsea: M. I. Lewis Publishers Inc.

McKim, J., P. Schmieder and G. Veith. 1985. Absorption of organic chemical transport across trout gills as related to octanol-water partition coefficient. *Toxicol. Appld. Pharmacol.* **77**: 1–10.

McKinney, J. D., K. Chae, E. E. McConnell and L. S. Birnbaum. 1985. Structure-induction versus structure-toxicity relationships for polychlorinated biphenyls and related aromatic hydrocarbons. *Environ. Health Perspect.* **60**: 57–68.

Nebert, D. W., M. Adesnik, M. J. Coon, R. W. Estabrook, F. J. Gonzalez, F. P. Guengerich, I. C. Gunsalus, E. F. Johnson, B. Kemper, W. Levin, I. R. Phillips, R. Sato and M. R.

Waterman. 1987. The P–450 gene superfamily: recommended nomenclature. *DNA* **6**: 1–11.

Nelson, A., N. Auffret and J. Borlakoglu. 1990. Interaction of hydrophobic organic compound with mercury adsorbed dioleoylphosphatidylcholine monolayers. *Biochem. Biophys. Acta* **1021**: 205–216.

Oliver, B. G. and A. J. Nliml. 1985. Bioconcentration factors of some halogenated organics for rainbow trout: limitations in their use for prediction of environmental residues. *Environ. Sci. Technol.* **19**: 842–849.

Opperhuizen A. and R. C. A. Stokkel. 1988. Influence of contaminated particles on the bioaccumulation of hydrophobic organic micropollutants in fish. *Env. Res.* **51**: 165–177.

Opperhuizen, A., E. W. van de Velde, F. A. P. C. Gobas, D. A. K. Llem, J. M. D. van de Steen and O. Hutzinger. 1985. Relationship between concentration in fish and steric factors of hydrophobic chemicals. *Chemosphere* **14**: 1871–1896.

Peakall, D. 1991. *Biomarkers in Environmental Health Assessment.* Chapman Hall.

Perdu-Durand, E. F. and J. P. Craven. 1989. Characterization of xenobiotic metabolizing enzymes in sturgeon (*Acipenser baeri*). *Comp. Biochem. Physiol.* **93B**: 921–928.

Roberts, E. A., N. H. Shear, A. B. Okey and D. K. Manchester. 1985. The *Ah* receptor and dioxin toxicity: from rodent to human tissues. *Chemosphere* **14**: 661–674.

Safe, S. 1990. Polychlorinated biphenyls (PCBs), dibenzo-*p*-dioxins (PCDDs), dibenzofurans (PCDFs), and related compounds: environmental and mechanistic considerations which support the development of toxic equivalency factors (TEFs). *Crit. Rev. Toxicol.* **21**: 51–88.

Scott, D. L. 1991. The physical chemistry of oxygen. In C. Bryant (ed.), *Metazoan Life Without Oxygen.* Chapman and Hall, London, pp. 1–18.

Schulz, T. W., G. W. Holcombe and G. L. Phipps. 1986. Relationships of quantitative structure-activity to comparative toxicity of selected phenols in the *Pimephales promelas* and *Tetrahymena pyriformis* test systems. *Ecotoxicol. Environ. Safety* **12**: 146–153.

Seitz, A. and H. T. Ratte. 1991. Aquatic ecotoxicology: on the problems of extrapolation from laboratory experiments with individuals and populations to community effects in the field. *Comp. Biochem. Physiol.* **100C**: 301–304.

Shaw, G. R. and D. W. Connell. 1984. Physicochemical properties controlling polychlorinated biphenyl (PCB) concentrations in aquatic organisms. *Environ. Sci. Technol.* **18**: 18–23.

Sijm, D. T. H. M. and A. Opperhuizen. 1989. Biotransformation of organic chemicals by fish: enzyme activities and reactions. In O. Hutzinger (ed.), *Handbook of Environmental Chemistry*, Vol. 2E. *Reactions and Processes*. Springer-Verlag, Berlin, pp. 163–235.

Silberhorn, E. M., H. P. Glauert and L. W. Robertson. 1990. Carcinogenicity of polychlorinated biphenyls: PCBs and PBBs. *Crit. Rev. Toxicol.* **20**: 439–496.

Somerville, L. and C. H. Walker. 1990. *Pesticide Effects on Terrestrial Wildlife*. Taylor and Francis, London, pp. 404.

Stegeman, J. J. 1981. Polynuclear aromatic hydrocarbons and their metabolism in the marine environment. In H. Gelboin and T. S'o. Pop (eds.), *Polycyclic Hydrocarbons and Cancer*, Vol. 3, Academic Press, New York, pp. 1–60.

Stegeman, J. J. 1989. Cytochrome P450 forms in fish: catalytic, immunological and sequence similarities. *Xenobiotica* **19**: 1093–1110.

Stegeman, J. J., R. L. Binder and A. Orren. 1979. Hepatic and extrahepatic microsomal electron transport components and mixed-function oxygenases in the marine fish *Stenotomus versicolor. Biochem. Pharmacol.* **28**: 3431–3439.

Stegeman, J. J., P. J. Kloepper-Sams and J. W. Farrington 1986. Monooxygenase induction and chlorobiphenyls in the deep-sea fish *Coryphaenoides armatus. Science* **231**: 1287–1289.

Thompson, H. M., M. L. Mackness, C. H. Walker and A. R. Hardy. 1991. Species differences in avian serum B esterases revealed by chromatofocussing and possible relationships of esterase activity to pesticide toxicity. *Biochem. Pharmacol.* **41**: 1235–1240.

Underwood, A. J. and C. H. Peterson. 1988. Towards an ecological framework for investigating pollution. *Mar. Ecol. Prog. Ser.* **46**: 227–234.

Van der Hoeven, N. 1991. Effects of toxicants on individuals and populations of *Daphnia*, a simulation study. *Comp. Biochem. Physiol.* **100C**: 283–286.

Van der Oost, R., H. Heida, A. Opperhuizen and N. P. E. Vermeulen. 1991. Interrelationships between bioaccumulation of organic trace pollutants (PCBs, organochlorine pesticides and PAHs), and MFO-induction in fish. *Comp. Biochem. Physiol.* **100C**: 43–47.

Van Veld, P. A. 1990. Absorption and metabolism of dietary xenobiotics by the intestine of fish. *Rev. Aquat. Sci.* **2**: 185–203.

Vighi and Calamari. 1990. Evaluative models and field work in estimating exposure to pesticides. In C. H. Walker and L. Somerville (eds.), *Pesticide Effects on Terrestrial Wildlife.* Taylor and Francis.

Walker, C. H. 1987. Kinetic models for predicting bioaccumulation of pollutants in ecosystems. *Environ. Pollut.* **44**: 227–240.

Walker, C. H. 1990. Kinetic models to predict bioaccumulation of pollutants. *Funct. Ecol.* **4**: 295–301.

Walker, C. H. and M. J. J. Ronis. 1989. The monooxygenases of birds, reptiles and amphibians. *Xenobiotica* **19**: 1111–1121.

Winston, G. W. 1991. Oxidants and antioxidants in aquatic animals. *Comp. Biochem. Physiol.* **100C**: 173–176.

Zaroogian, G. E., J. F. Heltshe and M. Johnson. 1985. Estimation of bioconcentration in marine species using structure-activity models. *Environ. Toxicol. Chem.* **4**: 3–12.

Index